SOLID WASTE MANAGEMENT ENGINEERING

PRENTICE HALL INTERNATIONAL SERIES IN CIVIL ENGINEERING AND ENGINEERING MECHANICS

William J. Hall, Editor

AU AND CHRISTIANO, *Structural Analysis*
BARSOM AND ROLFE, *Fracture and Fatigue Control in Structures, 2/E*
BATHE, *Finite Element Procedures in Engineering Analysis*
BERG, *Elements of Structural Dynamics*
BIGGS, *Introduction to Structural Engineering*
CHAJES, *Structural Analysis, 2/E*
COLLINS AND MITCHELL, *Prestressed Concrete Structures*
COOPER AND CHEN, *Designing Steel Structures*
CORDING, ET AL., *The Art and Science of Geotechnical Engineering*
GALLAGHER, *Finite Element Analysis*
HENDRICKSON AND AU, *Project Management for Construction*
HIGDON, ET AL., *Engineering Mechanics, 2nd Vector Edition*
HOLTZ AND KOVACS, *Introduction to Geotechnical Engineering*
HUMAR, *Dynamics of Structures*
JOHNSTON, LIN AND GALAMBOS, *Basic Steel Design, 3/E*
KELKAR AND SEWELL, *Fundamentals of Analysis and Design of Shell Structures*
MACGREGOR, *Reinforced Concrete: Mechanics and Design, 2/E*
MEHTA, *Concrete: Structure, Properties and Materials*
MELOSH, *Structural Engineering Analysis by Finite Elements*
MEREDITH, ET AL., *Design and Planning of Engineering Systems, 2/E*
MINDESS AND YOUNG, *Concrete*
NAWY, *Prestressed Concrete*
NAWY, *Reinforced Concrete: A Fundamental Approach, 2/E*
POPOV, *Engineering Mechanics of Solids*
POPOV, *Introduction to Mechanics of Solids*
POPOV, *Mechanics of Materials, 2/E*
SCHNEIDER AND DICKEY, *Reinforced Masonry Design, 2/E*
WANG AND SALMON, *Introductory Structural Analysis*
WEAVER AND JOHNSON, *Finite Elements for Structural Analysis*
WEAVER AND JOHNSON, *Structural Dynamics by Finite Elements*
WOLF, *Soil-Structure Interaction Analysis in Time Domain*
WRAY, *Measuring Engineering Properties of Soils*
YANG, *Finite Element Structural Analysis*

SOLID WASTE MANAGEMENT ENGINEERING

John T. Pfeffer
University of Illinois at Urbana-Champaign

PRENTICE HALL, *Englewood Cliffs, New Jersey 07632*

Library of Congress Cataloging-in-Publication Data

Pfeffer, John T.
Solid waste management engineering / John T. Pfeffer.
p. cm.
Includes bibliographical references and index.
ISBN 0-13-824905-9
1. Refuse and refuse disposal. I. Title.
TD791.P44 1992
628.4'4—dc20

91-25053
CIP

Acquisitions editor: Doug Humphrey
Production editor: Merrill Peterson
Supervisory editor and interior design: Joan L. Stone
Cover designer: Joe DiDomenico
Prepress buyer: Linda Behrens
Manufacturing buyer: Dave Dickey
Editorial assistant: Jaime Zampino

Printed in the United States of America

10 9 8 7 6 5 4 3 2 1

ISBN 0-13-824905-9

Prentice-Hall International (UK) Limited, *London*
Prentice-Hall of Australia Pty. Limited, *Sydney*
Prentice-Hall Canada Inc., *Toronto*
Prentice-Hall Hispanoamericana, S.A., *Mexico*
Prentice-Hall of India Private Limited, *New Delhi*
Prentice-Hall of Japan, Inc., *Tokyo*
Simon & Schuster Asia Pte. Ltd., *Singapore*
Editora Prentice-Hall do Brasil, Ltda., *Rio de Janeiro*

Contents

Preface

The management of the vast quantities of solid waste generated by urban communities is a very complex process. A variety of social, political, economic, and technical factors dictate the final solution for a specific community or region. This text focuses on the technology for managing urban solid waste. It encompasses the entire spectrum: the quantities of solid waste generated by residential, commercial, institutional, and municipal sources; trends in the quantity and characteristics of solid waste; techniques for storage, collection, and transport; processing technology for material and energy recovery; and the ultimate disposal of the residual materials. The text does not concentrate on or emphasize the social and political issues that have driven the field for the past two decades. It does address these issues to the extent that they have influenced the development and application of technology, both from a historical perspective and likely future impacts. This text is an introduction to the technology and engineering design of the systems employed in managing the urban solid waste of the United States. The current social, political, and regulatory constraints that control the application of the technology are also discussed. It assumes that students have a technical background and have had an introductory course in environmental engineering or technology that has familiarized them with air, water, and land pollution issues.

CHAPTER 1

Introduction / Definitions

One of the most pressing problems facing municipalities is the efficient and long-term disposal of urban solid waste. The disposal of solid wastes, once taken for granted, has become an issue of immense proportions. More and more these problems are in the news, with the political concerns overshadowing the technical and economic issues. Construction and operation of landfills, material recovery systems, and incineration systems have become costly. Because these systems are unpopular with the general public as a method for solid waste disposal, obtaining sites for new facilities has become very difficult. This resistance to the development of new facilities has imposed a capacity limitation on the existing facilities and caused a significant increase in the cost of refuse disposal in recent years.

It is important to note, however, that landfills will continue to be employed for the foreseeable future for the disposal of the majority of the urban solid wastes generated in this country. In 1970, material and energy recovery accounted for 7.1% of the 118.3 million tons of urban wastes generated, leaving 109.9 million tons to be landfilled. In 1984, the corresponding numbers were 14.6% recovered of the 148.1 million tons generated, leaving 126.5 million tons to be landfilled. It is expected that in the year 2000, 30.3% of the 182.2 million tons generated will be recovered in some form, leaving 126.8 million tons for landfills.[1] If these predictions are borne out employing current options for material and energy recovery, it is apparent that there is a large market for competitive technologies. It is also apparent that even with extensive employment of energy and materials recovery, landfills

[1] *Waste Age*, Nov. 1986, p. 27.

will still be necessary to dispose of a final, unusable residue, albeit in much smaller amounts.

To fully understand the problem faced by the state and local governments and private enterprise, it is necessary to have a detailed knowledge of the problem: What is solid waste? Why do we produce so much of it? Why is it so costly to manage? What are the effects on the environment? Why are residents so opposed to the current methods of refuse disposal? What are the regulatory agencies doing to control solid waste management? What are the opportunities for recycling practices to reduce the amount of solid waste generated? What are the technologies available for the management of solid waste? In the following chapters, these questions are discussed. Although it is extremely difficult to provide complete answers, it is intended that the reader obtain sufficient information to have a sounder base for decision making.

DEFINITION OF SOLID WASTE MANAGEMENT

Definition of Solid Waste/Refuse

Solid waste is any solid material in the material flow pattern that is rejected by society. What is a solid material? It is a material having a significant angle of repose. The angle of repose is a characteristic of the fluidity of a substance. A material that does not exhibit an angle of repose will assume a flat horizontal surface if allowed to stand unconstrained. The angle that the surface of the pile makes to the horizontal is the angle of repose. A significant angle of repose is subjective but can generally be viewed as that angle that will permit the material to be handled by solid handling equipment such as conveyors, front-end loaders, and shovels. If it has sufficient fluid properties that prevent forming a pile without containment walls, it generally is considered to be a liquid waste, not a solid waste. This is an important distinction since it is difficult to draw a clear line between what is solid and what is liquid when working at the interface between the two.

Definition of Management

Management can be defined as the judicious use of a means to achieve an end. "An end" is the removal of the rejected material from the material flow pattern. It is no longer of use to the population and it has no intrinsic value to the society. It is therefore discarded, and if not disposed of properly, will be a source of potential problems to the population that discarded it. In early societies, the material was discarded where used, and since many of the societies were nomadic, they moved away from their wastes. It is now no longer possible to move, so these materials must be removed from contact with the population.

Constraints Applied to Management

Protection of public health. The public health problems associated with the improper disposal of solid waste, especially food waste, have long been the driving motivation for the removal of these materials from human environs. The attraction of rats and other vermin made solid waste a source of significant health problems, especially with regard to diseases associated with rats and flies.

Minimum cost. It was generally accepted that the ''cost'' of solid waste management was the number of dollars required to eliminate the rejected material from contact with human populations. If this could be accomplished by dumping it in a used gravel pit, hauling it to sea, volatilizing it into the atmosphere, or whatever, that was the minimum cost. As the environmental impact of improper disposal of solid waste became known, it was necessary to impose more stringent constraints on disposal techniques. Also, it is no longer sufficient to consider only the current costs of the design, construction, and operation of a facility. Future economic liabilities, if not for the disposal site operator, then for society in general, are recognized and are now becoming a part of the cost evaluation.

Environmental acceptability. Initially the objective of solid waste disposal was to render the waste innocuous either by hauling it to remote locations where human contact was unlikely, or attempting to burn the combustible fraction. The resulting environmental insults were tolerated. The quantity of waste was limited both by the per capita production as well as by the total population. Also, the environmental insults from other societal activities were significantly greater than from solid waste disposal. As the concern for improving the quality of the environment began to be expressed in the 1960s and 1970s, it was clear that solid waste disposal practices were contributing to the deterioration of environmental quality. Since then there has been a steady increase in the control exercised over solid waste disposal.

Resource recovery and conservation. The concept of resource recovery and conservation has had a significant impact in defining what is and is not rejected by society. The oil embargo of the 1970s and the ensuing concerns for energy conservation resulted in the identification of materials having a high production energy requirement. (Perhaps the increase in the cost of energy was a stronger driving force than conservation.) These materials were the targets for recycling, and as more was learned about recycling, a variety of other materials has been added to the list.

This is discussed in detail later, but it is important to recognize that recycling has changed the definition of what is rejected by society.

SOLID WASTE SOURCES

Domestic/Residential Solid Waste

This category of waste includes the rejected solid material that originates from single- and multifamily household units.

Garbage. This type of residue results from food marketing, preparation, and consumption in relationship to residential units. It contains putrescible organic material that needs special consideration due to its nature of attracting vermin (rats and flies) and of producing very strong odors.

Rubbish/trash. This category consists of paper and paper products, plastics, cans, bottles, glass, metals, ceramics, dirt, dust, yard and garden wastes, and the like. Except for the yard and garden wastes, these materials are nonputrescible.

Ashes. This type of waste is the residue from any combustion process (i.e., fireplaces, wood or coal heating units, etc.) resulting from household activities.

Bulky wastes. This category includes furniture, appliances, mattresses and springs, and similar large items. Because of the size and weight of these items, it is usually not possible to collect them using normal collection equipment. They require special handling and collection.

Commercial and Institutional Solid Waste

The refuse that originates from offices, banks, retail stores, restaurants, schools, hospitals, and so on, is included in this category. It is subdivided similar to residential refuse. Garbage will originate primarily from restaurants and fast-food establishments. Rubbish may be generated in large quantities at certain commercial establishments because of the large amount of packaging materials. There is generally not much production of ashes, as incineration on a small scale in not practiced any more. The ash production used to be substantial since on-site incineration was a common means for reducing the volume of material for disposal.

There are two additional categories that are associated with the commercial and institutional wastes.

Construction and demolition waste. This class of refuse includes the lumber, bricks, concrete, plumbing, electrical wiring, and so on, associated with the destruction of old buildings and the construction of new buildings. The quantity of this material associated with building demolition and construction can be highly variable, due to the close correlation of the construction industry activity with the general economy of an area.

Special wastes. These wastes are the solid and semisolid materials generated by special facilities such as hospitals and research laboratories. These wastes may include explosive substances, toxic chemicals, radioactive materials, or pathological materials. Because of the hazardous nature of these materials, they are not permitted in the general waste stream, but require special collection, handling, and disposal, depending on the exact nature of the material.

Municipal Solid Waste

This category includes the solid residue that results from the municipal functions and services. These wastes are of a nature that require special collection, and in some cases, special processing.

Street refuse. This material results from normal street cleaning operations, including street sweepings and catch basin cleaning. It is primarily inorganic, containing a considerable percentage of sand, grit, and dirt. The quantity may be influenced by the season of the year and certain practices (i.e., the use of sand or cinders for ice and snow control). During the fall, leaves may be the primary component of this refuse stream, depending on the degree of urbanization and the policy regarding leaf pickup or burning. (There are still areas in this country that permit open burning of leaves in the fall. This may soon be banned totally.)

Dead animals. It is the responsibility of the municipality to remove large dead animals, either wild or domesticated, from the streets. It is a major problem in communities that are located adjacent to habitats that support a significant population of large wild animals, or communities that do not have animal control laws that keep animals (dogs in particular) from running loose.

Abandoned vehicles. It is common practice in some areas of a community simply to remove the licence plate from an old automobile and abandon it on the street. Removal and disposal then become a problem for the city. Title clearance is a major problem to consider when developing an abandoned automobile disposal system.

Water and sewage plant residues. In all urban areas, disposal of sludge resulting from the treatment of water and wastewater must be properly conducted. In some cases the municipality may own these systems, or they may be owned by a public utility or sanitary district. In all cases, the disposal problem must be addressed.

Park and beach refuse. This is the typical refuse generated by users of these facilities: cans, bottles, paper, and plastic. Also, there may be quantities of landscape waste resulting from maintenance of the vegetation and trees. This can be a significant problem when storms strike and damage a large number of trees in the park or beach area.

Landscape waste. In most cities, an arbor department has the responsibility for maintaining trees and areas in the parkways of streets and other public lands. This may also be a function of the park district. The removal of debris resulting from storms (i.e., wind or ice) can present a major task to the municipality.

Industrial Solid Waste

There are two general sources of refuse generated at industrial sites, the commercial/institutional component and the process solid waste. It is important to distinguish between the two. The quantities and characteristics of these residues are markedly different.

Commercial/institutional. This type of refuse is associated with the activities of the support personnel for the plant. The refuse is produced by the office staff, the cafeteria, and personnel-related activities, as well as quality control laboratories.

Process wastes. This is the residue remaining from manufacturing processes. All plants are less than 100% efficient and a percentage of the raw material as well as the product becomes a waste. In industries where the primary product accounts for a limited percentage of the input material, secondary products have been developed to improve the economics by increasing the income stream as well as reducing the cost of residue disposal. For example, lumber mills have developed a use for the chips from the planing operation (chipboard) and for the sawdust (either as a fuel for power production or for use in particleboard). These residues are unique for each industry and unique for each plant within a given industry.

Agricultural Residues

In rural areas the disposal of solid residues resulting from agricultural activities poses significant and unique problems. These residues will not be addressed except to indicate that problems do exist.

Confined animal feeding. It is not uncommon to concentrate 10,000 to 50,000 head of beef cattle in a confined feeding operation. The manure generation from these animals during their residence on the lots is substantial. Similar problems are associated with swine feeding operations with 30,000 pigs on site or from a million chickens or turkeys. Manure management becomes a major expense for operators of such facilities.

Crop residues. The residues from many crops, such as corn, wheat, and soybeans, are left on the fields and reintroduced into the soil. However, the cuttings or prunings from vineyards and orchards present disposal problems. There is both the volume of the plant material to consider and the potential for this plant material to harbor diseases and insects. There are many other examples of disposal problems associated with agricultural production.

FUNCTIONAL ELEMENTS OF A SOLID WASTE MANAGEMENT SYSTEM

There are a number of different operations associated with a solid waste management system. Each operation accomplishes a specific purpose in the chain of actions required to manage the solid waste satisfactorily. Understanding each of these steps is necessary in order to develop an efficient management system. Figure 1.1 is a representation of the sequence in which these operations occur. Our presentation of the material follows this flow diagram.

The operations in the highlighted boxes are mandatory for any management system; this represents the absolute minimum number of operations required. Refuse is produced at the source. The quantity and composition of refuse are determined by the characteristics of the source. Most refuse is produced over a period of time, requiring on-site storage until such time as it can be collected. The storage may be conventional trash cans for single-family residences or a large storage bin for large producers. Frequency of collection will depend on a number of factors, the most important being the production rate.

Collection will be accomplished with a special vehicle that is mechanically compatible with the storage systems. Collection in a single-family residential area will require a side- or rear-loading packer truck into which the cans can be emptied manually. Areas served with a large container will require a vehicle with mechanical mechanisms to transfer the refuse from the container to the vehicle. After the refuse has been collected, it must be transported to either a processing system or a sanitary landfill. For economic reasons, small solid waste management systems generally use the collection vehicle to haul the refuse to the landfill.

As the system becomes larger or more heavily regulated, additional steps may be warranted. These steps may involve a more efficient means of transport, including such methods as rail or barge haul, or processing for mass and volume reduction through a resource recovery system. Resource recovery may entail separation of specific components for recycle. Paper, aluminum, plastic, ferrous metals, and glass are components that have been identified for reuse. The organic portion of the refuse can be incinerated to generate steam or electricity. This fraction can also be processed biologically to produce fuel gas or composted for use as a soil conditioner.

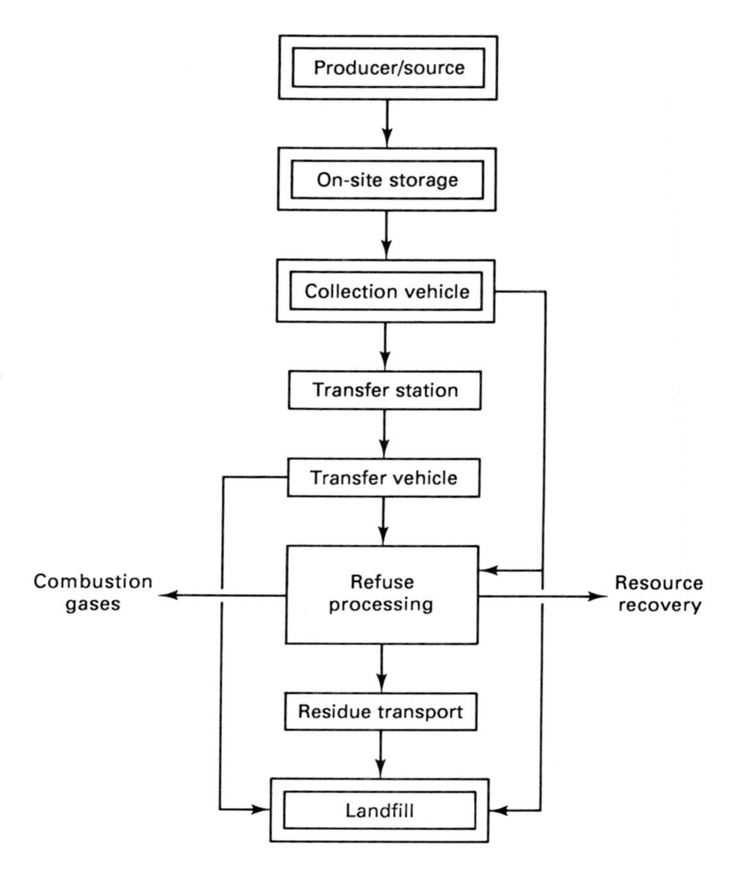

Figure 1.1 Functional elements of a solid waste management system.

There will always be a certain quantity of rejected material that is noncombustible (ash) or that has no recoverable value. This material will necessarily require a sanitary landfill for final disposition.

SOLID WASTE MANAGEMENT AGENCIES

The collection and disposal of the refuse generated in metropolitan areas, small towns, and rural areas may be conducted by a unit of local government such as the village or city, the county, or a special management district; by a private contractor; or by a combination of public and private agencies. In larger cities it is common for residential collection to be the responsibility of the city. The larger producers—apartment complexes and commercial and industrial sources—often contract with private haulers to collect their refuse. Disposal may be at a municipally operated

TABLE 1.1 COMPARATIVE REFUSE COLLECTION AND DISPOSAL COSTS

City	Average annual tonnage	Total Cost (10^6 $)	Tons collected per worker	Disposal cost ($/ton)
Philadelphia	820,000	128.5	323	157
Baltimore	335,000	48.9	332	146
New York	6,620,000	622.0	551	94
San Diego	340,000	12.4	1471	37
Los Angeles	1,500,000	33.0	1599	22

Source: Waste Age, Nov. 1986, p. 6.

site or at a privately owned site. It is difficult to generalize, however, since there are a large variety of combinations of public and private refuse collection and disposal operations.

Management of solid waste is a big business in the United States and in other industrialized countries. Municipal systems consume a significant portion of the municipal budget. There is a great difference in the efficiency (cost) of solid waste management in different communities. Table 1.1 illustrates this difference by comparing the 1986 cost of operation for several large cities. The table indicates several differences in the refuse management practices of these cities. First, not all of the refuse is collected by municipal workers. For example, New York has about 2.3 times the population in their service area than that of Los Angeles, but collects 4.4 times the tonnage of refuse. The role of private collectors is not reflected in the cost of the municipal systems. The number of workers associated with the management system has a major impact on the cost. Clearly, the operational efficiency in San Diego and Los Angeles is much higher, as reflected by the tons collected per worker.

An additional factor that determines the cost is the disposal method. In the cities that have adequate sanitary landfill capacity, the disposal cost is not as great as for those cities that incinerate all or part of the refuse. The data in Table 1.2 give the percentage of the refuse landfilled or incinerated prior to landfill of the ash. The high percentage of the refuse from Philadelphia and Baltimore that is incinerated increases the cost significantly. A typical landfill tipping fee (the charge for unloading the refuse at a specific site) will be in the range of $15 to $20 per ton, while the tipping fee at an incineration facility will be between $75 and $100 per ton. In the next decade there will probably be a significant increase in tipping fees associated with all disposal facilities.

The contribution to solid waste management made by private companies is difficult to quantify, in part because this segment of the industry is comprised of a large number of companies, ranging from a single truck owner to multibillion-dollar corporations. The record keeping and accessibility of the records for the small companies presents a significant problem in generating statistics on the contribution of private enterprise to the solid waste management problem. It has been estimated

TABLE 1.2 WASTE DISPOSAL PRACTICES OF SEVERAL LARGE CITIES

	Sanitary landfill		Incineration	
City	Number	Percent	Number	Percent
Philadelphia	2	50	2	50
Baltimore	3	50	1	50
New York	3	91	3	9
San Diego	2	100	0	0
Los Angeles	4	100	0	0

Source: Waste Age, Aug. 1984, p. 23.

that approximately half of the solid waste management business in the United States is conducted by private companies.

The data in Table 1.3 present some information on the two largest waste management companies in the United States, Waste Management, Inc. (WMI) and Browning Ferris Industries (BFI). In 1987, WMI's gross sales increased to \$2.757 billion and BFI's to \$1.656 billion.[2] These companies had about 8.2 million residential accounts, serving a population of approximately 40 million people. This leaves almost 200 million people served by municipal systems or other private companies.

In 1987, the third largest North American company in this business was Laidlaw Industries, with a gross revenue of \$481 million. It was followed by Western Wastes, with a gross revenue of \$98 million. When one considers the volume of business by these four companies relative to the total population and the rapid rate at which the size of the companies decrease, it is not surprising to find that there are a large number of small operators collecting and disposing of solid waste. In the late 1960s, WMI was a small local company that initiated an acquisition program that has been responsible for much of the company growth since it was formed.

TABLE 1.3 1985 STATISTICS FOR THE TWO LARGEST WASTE MANAGEMENT COMPANIES

Parameter	WMI	BFI
Number of employees	19,800	17,300
Number of vehicles	8,600	5,400
Residential accounts	4,900,000	3,300,000
Commercial and industrial	350,000	425,000
Number of landfills	97	80
Number of transfer stations	20	25
Number of incinerators	0	0
1985 Gross sales	$\$1.625 \times 10^9$	$\$1.445 \times 100^9$

Source: Waste Age, May 1986.

[2]*Waste Age*, June 1988.

WMI's revenues for 1989 were $4,458,904,000. This has been a real growth industry for a few companies. This trend continues as the smaller companies are acquired by the few large firms.

The presence of a multitude of private and public entities in the solid waste management business presents an interesting challenge. Private companies are profit oriented, whereas public agencies are politically oriented. One generally observes that the private operations are more economical when all appropriate costs are counted. However, many public agencies that are tax supported may have hidden costs. Equipment may serve two purposes; for example, garbage trucks can be fitted with a snow plow and used for snow removal in communities that have limited snow fall. It is very difficult to determine the true cost of public solid waste management agencies. As revenue bonds are used to fund more of the solid waste management facilities operated by public agencies, it will be possible to develop better cost data for comparing private versus public solid waste management agencies. These bonds are retired from income from the facility. Consequently, better cost accounting practices are necessary.

STUDY QUESTIONS

1. Define the following terms: (a) solid waste management; (b) commercial refuse; (c) garbage; (d) institutional refuse; (e) rubbish; (f) refuse/solid waste; (g) municipal solid waste; (h) domestic/residential refuse.
2. What constraints are imposed on the management of solid waste?
3. What role do sanitary landfills play in solid waste disposal today, and what is their likely future role?
4. What are the functional elements of a solid waste management system?
5. What is the role of the municipal government in solid waste management?
6. What is the role of private industry in solid waste management?
7. Do small, privately owned collection and disposal companies play a significant role in the collection and disposal of solid waste?

CHAPTER 2

Environmental Effects

Historically, public health was the motivation for removing solid waste from the human habitat. The waste material frequently contained items that were potentially dangerous, such as broken glass and ceramics. The food residues and human waste were attractions for rats, flies, mosquitoes, and other vermin. These vermin were vectors for the transfer of human and animal diseases to the human population. This issue was responsible for the efforts of municipalities during the late nineteenth and early twentieth century in the control of solid wastes. This responsibility was delegated to the ''Department of Sanitation'' or ''Department of Streets and Sanitation.'' Even today, solid waste collection and disposal in the larger cities is frequently the responsibility of such municipal departments.

The primary objective was the removal of the refuse from contact with the human population. In most cases this only required hauling these solid residues to a ''dump'' located on the edge of the city. Garbage was commonly collected from large generators and fed to swine. Garbage was also hauled to the dump. As the material accumulated at the disposal site, fires set deliberately or accidentally would burn a large portion of the combustible material and create new space for more refuse. It was possible to extend the life of these sites for decades by this technique, and such was the case until as recently as the early 1970s.

After World War II, the larger cities were faced with ever-increasing quantities of rejected material. The cost of hauling these large tonnages to remote areas increased substantially. To control costs it was necessary to reduce the tonnage for disposal. In most cities, single-family homes that had space in their yards employed a ''burn barrel'' to eliminate the combustible fraction. In the more congested residential areas and central business districts, this was not possible. Large generators of refuse, such as department stores and large apartment complexes, employed on-site incineration to eliminate the material that would burn. In areas where neither the burn barrel nor on-site incineration was possible, the refuse was collected and hauled to a central location for burning. The central incineration system was instituted to burn the combustible fraction of this refuse. In some cities, separation of combustible from noncombustible refuse was required. Separate collection of these streams resulted in a better fuel for incineration. As collection costs increased in more recent times, separate collection was discontinued and the entire refuse stream was incinerated.

The foregoing scenario was common throughout the United States during this period in history. The refuse was removed from human contact, but it was only a matter of time before these ''disposal'' techniques created major environmental problems in metropolitan areas. In this chapter we discuss the problems associated with the improper management of solid waste, from the public health aspects to the pollution of the air and water resources.

PUBLIC HEALTH ASPECTS OF SOLID WASTE MANAGEMENT

The public health problems associated with solid waste fall into two categories: diseases carried by vermin, and the physical and chemical hazards resulting from certain components in solid waste. Discarded solid waste provides the food and harborage for rats, flies, and mosquitoes. Food wastes attract these insects and rodents. The debris associated with the refuse provides shelter as well as breeding environments, especially for insects. Therefore, the population of vermin can be expected to increase if the refuse is not managed. Broken glass, rusty metals, household pesticides, solvents, and so on, are the sources of chemical and physical hazards associated with solid waste.

Disease Sources and Pathways

The diseases that can be attributed to the vermin associated with solid waste are varied, depending on the contaminated material that may be present in the refuse and the endemic disease reservoir of the human and animal population of the area. Refuse can contain human and animal fecal material. Any disease present in the population producing this waste can be transmitted by flies. Diseased animal remains may be present. Flies contacting this material can transmit certain diseases to other populations. Mosquitoes that breed in the water retained by the refuse can be the carrier of a number of diseases that may be endemic in the human or animal population. Finally, there is the disease reservoir in the rat population, which can be transmitted to human beings in contact with this population.

Diseases associated with the direct carrier pathway (i.e., carried by flies and mosquitoes) include typhoid fever, paratyphoid fever, gastroenteritis, dysentery, cholera, yellow fever, hepatitis, encephalitis, and malaria. Two conditions are necessary for these diseases to be a public health problem: (1) they must be endemic in either the human or the animal population of an area, and (2) there must be a carrier to transmit the etiological agent from the original host to the receptor. Consequently, control of the conditions that foster the growth of flies and mosquitoes is a critical component in the control of many public health diseases. Because of the progress in control of the insect populations and the reduction in the number of persons infected, many of these diseases are no longer a problem in the industrialized world. Gastroenteritis, dysentery, hepatitis, and encephalitis are currently the primary public health problems in the United States associated with flies and mosquitoes.

Rats also contribute to transmission of diseases. They are a reservoir for plague and typhus fever. The etiological agent is present in the rodent population and is transmitted to human beings by lice and fleas that live on the rats. So if rats inhabit the same environs as people, the probability for transmitting these diseases increases. Rat bite fever is also transmitted by rats. A rat bite may become infected by microorganisms carried in the rat's mouth, causing a fever to develop that can be fatal if not treated promptly and properly.

Role of Solid Waste in Disease Prevention

In addition to being a potential source of pathogenic (disease-causing) organisms, solid waste has two other roles in the increased incidence of common public health diseases. The garbage (food waste) is a source of food for vermin. Flies find moist garbage an ideal environment for reproduction and, as discussed later, the fly population correlates closely with the way in which garbage is managed. In a similar manner, the organic matter in garbage is also a food source for the mosquito larvae. Of course, rats are scavengers and are attracted by exposed garbage. The populations of both mosquitoes and rats are increased by the presence of food wastes, but careful management of garbage will not eliminate these vermin.

Discarded rubbish may retain rainwater that provides a breeding place for mosquitoes. One area that continues to receive attention is the growth of mosquitoes in discarded tires. It has been common practice to collect used tires at a central location so that when a sufficient quantity is obtained, some use may be found for them. Unfortunately, uses have been slow in developing and the number of tires in a single location may be in the many millions. It is impossible to place tires so that water does not collect in them. This is a perfect breeding place for mosquitoes. Of particular public health concern is the *Culex* mosquito, which transmits encephalitis. If rubbish is allowed to accumulate, it is a potential breeding habitat for mosquitoes.

Rats are generally found in all rural and urban areas. They are very secretive animals that have learned to live in close proximity with human beings. They are commonly found in sewers and similar protected environments. Consequently, the presence of a food source (discarded garbage) will attract them. If the food supply is adequate and there is adequate habitat, the rat population will increase significantly. Of course, discarded rubbish that accumulates also provides an ideal harborage for these animals. A concurrent increase in diseases carried by the rat can be expected.

Physical and Chemical Hazards of Solid Waste

A problem frequently overlooked is that of the physical and chemical hazards that can result from certain components in the refuse stream. The general public faces the following hazards:

1. Explosive hazard of solvents, gasoline, and so on
2. Toxic chemicals used as pesticides, cleaning solutions, solvents, and so on
3. Direct contact: infected cuts and abrasions resulting from contact with broken glass, metals, and other sharp objects that are heavily contaminated with bacteria

It is common practice to dispose of "empty" gasoline, paint, and solvent containers by putting them in the refuse container. Sufficient vapors may remain in these containers to create an explosive mix. With a spark or match, an explosion is possible. Because of the quantity of vapors, the explosion will not be large but may be sufficient to cause injury to a person close to the container. More esoteric sources of explosive materials may be a military souvenir brought home from a war. Since the item has been sitting on a shelf for several decades, it is now time to discard it. What was thought to be a disarmed shell or grenade has more than once been found to be live.

Toxic household chemicals are used in substantial quantity in the modern household. Every chemical, including toxic metal compounds of arsenic or lead, insecticides, herbicides, cleaning ammonia, caustic solvents (sodium hydroxide or lime) and organic solvents, may be found in households. The quantities of these materials are small, but they may be very toxic if ingested or absorbed through the skin. Containers that retain quantities of unused chemical may be placed in the trash can for disposal. Contact with such chemicals can be dangerous to people, especially if they do not know what is in the container.

Finally, the injury that can result from cuts and abrasions from the broken glass and ceramics, sharp metal edges, and pointed objects can be severe. Cuts easily become infected because the objects will probably be heavily contaminated with bacteria. Blood poisoning and tetanus are common problems associated with these injuries.

Workers in the refuse collection industry are more susceptible than the general public to the foregoing physical and chemical hazards. Continued exposure to the refuse on a daily basis leads to frequent injuries. Additional injuries result from equipment-related incidents where the mechanical mechanisms may injure the careless worker. The collection vehicles are cumbersome and visibility is limited, so it is not infrequent for a worker to be run over by a collection truck. Muscle strains and back injuries are frequent. The physical effort required to lift, carry, and load the contents of 500 or more trash cans in a day is substantial. Most communities limit the weight of the can contents, but this limit is not always followed. Attempting to lift an overweight can is a frequent cause of back injury.

A survey conducted by the National Safety Council in 1974[1] found that the injury rate in the refuse collection industry is the highest in the work force. The survey found 60.77 disabling injuries and 2012 days lost per million person-hours worked. The days lost were almost four times the national average for all industries. Improvements in the design of equipment and more safety training has reduced this injury rate, but it is still among the highest in the work force. The job specifications in many communities require the applicant to pass a rather strenuous physical test. This requirement is necessary to reduce muscle strain-related injuries.

CONTROL OF FLY PRODUCTION

Understanding the life cycle of the fly is necessary to develop solid waste management techniques that can control the population. Flies have four stages in their life cycle:

$$\text{adult} \rightarrow \text{egg} \rightarrow \text{larva} \rightarrow \text{pupa} \rightarrow \text{adult}$$

The adult fly will deposit the eggs in a moist environment that has an acceptable food for the larvae that develop when the eggs hatch. Garbage is an excellent environment. When the eggs hatch, the larvae feed on the organic material until they reach a certain maturity. The larvae are then transformed into pupae, which is an inactive state. Before pupating, the larvae migrate from the moist garbage to a dry loose material such as sand or soil. The pupae that are formed then mature until the adult flies emerges. This migration is important since the larvae will no longer be in the refuse. To break the life cycle of the fly, the refuse should be removed before this migration occurs.

Field studies were conducted during the 1950s in California to evaluate the role of solid waste management in control of fly population. One program was established to determine the effect of collection frequency on the fly population.[2] The evaluation simply used a larvae trap to collect the larvae that migrated from the refuse cans after different periods of time. The significant data show that only 7.8% of the larvae migrated from the can in a 4-day period. However, in a 7-day period, 93.1% of the larvae had left the can. Consequently, during the warm

[1]*Solid Waste Management,* Jan. 1975, p. 10.

[2]Campbell and Black, *California Vector Views,* Vol. 7, Feb. 1960; Ecke and Linsdale, *California Vector Views,* Vol. 14, 1967.

months it will be necessary to collect the refuse twice weekly if fly control is an objective.

The condition of the refuse storage container has been found to be a significant factor in the fly population. Only 26% of new containers that were in good condition and restricted access of the adult fly to the refuse were found to generate flies. Larvae production was 92 per can per week. However, 46% of the containers that were in poor condition and made it easy for the adult flies to deposit eggs in the refuse produced larvae at a rate of 877 per can per week. Can condition was responsible for about a 20-fold increase in the fly population.[3] Additional factors that restricted access of the adult fly to the garbage as well as the quality of the garbage were effective in reducing the population. These included wrapped and well-drained garbage, the type and quality of food scraps, exposure time of garbage to adult flies, can cleanliness, and spillage of garbage around the container.

The number of flies generated from refuse is also a function of other conditions. Temperature is a major factor in the reproduction rate of all insects. Consequently, the warm months are when fly problems will be experienced. If not properly managed, household refuse has the potential of producing about 70,000 adult flies per cubic foot. It is clear that general cleanliness and good housekeeping at the on-site storage location along with frequent collection of the refuse can be expected to minimize the population of flies generated by solid wastes.

CONTROL OF THE POPULATION OF OTHER VERMIN

The rat population is controlled by keeping the refuse in containers so that the rat cannot get to it. The use of metal and plastic containers prevents the rat from accessing the garbage. Consequently, there is no attraction. Also, the collection frequency and the general cleanliness of the on-site storage area are important in keeping the rat from being attracted. If garbage is allowed to accumulate, even if it is in good containers, the rat will find a way to get to it.

Mosquito control simply depends on the elimination of the breeding places. If the refuse is not allowed to accumulate, there will be no accumulated water. This will eliminate the mosquito population associated with solid waste.

AIR POLLUTION FROM SOLID WASTE MANAGEMENT

Historical Perspective

Improper disposal of solid waste has a history of polluting the atmosphere. Early dumps were located away from human habitats because of the air pollution problems ranging from odors of decaying garbage to smoke and fumes caused by the

[3]Black and Magy, *California Vector Views*, Vol. 9, Nov. 1962.

TABLE 2.1 SUMMARY OF AIR POLLUTION EMISSIONS FROM SOLID WASTE DISPOSAL IN THE ST. LOUIS METROPOLITAN AREA, 1963

	Emissions (tons/yr)				
Source	Aldehydes	HC	SO_2	Particles	B(a)P[a]
Refuse incineration (285,600 tons per year—21% of total refuse collected)					
Total	150	396	226	1,486	14
Municipal	39	33	201	1,001	3
Residential	25	19	25	80	
Industrial	52	208		262	6
Other	34	136		140	5
Open burning of refuse (602,000 tons per year—40% of total refuse collected)					
Total	1,140	84,300	297	14,147	402
On-site	572	44,500	127	7,473	245
Dumps	528	39,600	170	6,674	157
Refuse total	1,290	84,696	523	15,633	416
All other sources	2,400	289,000	455,000	130,500	1,130

[a]B(a)P, Benzo[a]pyrene in lb/yr.

Source: Bureau of Solid Waste Management, U.S. Public Health Service, Washington, DC, 1964.

frequent fires that occurred at the site. The effect of on-site incineration was reduced by having a stack that dispersed the combustion gases above the ground. The ever-present backyard burn barrel was a serious nuisance and, in some cases, an acute health threat to asthmatics. The environmental revolution that was gestating in the 1960s began the attempt to determine the role of various societal activities in environmental degradation.

It was obvious that the uncontrolled burning of refuse was creating air pollution, but the contribution to the total air pollution problem was not documented until studies such as the one shown in Table 2.1 were completed. Sixty-one percent (887,600 tons) of the refuse produced in 1963 in the St. Louis area was subjected to some type of combustion. The incinerated refuse produced a measurable quantity of air pollution, but it was insignificant compared to the quantity generated by open burning. The true impact of uncontrolled combustion of refuse is apparent compared to the total air pollutants from all other sources in the area.

Aldehydes represent a class of compounds that may be associated with incomplete combustion of any fuel. Open burning of refuse contributes an amount equal to half that from all other sources. Hydrocarbons (HC) are another indicator of poor combustion as well as a measure of volatile organic compounds such as gasoline and organic solvents. Open burning of refuse contributed about one-third the amount produced by all other sources. The production of benzo[a]pyrene, a suspected carcinogen, by open burning was about one-third the amount produced from other sources.

Refuse combustion was not a significant contributor to sulfur oxide (SO_2) since the sulfur content of refuse is low. The particulate emission from refuse burning was also a smaller portion of the total particulate load on the atmosphere. The ratios are no longer applicable since these data were collected when there was little or no control of air pollution from any source. As a result of vigorous enforcement of air pollution control standards, many of the air pollutants have been reduced. Data of this type clearly identified the open burning dump as a major contributor to the air pollution problem. As discussed later, legislation soon prohibited this practice.

In the 1960s it was common practice for the refuse produced at large complexes such as apartments, shopping centers, and commercial buildings to use on-site incineration. In large cities, multistory apartment units used what was termed a "flue-fed" incinerator. The incinerator was located in the basement with the flue rising through the building to the roof. On each floor, refuse disposal doors were located in the incinerator flue. The refuse was dropped to the incinerator in the basement. When the pile was large enough, the building maintenance personnel would ignite and burn the refuse.

An evaluation of the flue-fed incinerators in New York City found that there were 11,000 units that burned about 600,000 tons per year. The combustion temperatures were very low and most of the organic material would volatilize rather than be oxidized to carbon dioxide and water. The average temperature in these incinerators ranged between 460 and 670°F, while the maximum was found to range between 970 and 1200°F. A temperature of 1200°F is the bare minimum for efficient combustion. In a modern incinerator, to ensure complete combustion, the temperatures are maintained at 1600°F or greater. Obviously, the combustion conditions in these units were very poor, resulting in massive air pollution. The mass of particles emitted to the atmosphere was estimated to be between 0.85 and 1.55% of the weight of the refuse fed to the incinerator. The quantity of "noxious gases" was estimated to vary from 0.9 to as much as 30% of the tons of refuse fed. A major environmental insult has been corrected with tight air pollution control standards for on-site incineration. The standards have effectively made this process so expensive that it is no longer practiced.

Current Air Pollution Concerns

The ban on open burning of all refuse in urbanized areas has shifted the concern for air pollution to the central incineration or mass-burn systems. These units are large, processing up to 3000 tons per day of refuse and, because of their size, can present a significant burden on the local atmosphere if proper air pollution control is not exercised. Table 2.2 shows the expected emissions from an uncontrolled incineration system. The volume of combustion gases associated with these weights would be about 300,000 standard cubic feet (scf) per ton (dry solids) of refuse.

TABLE 2.2 MODERN INCINERATION EMISSION FACTORS

Pollutant	Emissions (lb/ton)
Particulate	260
Sulfur oxides	8
Nitrogen oxides	5
Hydrogen chloride	10
Carbon monoxide	3
Total hydrocarbons	0.6
Lead	0.7
Mercury	0.002
Hydrogen fluoride	0.04

In addition to the metals listed in Table 2.2, a variety of other metal particles may be in the incinerator stack gases. All of the metals that are present in the refuse have an opportunity to be discharged as particles. The most important metals in addition to those listed in Table 2.2 are arsenic, cadmium, and selenium. These metals are toxic at relatively low exposure levels and their discharge should be controlled.

The technology associated with these systems is capable of producing a high-quality stack gas. Table 2.3 presents the standards proposed for the European Economic Community (EEC) as well as the stack emissions reported from operating incineration plants that employ the appropriate air pollution control technology (see Chapter 9). The emission data are within the standards that are proposed for the EEC, showing that the control technology is adequate for these pollutants.

Table 2.3 introduces a new pollutant, dioxin. There are two general classes of organic compounds that are of concern in this category: the polychlorinated dibenzo-*p*-dioxins (PCDDs) and the polychlorinated dibenzofurans (PCDFs). They are commonly called dioxins and furans. Because of their chemical structure there are a large number of isomers, some are more toxic than others. However, all are of great concern because of their toxicity, carcinogenicity, and probable mutagenicity.

The structures of the dioxin and furan are presented below. They consist of two benzene rings with a double (dioxin) or single oxygen (furan) linkage between the rings. The isomers are formed by the substitution of chlorine atoms at any or all

TABLE 2.3 EEC STANDARDS AND EMISSIONS FROM OPERATING UNITS

Parameter	Proposed standard	Measured levels
Carbon monoxide	80 ppm	70 ppm
Sulfur oxide	70 ppm	60 ppm
Nitrogen oxide	390 ppm	150 ppm
Hydrogen chloride	62 ppm	25 ppm
Particulate	0.08 gr/dscf*	<0.005 gr/dscf
Total dioxin	?	<minimum detectible

*Grains/dry standard cubic foot.

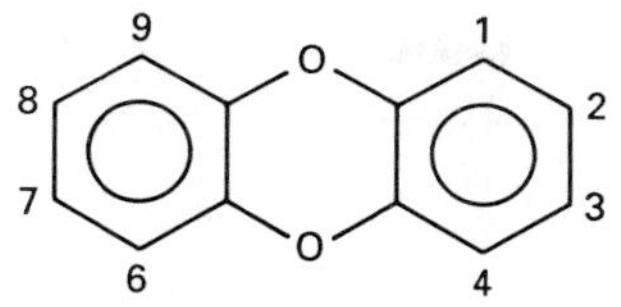

9 1 8 2 7 3 6 4 O O

Dioxin

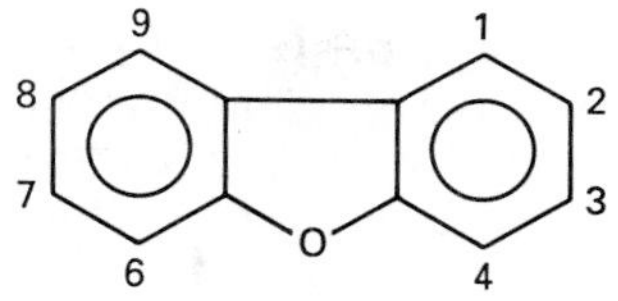

9 1 8 2 7 3 6 4 O

Furan

of the numbered carbons in the benzene rings. There are five dioxins and seven furans that are considered as the most toxic. One is the dioxin 2,3,7,8-TCDD (tetrachlorinated dibenzo-*p*-dioxin). This compound is frequently used as a reference for the determination of the concentration of these compounds in the environment. The toxicity threshold is very low; the LD_{50} (dosage at which 50% of the test population dies in a specified time period) for guinea pigs is less than 10 μg/kg of body weight.[4] Environmental concentrations in the range of ppt (parts per trillion) are of interest. These levels are very near the limits of the sensitivity of the analytical techniques.

The presence of compounds such as dioxins cannot be taken lightly because of the very low toxic threshold. As new analytical techniques are developed, it will be possible to detect more materials that are extremely toxic. The dilemma for regulation is evaluation of the risk associated with the levels that are measured. When it was not possible to detect the presence of a toxic material, there was no concern about the possibility of it being generated in the processing of solid waste. As the analytical capabilities improve, it will be possible to detect more and more potentially toxic substances. However, it is not possible to determine the long-term exposure effect of these concentrations until many years after they are detected. Risk evaluation has been and will be a major component in establishing rational standards for all discharges to the environment.

Standards for the emissions from the solid waste incineration systems are in an evolutionary stage. It will probably be the year 2000 before the standards solidify. Therefore, it will be necessary to monitor the technical literature and the regulatory activity to keep current on emission standards.

WATER POLLUTION FROM SOLID WASTE MANAGEMENT

Historical Perspective

Sanitary landfills are a recent technology that has been applied nationwide only during the past two decades. Prior land disposal sites were better defined as dumps. Little attention was given to the manner in which the refuse was placed and soil cover was used only when convenient. California has been leading the nation in the development of this technology. Because of the unique air sheds in many of the

[4]*Environmental Science and Technology*, Vol. 17, No. 3, 1983, p. 124a.

populated areas of southern California, air pollution has been a major problem since the rapid population growth that started after World War II. Open burning of any solid waste in the urban areas was quickly prohibited. Central incineration was also not permitted, so landfills were the only option.

The population density and the cost of hauling the refuse long distances required many of these disposal sites to be located close to populated areas. It was necessary to operate the landfill in such a manner that it did not create problems for the neighbors. An acceptable technology for this climate developed rapidly and until recently it was the only method considered for the disposal of solid waste in southern California. Since this region is relatively dry, leachate production was not a problem except under unusual circumstances.

In the larger urban areas in the rest of the country, the refuse was generally incinerated. In the smaller communities, the lower population densities made it possible to find land areas relatively close by that could be used for refuse disposal. Unfortunately, there was no local pressure that required the design and operation of an engineered disposal site. State and federal control did not yet exist. Consequently, most of the refuse was simply dumped at an innocuous location. No effort was made to control the access of water to the sites. In fact, it was not uncommon to dump the material into an excavation such as a sand or gravel pit that was full of water. Refuse was considered a good fill material to use to reclaim these excavations.

Also, control of surface water was not practiced. Frequently, drainage from surrounding land areas was permitted to drain into the excavation for the landfill. Any precipitation that fell on the site percolated into the refuse. Consequently, substantial volumes of water entered the fill. As this water percolated through the refuse, it dissolved large quantities of soluble inorganic and organic pollutants. The water generally found its way into the groundwater. As long as the refuse contained only natural materials, the subsurface soils were effective in removing most of the pollutants by the time the leachate-polluted water reached the point of use. There was some concern about the presence of heavy metals, but most of the wells that were contaminated with this leachate had a quality of water that masked the potential hazard from the leachate contamination.

The introduction of anthropogenic (human-made) chemicals into the material flow created a new problem. The blatant dumping by industrial sources of large quantities of these toxic compounds into the landfills created a source of new contamination in the leachate. These compounds, being synthetic, were frequently not biodegradable. The presence of these substances in groundwater in increasing quantities created new concern for leachate pollution. In many cases, natural processes were not effective in containing these materials. Therefore, the pollution of ground and surface water with leachate became a problem of major importance.

Leachate Production

Water pollution can result from the improper design and/or operation of a sanitary landfill. Control of infiltration that results from precipitation and surface runoff from the adjacent land areas is an essential component of any satisfactory landfill

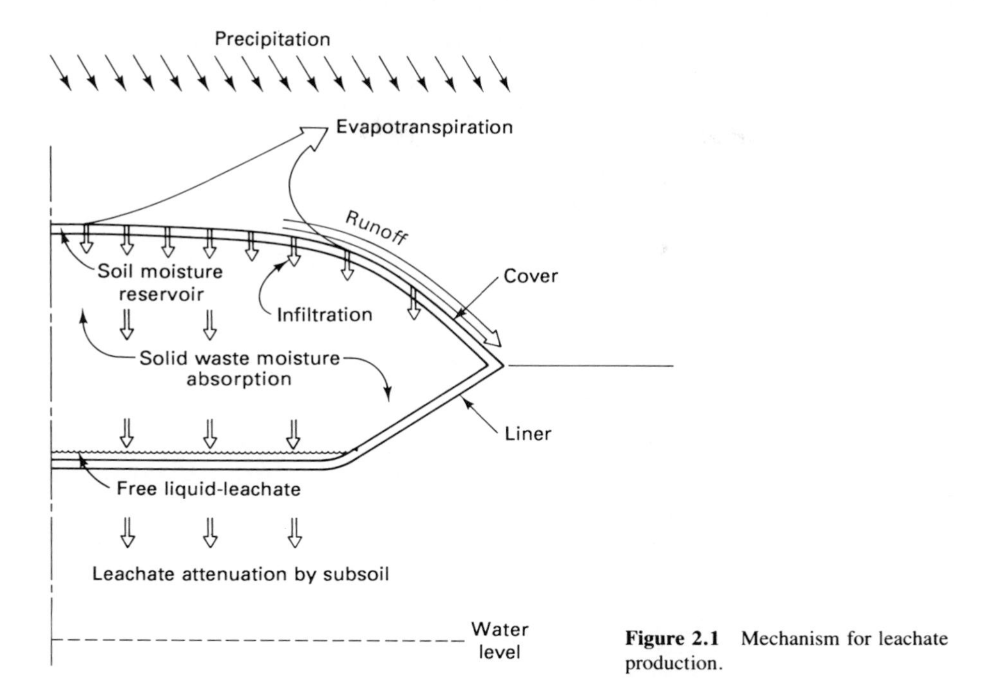

Figure 2.1 Mechanism for leachate production.

operation. If this water is allowed to infiltrate into the landfill, it will increase the water content of the fill until excess moisture exists. The water will slowly move down through the refuse and absorb the soluble materials present in the solid waste. Figure 2.1 illustrates how leachate is generated. The quality of the leachate is dependent on a number of factors, and the composition, as illustrated in Table 2.4, can be highly variable, depending on the source. These data represent the range of leachate characteristics that have been reported in the literature. The characteristics of the landfills producing this material are equally variable, with operating ages ranging from 1 to 16 years.

The volume of the leachate generated will depend on the quantity of water infiltrated. One acre-foot of infiltration (12 in. of water over 1 acre) equals 325,000 gallons. Without control of surface water, this would be a typical amount of infiltration in an area with an annual precipitation of 30 to 40 in. For example, each foot of water entering the surface of a 100-acre landfill per year will generate 32,500,000 gallons of leachate per year. The effect of this volume of leachate of the quality indicated in Table 2.4 can have a major impact on either the surface or the groundwater system.

The biological activity in the landfill significantly changes the chemical environment. The biodegradable organic solids are hydrolyzed, and in many cases the product is an organic acid. The pH can become acidic, less than 5.0. Also, the oxidation-reduction potential is lowered into the range where many metals are soluble. The combination of low pH, high salt concentration, and reduced environment

TABLE 2.4. LANDFILL LEACHATE CHARACTERISTICS

Parameter	Range of values (mg/L except pH)
BOD	40–89,500
COD	81–33,360
pH	3.7–8.5
Dissolved solids	584–44,900
Suspended solids	10–700
Alkalinity ($CaCO_3$)	240–20,500
Hardness ($CaCO_3$)	540–22,800
Total P	0–130
NH_3-N	0–1106
NO_2 + NO_3-N	0–10.3
Calcium	60–7200
Chloride	5–2467
Sodium	34–7700
Potassium	28–3770
Sulfate	1–1558
Manganese	0.1–125
Magnesium	17–15,600
Iron	0–2820
Zinc	0–370
Copper	0–10
Cadmium	0–17
Lead	0–2.0

Source: R. Gardner and E. Conrad, The Use of "HELP" Model in Evaluating Alternative Leachate Management Plans for Three New York City Landfills, Proceedings Waste Tech 86, Chicago, IL. Oct 1986

create a very corrosive environment that is responsible for converting many of the metals in the refuse into soluble ions.

Surface Water Pollution

The pollution of surface waters (streams and lakes) by leachate is not a significant problem. It is an unusual site design that causes the leachate to exit the landfill above ground. If this should happen, the concentrated leachate causes severe pollution of the surface water. The high BOD rapidly depletes the dissolved oxygen in the water and fish kills occur in a very short time. A leachate with a BOD of 10,000 mg/L is approximately 500 times as strong as domestic sewage. The 32.5 million gallons per year generated from the 100-acre site discussed above would be the equivalent of the untreated sewage discharge from a city with the population of 45,000.

Additional contamination is related to the various metals that may be present in the leachate. Table 2.5 lists the general-use water quality for the state of Illinois.

TABLE 2.5 STATE OF ILLINOIS WATER QUALITY STANDARDS

Constituent	Chronic Standard (μg/L)
Arsenic	190
Barium	2,000
Cadmium	$e\{0.7852[\ln(\text{hardness})] - 3.490\}$
Chromium (total)	11
Copper	$e\{0.8545[\ln(\text{hardness})] - 1.455\}$
Cyanide	100
Fluoride	15,000
Iron (total)	1,000
Lead	$e\{1.273[\ln(\text{hardness})] - 1.46\}$
Manganese	1,000
Mercury	0.5
Nickel	1,000
Silver	100
Zinc	1,000

Source: Illinois Environmental Register 366, Illinois Pollution Control Board, Chicago, Sept. 1, 1988.

Manganese, iron, zinc, copper, cadmium, and lead have potential concentrations that could cause violation of these stream standards even with a large dilution factor. If the leachate does exit the fill above the elevation of the surrounding ground, it can be expected to cause severe water pollution when it enters a surface water.

Groundwater Pollution

The normal path for the leachate is to percolate downward into the groundwater. The rate at which the water moves is controlled by the permeability of the material underlying the site. A very tight clay will restrict the movement to inches per year. Conversely, sand and gravel deposits permit several feet of movement per day. As the contaminants move through the subsoils, there is significant biological, chemical, and physical interaction between the contaminants and the soil matrix. This interaction depends on the characteristics of the soil.

The biodegradable organic material (BOD) present in the leachate provides a substrate for a diverse population of soil bacteria. A substantial population of microorganisms will develop when this organic material is present. The organisms will stabilize the BOD if the concentration of toxic materials is low. The supply of oxygen is restricted due to the surface soils. Consequently, anaerobic degradation generally occurs. The carbon dioxide produced by the biological activity is dissolved in the water and will keep the pH depressed, causing the water to dissolve the minerals as it passes through the aquifers.

The biodegradable material does not travel far from the landfill before it is stabilized by the microorganisms in the soil. The change in the chemical quality of the groundwater may be significant. The acidity associated with the carbon dioxide will cause an increase in the dissolved solids. The effect of these dissolved materials will depend on the characteristics of the aquifer. Typically, one can expect an increase in hardness and probably an increase in the iron and manganese. In addition, the chemical characteristics of this water will keep any metal ions leached from the refuse in solution. If this flow is in a permeable sand and gravel aquifer, it is possible to find contaminated groundwater considerable distances from the landfill.

If the subsoils are tight clays, the rate of movement is greatly reduced. In addition, the clay has a cation-exchange capacity that will prevent the metals leached from the landfill from moving. As the metal ions pass through the clay, they are exchanged for the monovalent ions present in the clay matrix. With the proper subsoil, it is possible to greatly reduce the rate at which the leachate moves, providing sufficient time for the biological activity to stabilize the BOD. The ion-exchange capacity of the clay will attenuate the movement of metal ions by capturing them in the soil matrix.

The introduction of the refractory organic compounds creates a different scenario. Since these compounds do not degrade, it is only a matter of time before they will reach a point of use. It may be 10 years, or 100 years, but sooner or later they will move from the landfill unless the site is designed to minimize leachate production and to capture the leachate that is produced. If this material had not been dumped into the landfills in the past, the current public hysteria about toxic materials in leachate would not exist.

Past practice did not separate the normal urban solid waste from industrial hazardous waste. Drums of a variety of toxic chemicals were routinely hauled to the most convenient landfill. Since these sites were not designed to contain leachate, numerous cases of groundwater pollution have been documented. The contaminants consist of a variety of organic chemicals, including organic solvents (volatile organic compounds), herbicides, insecticides, oils, polychlorinated biphenyl (PCB), and heavy metal sludge. It will be many decades before the existing problem sites are finally cleaned and the public resistance to sanitary landfill disposal of nonhazardous urban waste is eliminated.

STUDY QUESTIONS

1. What was the objective of solid waste disposal prior to the 1970s?
2. What is the potential impact of improperly managed solid waste on the traditional public health diseases?
3. What is the role of solid waste in controlling the population of flies, mosquitoes, and rats?
4. What chemical and physical hazards are associated with solid wastes?

5. What is the life cycle of the fly, and how does this relate to the collection of solid waste?
6. What were the major environmental problems associated with the improper disposal of solid waste during the 1960s and 1970s?
7. What are the current environmental concerns with solid waste disposal?
8. Why is the dioxin and furan class of chemical compounds of such concern to the public?
9. What is the mechanism for the production of leachate from a sanitary landfill?
10. What are the general characteristics of leachate?
11. Differentiate between the pollution of surface water and groundwater by sanitary landfills.

CHAPTER 3

Legal Issues and Authority

POWER OF THE LAW

The original power to regulate environmental pollution was founded in the police power that has been vested in the state government. The Constitution guarantees that the rights of the individual states shall be supreme and all rights assigned to the states are to be enforced by the states. (See Amendment X to the U.S. Constitution, which was approved in 1791.) This has special significance in the area of pollution control because all control actions for intrastate pollution problems rest with state government.

However, the U.S. Constitution reserves certain rights for the federal government (see Article 1, Sections 8 and 10). These include the power to regulate interstate and international problems as well as those problems that pertain to the welfare of all of the people of the United States. Also, the Constitution does not permit treaties between states. When such an arrangement is desired, it requires action by the federal government to legalize the agreement. This means that any cooperative arrangements between two or more states for the purpose of environmental quality control must be authorized by the federal government.

The state, in its supreme position, can delegate its powers to other units of government. This power can be delegated not only to the federal government but also to lesser governmental units. This is seen in the regulatory power of many political entities, such as county and municipal government. The police power of these local units of government is never inherent, but it is a delegated power from a higher source by constitutional, statutory, or chartered positions. The state government retains the right to discharge or cancel the powers that they grant.

It soon becomes obvious that the residence of power can be a highly complex problem. The states guard their power carefully and are reluctant to yield it to a higher government. However, the federal power over interstate matters offers an inroad into environmental pollution control. This is basically because the environment does not confine itself to political boundaries but tends to follow natural boundaries such as drainage basins or air sheds. It is generally true that most environmental pollution problems cross state lines and, consequently, open the door for federal action.

HISTORICAL PERSPECTIVE

Control of environmental pollution has a long and complicated legal background. The basis for most of the legal activities in the United States prior to the explosion of legislation in the past 40 years was the common law doctrine. This doctrine was brought from England by the early settlers of this country. It has as its basis the centuries of decisions that were made by various bodies. In essence, it reflects what the society considers to be fair and reasonable. There is no specific written document, so the interpretation of what is fair and reasonable tends to change as society's perception changes. Consequently, as the quality of life improves, what was considered fair and reasonable 50 years ago may not be today.

Historically, the primary mechanism for the control of pollution was the public health powers. These powers were central to elimination of the gross pollution that occurred up until perhaps the 1930s. During the 1930s, significant funding was provided by the federal government for the construction of sewage treatment plants. This was part of the public works programs instituted to increase employment during the depression era. Funds were made available for the construction of treatment facilities for communities that discharged into small streams. In addition to increased employment, construction of these treatment plants reduced the health hazard associated with raw sewage discharges. The second mechanism that was occasionally used to correct some environmental problems was legal action initiated under the common law doctrine of nuisance.

The public health powers arise from the police powers of the state, which are "to assure safety, comfort, health and convenience in the State by preserving public order and insuring to each citizen an uninterrupted enjoyment of all rights and privileges." This power has been delegated to local agencies, at both the county and municipal level. This is the basis for the municipal police departments and the county sheriff departments, as well as other agencies, such as city or county health departments.

Since public health falls under the police power umbrella, any activity that causes a defined public health risk can be terminated by the appropriate state or local health agency. Perhaps the most common use of this power at present is the closing of eating establishments that have had cases of food poisoning. Rapid action on the part of the health department in closing an establishment is necessary to prevent the exposure of a large number of people to potential health problems. The state, county, or municipal health department has the power to effect such closures in the interest of public health.

The intentional or accidental dumping of a large quantity of a toxic chemical into the water supply of a community would be the type of event that could be considered as a public health problem and immediate action could be taken by the appropriate health department. The release of toxic gases into the atmosphere in quantities that are known to be toxic can be declared an emergency and immediate action taken to protect the population and eliminate the discharge. The acute health effect must be identifiable before the health officer can exercise the power of the office. This mechanism works well for well defined cause–effect scenarios. It is, however, not effective in addressing the multitude of pollution problems that do not have such a clearly identifiable effect.

A similar situation exists with the common law doctrine of nuisance. *Black's Law Dictionary* defines "nuisance" as that which annoys or disturbs one in the possession of one's property, rendering its ordinary use or occupation physically uncomfortable to her/him. A "nuisance per se" is defined by the same authority as "an act, occupation, or structure which is a nuisance at all times and under all circumstances, regardless of location or surroundings." Therefore, if some activity generated a substance that had been legally or judicially defined as a "nuisance per se," it is unnecessary to demonstrate that it caused injury to a particular person or persons or the property of said persons. For example, in many jurisdictions, air pollution by a thick black smoke has been declared a nuisance per se by legislative fiat.

Under the nuisance doctrine, it is possible for any citizen to bring suit against someone causing pollution. The basis for the suit is not necessarily health related but can be any adverse effect the pollution event has on "the rightful use and enjoyment of one's property." A frequent basis for such suits in recent times is odor: for example, from industrial processes or animal feed lots. The suit will generally be heard in circuit court before a jury. If the plaintiff demonstrates successfully that the pollution interferes with the rightful use of the property in question, there are three possible results. There could be a cease-and-desist order, a money penalty, or a combination of the two. The severity of the penalty will depend on how the case is presented and the mood of the jury.

If the plaintiff is successful in establishing great mental anguish or some other equally ill defined malady, the monetary penalty may be substantial. Also, the cost of eliminating the source of pollution could be significant. Consequently, if the financial cost to the polluter is great, an appeal will probably be filed. The time frame for hearing the suit in the circuit court may have already been 1 to 2 years. The appeals process can extend the resolution of the case for many years. In some suits initiated in the 1950s, up to 15 years passed before resolution was achieved. The legal fees on both sides increase in proportion to the time required to settle the case. This is not a very efficient mechanism for correcting pollution. It is not timely, it is expensive, and it frequently addresses only one pollution event rather than the general question of environmental deterioration.

Common law has little preventive power, as the practice of common law was to settle for monetary damages only after the act had been committed. In common law and even in some states today, there is a doctrine that one who moves near an existing nuisance cannot complain. This has proved to be a considerable hindrance to environmental pollution control. Fortunately, this doctrine has been repudiated by most states.

Because of the shortcomings of common law procedures for combating environmental quality degradation, the federal and state governments began to search for more efficient mechanisms for controlling pollution. This was achieved by the passage of statute laws directed specifically at environmental protection.

STATUTE LAW

A statute law is a law enacted by the legislative body of a state or nation and recorded in a formal document. It addresses a specific problem and establishes a mechanism for defining what is considered to be a violation of the law. By enacting the statute, the legislative body determines what is acceptable for the citizens of the state. For example, the speed limit on the highway is set at a specific level. It is not necessary to show that the speeder endangered someone. The legislative body has determined that this is a speed at which a vehicle can be operated with a reasonable degree of safety. Exceeding that speed is a violation of the law. Because of the unique status of the states in the United States, the majority of the statute laws relating to the health, safety, and well-being of the public is legislated at the state level.

After a statute law is passed, it must be able to withstand any court challenges. If the law is infringing upon rights guaranteed by the state or U.S. Constitution, the law is subject to being declared unconstitutional. The courts hearing the challenge will range from the state courts to eventually the U.S. Supreme Court if they decide that the issue is important enough for them to consider. When these laws survive the court challenge, they become very powerful tools for regulatory agencies.

The current state environmental regulatory activity is based on statute laws. As with speed limits, the legislation is specific. It does not just say that one shall not pollute the environment, but sets up specific mechanisms to ensure that anyone operating a system that can pollute operates it in such a way that pollution does not occur. This is accomplished through enabling legislation that establishes the necessary administrative organization to carry out the intent of the legislation. This organization generally includes a rule-making body as well as an enforcement body. An example of such legislation is the 1970 Illinois Environmental Protection Act. This legislation contains the 13 titles listed below. A brief summary of the contents of each title is presented, with more emphasis placed on the titles pertaining to solid waste.

Title I: General provisions. This title presents the legislative declaration indicating the intent of the legislature in passing this act, as well as the definitions of the various terms used in the document. Most important is the establishment of the agencies responsible for administrating the act. The Environmental Protection Agency (EPA) (this is the state agency, not to be confused with the U.S. EPA) is the primary administrative agency. They are responsible for environmental monitoring, initiating enforcement action, managing the federal programs delegated to the state, defining technical standards, and issuing of development, construction, and operating permits. The director of the EPA is appointed by the governor with Senate approval. There is a large professional staff—accountants, engineers, lawyers, scientists, and so on—that is responsible for the conduct of the agency business.

The Pollution Control Board (PCB) is the rule-making body created by this act. They have the responsibility for establishing the standards necessary for achieving the desired environmental quality. The act established well-defined procedures for the PCB to follow when developing standards. The PCB has set both general environmental standards such as dissolved oxygen levels in the streams or ozone level in the atmosphere, as well as discharge standards required to achieve these general standards. The PCB also acts as a judicial board by hearing any enforcement action initiated by the EPA or a citizen. A concise record is developed when hearing a case, and an appeal of a PCB decision goes directly to the appellate court. The record of the PCB hearing is the basis for any appeal. The PCB also imposes any penalties they deem appropriate. The PCB has seven members appointed by the governor with Senate approval. A limited professional staff is also available to the PCB.

The Institute for Environmental Quality was established as the research and data collection entity for the Illinois environmental program. Funds were provided for contract work in support of the PCB and EPA mission. Legislative action in 1978 transferred the mission of the Institute to the Department of Energy and Natural Resources (DENR). A primary responsibility of DENR in support of the state environmental program is the preparation of economic impact statements for any new substantive regulations. The economic effects of regulations have been a significant concern of the legislature since the late 1970s. Additional sections in this article provide descriptions for the public inspection of the records and the protection of trade secrets.

Title II: Air pollution. This title addresses the air environment and authorizes the PCB to set air quality standards and discharge standards necessary to achieve the air quality standards. It also defines how the state agencies should interact with the federal agency in setting and enforcing air quality standards. This title contains two sections of special interest. Section 9.b says: "No person shall construct, install, or operate any equipment, facility, vehicle, vessel, or air craft capable of causing or contributing to air pollution in Illinois, or designed to prevent air pollution, of any type designated by Board regulations, *without a permit* granted by the Agency, or in violation of any condition imposed by such a permit.''

This section introduces the permit mechanism for the control of air pollution and the same concept is applied to the other pollution problems in subsequent titles. A permit is required for any discharge to the atmosphere, and in order to obtain the permit, certain agency stipulations must be satisfied. Failure to have a permit is a violation of the law regardless of the lack of any impact on the environment. The requirements for obtaining the permit can change, without new legislation, as the conditions warrant. The PCB and, to some extent, the EPA determine what these conditions are, and as the quality of the environment changes, so may the discharge standards and the permit requirements.

One section that is especially germane to solid waste management is Section 9.c, which stipulates: ''No person shall cause or allow the open burning of refuse, conduct any salvage operation by open burning, or cause or allow the burning of refuse in any chamber not specifically designed for the purpose and approved by the Agency pursuant to regulations adopted by the Board under this Act. . . .'' This section tied the improper disposal of refuse to air pollution problems and essentially eliminates open burning.

Title III: Water pollution/Title IV: Public water supplies. These titles address the water environment and contain essentially the same provisions as Title II: Air Pollution, except for the reference to refuse disposal.

Title V: Land pollution and refuse disposal. This title addresses the issue of proper disposal of refuse, including hazardous wastes. It contains the general pronouncement of legislative intent plus several prohibitions regarding the handling of solid waste. In particular, Section 21.a stipulates: ''No person shall cause or allow the open dumping of any waste.'' Section 21.b requires that ''no person shall abandon, dump, or deposit any waste upon the public highways or other public property. . . .'' Finally, Section 21.c prohibits ''the abandonment of any vehicle in violation of the Abandoned Vehicles Amendment to the Illinois Vehicle Code.'' These three subsections identify specific problems that were presented to the legislature at the time the legislation was passed.

The substance of this title is contained in Section 21.d and f. Specifically, subsection d specifies: ''No person shall conduct any waste-storage, waste-treatment, waste-disposal, or special waste-transportation operation: 1. *without a permit* granted by the Agency or in violation of any condition imposed by such a permit . . . or 2. In violation of any regulations or standards adopted by the Board under this Act.'' Subsection 21.f addresses the management of hazardous waste. This has all of the requirements of the above plus additional requirements imposed by the nature of the material.

Section 22 permits the PCB to set standards for the location, design, construction, sanitation, operation, maintenance, and discontinuance of the operation of refuse collection and disposal, storage and treatment sites and facilities, and resource conservation and recovery sites and facilities; standards for the certification of personnel to operate refuse disposal facilities or sites; standards for the dumping of any refuse; and standards for the handling, storing, processing, transporting, and disposal of any hazardous waste. In addition, the PCB can set standards for record keeping, standards for equipment used, procedures for monitoring contaminant discharge, land pollution emergency alerts, closure and postclosure requirements, and the restriction on the disposal of hazardous wastes in sanitary landfills.

This title also contains considerably more stipulations regarding hazardous waste management and disposal. A number of these requirements for hazardous waste were added in amendments to the original act. In some cases the amendments were in response to federal legislative requirements.

Title VI: Noise. This title recognized that noise is also a means for environmental deterioration. The title is structured similar to the other titles.

Title VII: Regulations. This title establishes the operational philosophy for setting the environmental regulations. It gives the PCB the power to define the procedural rules for establishing a record for any regulations or enforcement action. It provides for public hearings and debate for enacting, amending, or repealing any substantive regulation. A substantive regulation refers to the ambient standards, emission standards, standards for issuance of permits, and so on. The purpose of this title is to establish uniform procedures and policies for setting regulations and enforcing environmental quality control.

Title VIII: Enforcement. Under this title the EPA is given the responsibility for investigating any complaint made by the PCB, a complaint filed by a citizen, or any other condition deemed worthy of investigation. The EPA is essentially defined as the policing agency. Certain procedures that must be followed by the PCB in an enforcement hearing are also defined in this title. After a duly constituted hearing before the PCB, orders to cease and desist may be issued. These orders may be enforced by monetary penalties. Other actions include the revocation of operating permits.

This title also recognizes that episode or emergency conditions may occur. When such conditions occur as specified in the PCB regulations, the EPA may so declare and take any action deemed necessary to protect the health and security of the population. This action may include sealing any vehicle, equipment, vessel, aircraft, or other facility that is in violation of such regulations.

Title IX: Variances. Variances from any PCB regulation may be granted when the regulation is found to cause undue hardship upon the discharger. A variance relieves the discharger of the responsibility for meeting the standards. There are usually time limits for the variance and there may be additional stipulations imposed. This title defines the intent of variances and sets some guidelines for issuance of variances.

Title X: Permits. This is the heart of the state mechanism for environmental quality control. The requirement that every discharger, operator of pollution control equipment, refuse disposal operation, and so on, must have a permit gives the state control over the manner in which the system functions. This control emanates from the technical standards and guidelines set by the EPA for the design, installation, and operation of any pollution control system or refuse disposal operation. This is in addition to the general environmental quality standards and the discharge standards set by the PCB. Any discharger of air or water or supplier of water must have a permit to function. Lack of a permit is a violation of the law and subject to penalty. The permit can be obtained only if the applicable standards and guidelines will be satisfied.

This title specifies a number of considerations necessary for awarding a permit. Additional permitting requirements are a result of the delegation of federal permits. One such delegation is the National Pollution Discharge Elimination System Permits (NPDES) required by the federal government for any point-source discharge to a surface water. The state has been delegated the responsibility for these permits, which define the discharge limits for an effluent. Other delegated permitting authority relates to the Resource Recovery and Conservation Act (RCRA) Permits for persons owning or operating a facility for the treatment, storage, or disposal of hazardous waste. As new federal permitting requirements are developed and delegated to the states, these activities would be covered under this title.

Section 39.2 of Title X is particularly relevant to refuse disposal. This section was added by the Illinois Legislature in 1982 and was known as Illinois Senate Bill 172 in recognition of its being introduced in the Illinois Senate. While the act was not specific for solid waste disposal facilities, the impact of this legislation on the development of new sites for sanitary landfills was devastating. Passage of this bill effectively stopped all new sanitary landfill construction in Illinois until 1989. The following are the relevant sections:

a. The county board of the county or the governing body of the municipality, as determined by paragraph (c) of Section 39 of this act, shall approve the site location suitability for such new regional pollution control facilities only in accordance with the following criteria:
   1. The facility is necessary to accommodate the waste needs of the area it is intended to serve.
   2. The facility is so designed, located, and proposed to be operated that the public health, safety, and welfare will be protected.
   3. The facility is located so as to minimize incompatibility with the character of the surrounding area and to minimize the effect on the value of the surrounding property.

4. The facility is located outside the boundary of the 100-year floodplain as determined by the Illinois Department of Transportation, or the site is flood-proofed to meet the standards and the requirements of the Illinois Department of Transportation and is approved by the department.
5. The plan of operation for the facility is designed to minimize the danger to the surrounding area from fires, spills, or other operational accidents.
6. The traffic patterns to and from the facility are designed as to minimize the impact on existing traffic flows.
7. If the facility will be treating, storing, or disposing of hazardous waste, an emergency response plan exists for the facility, which includes notification, containment, and evacuation procedures to be used in case of an accidental release.
8. If the facility is to be located in a county where the county board has adopted a solid waste management plan, the facility is consistent with the plan.
9. If the facility will be located within a regulated recharge area, any applicable requirements specified by the board for such areas have been met.

The section continues with specific requirements for filing of applications and the public hearings associated with the application for a regional facility permit.

This legislation provided the perfect vehicle for the neighborhood opponents of a site. It was possible to exert sufficient political pressure on the county board members, especially the one representing the district in question, that receiving site approval became almost impossible. The "not in my back yard" (NIMBY) syndrome dominated the decision-making process. Locating a site for a new regional pollution control facility has been taken out of the realm of economic, engineering, and scientific judgment and has become almost totally emotional and political. If a site does not have perfect economic, engineering, and scientific characteristics, it has absolutely no chance of being approved. Even if perfect from every technical perspective, approval is not assured.

Because of the difficulty in obtaining new sites for sanitary landfills, the Illinois legislature was considering (1990) the establishment of a statewide body that has the responsibility for the approval of any regional waste disposal site. This board would be similar to the state public utilities commission and would take control from the local politicians and place it in the board, which is more isolated from public pressure.

Title XI: Judicial review. Any appeals from PCB decisions are relegated directly to the appellate court for the district in which the cause for the action arose and not in the circuit court.

Title XII: Penalties. The general penalty statement is contained in Section 42.a and is: "Any person that violates any provisions of this Act or any regulations adopted by the Board, or any permit or term or condition thereof, or that violates any determination or order of the Board pursuant to this Act, shall be liable to a civil penalty not to exceed $10,000 for said violation and an additional penalty may not exceed $1,000 each day for which the violation continues.'' There are additional sections that increase the maximum penalty. For example, violation of the NPDES permit or terms or conditions thereof is liable for a civil penalty not to exceed $10,000 per day of violation. Violation of a RCRA permit carries a possible fine of $25,000 per day of violation.

Because of the seriousness of improper management of hazardous waste, the act sets some very stiff penalties. Section 44.c specifies that any person knowingly transporting, treating, storing, or disposing of hazardous waste without a permit, or who obtained the permit with falsified information, commits a class 4 felony and is liable to a fine of not more than $250,000, except that a defendant that is an organization is liable to a fine of $1,000,000. If these acts place any person or persons in imminent danger of death or serious bodily injury, a class 3 felony is committed. Felonies can result in a jail sentence.

In most cases, the penalties are sufficient to encourage compliance with the PCB orders. The ultimate course of action for those violations that carry a fine only and the fine fails to generate compliance is a court order requiring compliance. Failure to obey the court order can be considered contempt of court, which can result in a jail sentence.

Title XIII: Miscellaneous provisions. This title contains what could be considered ''housekeeping'' provisions. They do not fall under any of the previous categories, or refer only to actions taken during the transition from the existing agencies to the new agencies created by the act.

FEDERAL LEGISLATION FOR THE CONTROL OF SOLID WASTE

Federal legislation that has been used for pollution control dates back to the Rivers and Harbors Act in 1899. This was the first attempt to control the dumping of refuse into surface waters. Its Section 13, known as the Refuse Act, prohibited the discharge or deposit into navigable waters of any solid refuse. The act was designed to control the discharge of materials that interfered with shipping. This act had little or no impact on the quality of the environment as long as the refuse was not interfering with shipping. It was, however, very important in the initial attempts to control pollution during the 1950s and 1960s. It was an existing law and it was interpreted in a manner that assisted the pollution control agencies in control of some of the pollution problems before the needed legislation was passed.

The Clean Air Act Amendments and Solid Waste Disposal Act of 1965

The first legislation that specifically recognized the role of solid waste in environmental degradation was the 1965 amendments to the 1963 Clean Air Act. These revisions became known as the Solid Waste Disposal Act of 1965. This act was a result of a 1964 congressional subcommittee investigation into air pollution problems. The subcommittee report stated: "No more obvious and disgraceful illustration of the need for applied technology can be found than the appalling state in which the nation's waste disposal practices exist." The act specifically recognized that the disposal of solid waste in open burning dumps was a major contributor to air pollution. The "backyard" burn barrels and the low-technology incinerators used for on-site incineration were also making significant contributions to a degradation of the air quality. At this time, air pollution was the primary problem being associated with solid waste disposal.

This act called for a national research program to find and perfect methods of disposal that would avoid environmental contamination. Little or no information existed on the quantity and characteristics of solid waste. The larger cities had information on the tonnages that they processed through their incinerators and landfills. However, few data were available on the quantities collected by private industry, even in the metropolitan areas. This was compounded by the tens of thousands of disposal systems in small communities, some of which were city operated while others were private. The large private companies were not yet in existence, so the private sector was represented by literally thousands of small operators.

Funding was provided by this act to conduct studies that would define the problem. Surveys were conducted of the industry in an attempt to determine the quantity of refuse generated. What were the characteristics of the discarded material? Surveys of the manufacturing industry attempted to establish the quantity of material used in various consumer products such as glass containers, metal containers, paper and plastic packaging, and so on. Additional information was obtained on the current methods for disposal of the refuse. In particular, what was the environmental impact of refuse disposal? The results of these studies and surveys provide a much needed data base for future legislation and control strategies.

One fact that surfaced from these studies was the large quantity of urban waste material generated (estimated at about 125 million tons per year). Analysis of the composition of the refuse indicated that it was potentially a valuable resource being wasted. The paper, glass, and metals had potential for recovery and reuse. Also, the energy content of the refuse was significant, equivalent to about 50 million tons of coal per year. If techniques could be developed to use the refuse as a raw material, it would cease to be a waste. Consequently, the next piece of legislation emphasized resource recovery.

One additional concern surfaced during these investigations. The publication of *Silent Spring* by Rachel Carson in 1962 raised the consciousness of society concerning the hazardous and toxic chemicals being used. Data collected on waste materials generated by the production and use of these hazardous materials indicated that the quantity was significant. More important was the lack of information as to the ultimate disposition of these chemicals. The national concern for hazardous waste was in an embryonic stage at this time in the development of solid waste control legislation.

Resource Recovery Act of 1970

This legislation was an extension of the Solid Waste Disposal Act of 1965 designed to foster the development of techniques for reuse of the refuse constituents. The specter of a poisoned environment had been raised, and the sparsity of data on the management of hazardous materials was a clear signal that this area must be better defined. In addition to the general pronouncements, this act contained five provisions that were intended to address the solid waste management issues.

The first provision made available funds for the demonstration of new technology for energy and material recovery. Federal funding up to 75% of the capital costs was available for the construction of full-scale demonstration plants that recovered energy and/or materials from solid waste. It is very difficult to convince public officials to invest in new technology unless they can see a full-scale operating plant. These funds were to be used to provide the operating systems that would encourage the use of the new technology. The success of this funding was very limited. The only technology that was successfully demonstrated and utilized to any extent was refuse-derived fuel (RDF). The refuse is processed to remove a substantial portion of the noncombustible components. The combustible fraction is burned in a coal-fired central power generating plant.

Solid waste management was chaotic at all levels of government. The problem was poorly defined at the state and local level. Regulations were nonexistent and the states did not have any coordinated plan for addressing the problem. A second provision in this act funded studies for the development of local, state, and interstate solid waste management plans. These funds assisted the various governmental units in defining the problem and developing a plan for managing the refuse stream.

Availability of funds for capital improvements is frequently an impediment to the local governments in solving environmental problems. They may be near their bonding limit, or the voters will not approve bond issues for capital projects. At this period of time, the federal government was in the midst of a construction grants program for the construction of water pollution control plants. This section of the act provided grants to local agencies for the construction of resource recovery plants. However, money was never appropriated for these grants.

Because of the uncoordinated approach throughout the country and the lack of any uniform standards and guidelines for solid waste management systems, a section of the act was designed to support the development of guidelines for the collection, disposal, and recycle of solid waste. This charge was given to the Office of Solid Waste Management of the U.S. Public Health Service, the federal agency that was responsible for solid waste management at that time. It also provided funds for training programs for personnel involved in the design, operation, and maintenance of disposal systems.

Finally, the act recognized that there was a serious problem in the way that toxic and hazardous materials were managed. Because the issue was so poorly defined, the act concentrated on providing a mechanism for better defining the problem. This was the beginning of a still intensive effort to bring the hazardous waste problem under control. It is also the beginning of the confusion between solid waste management and hazardous waste management issues. This was, in part, a result of the indiscriminate dumping of hazardous waste along with regular non-hazardous refuse. The philosophy was "out of sight, out of mind," and landfills were very convenient places to achieve the "out of sight" component of this philosophy. Unfortunately, the second component did not apply, as these materials found their way back into the environment.

1976 Resource Conservation and Recovery Act (RCRA)

Public Law 94-580, signed into law on October 21, 1976 by the president, changed the nature of solid waste management dramatically. This legislation initiated the separation of hazardous waste from nonhazardous waste. It is essential to have an appreciation of the impact that the hazardous waste management legislation had on solid waste management. As will be seen, the regulation of hazardous waste was much more intense than normal solid waste. As long as the waste is nonhazardous, the regulation is less severe. Unfortunately, the indiscriminate mixing of hazardous and nonhazardous waste in prior years has left a legacy that will take many decades to eliminate.

Subtitle A contains the statements referring to the general intent of the legislation, but it also includes some salient points that shaped the response of the state regulatory agencies for many years into the future. One such statement indicated that one object of the legislation is the "protection of health and the environment . . . by prohibiting future open dumping on the land and requiring the conversion of existing open dumps to facilities which do not pose a danger to the environment or health." This statement was a directive to close all open dumps.

Following the statement of findings and objectives, Subtitle A of the new law also provides some new definitions. Vastly expanded over previous law, the definition of solid waste includes "any garbage, refuse, sludge from a waste treatment plant, water supply treatment plant, or air pollution control facility, or other discarded material, including solid, liquid, semisolid, or contained gaseous material resulting from industrial, commercial, mining, and agriculture activities. . . ." Only solid or dissolved material in domestic sewage, irrigation return flows, and otherwise regulated industrial discharges do not fall under this definition.

Also. it contained a much broader definition of disposal; "the discharge, deposit, injection, dumping, spilling, leaking, or placing of any solid waste or hazardous waste into or on any land or water so that such solid waste or hazardous waste or any constituent thereof may enter the environment or be emitted into the air or discharged into any water, including groundwater."

Subtitle A also requires the U.S. EPA to publish *suggested* solid waste management guidelines describing the "level of performance" that can be attained by various available solid waste management practices which protect public health and the environment. These guidelines must be published by October 21, 1977, and from time to time thereafter. It is interesting to note that some of the required guidelines were still in the draft stage in 1990. This indicated the complexity of the process. While the word *suggested* is used, the act makes it mandatory for these guidelines to be used by federal agencies. Also, use of these guidelines is a precondition for any federal funding to the states, so the guidelines are essentially mandatory.

Subtitle C addresses the topic of hazardous waste by providing the following definition: "A hazardous waste is a solid waste or combination of solid wastes, which, because of its quality, concentration, or chemical, physical, or infectious characteristics, may (A) cause, or significantly contribute to an increase in the mortality or an increase in serious irreversible, or incapacitating reversible, illness; or (B) pose a substantial present or potential hazard to human health or the environment when improperly treated, stored, transported, or disposed of, or otherwise managed." This was the first legislation that provided a detailed definition of what was to be considered hazardous waste.

The task of developing specific criteria for identifying hazardous waste was given to the U.S. EPA. As soon as these criteria were available, they were to publish a list of waste materials that were subject to federal regulation. Because of the requirements imposed on the management of hazardous waste as described below, it is essential that care be taken to exclude these materials completely from the normal solid waste stream.

For generators of hazardous waste, standards were to be (and have been) developed that cover record keeping, labeling practices, use of appropriate containers, chemical analysis, use of a manifest system, and reporting to regulatory authorities as to the quantities of hazardous waste generated and their disposition. The manifest system was designed to assure that all hazardous waste goes to a permitted treatment, storage, or disposal facility.

Standards were developed for transporters of hazardous waste to ensure that only properly labeled wastes are transported and only to permitted facilities designated on the manifest. Record-keeping requirements were imposed, and all regulations must be consistent with the Department of Transportation rules under the Hazardous Materials Transportation Act.

Facility owners and operators must have permits, and performance standards governing operating practices, contingency plans, and location, design, and construction of facilities were developed. Detailed record keeping and satisfactory reporting, monitoring, and inspection, as well as compliance with the manifest system were required. The owner is required to ensure maintenance of operation, which includes continuity of operation, personnel training, and financial responsibility.

The EPA was given the responsibility for developing the rules and regulations for implementation of this legislation. The deadline was April 21, 1978, but slippage in this date was expected and new deadlines were established to match the progress made by the EPA.

Subtitle D of the RCRA was designed to assist in developing and encouraging methods for the disposal of solid waste which are environmentally sound and which maximize the utilization of valuable resources and to encourage resource conservation. Such objectives are to be accomplished through federal technical and financial assistance to states or regional authorities for comprehensive planning pursuant to federal guidelines designed to foster cooperation among federal, state, and local governments and private industry. Guidelines developed by the EPA were to be available within 2 years.

Contained in this subtitle were references to six specific requirements that must be satisfied by the states in order to qualify for approval of funding from the federal government for the development of these plans.

1. The state must identify the responsibilities of state, regional, and local authorities relative to the distribution of federal funding among them and the means for coordination and implementation of regional planning.
2. The plan must prohibit the establishment of new open dumps within the state and require that all solid waste be processed for recovery or be disposed of in sanitary landfills or in some other environmentally sound manner.
3. The plan must provide for the closing or upgrading of all existing open dumps according to a predefined timetable, but not later than October 21, 1983.
4. The plan must provide for whatever regulatory powers the state may need to carry out the plan.
5. The plan must permit local governments to enter into long-term contracts for the supply of solid waste to resource recovery facilities.
6. The plan must provide for whatever waste management practices—resource recovery, sanitary landfill, and so on—may be necessary to use or dispose of solid waste in an environmentally sound manner.

The primary importance of these actions was the elimination of open dumping. However, it was necessary for rules and regulations to be developed that define a sanitary landfill relative to the design, construction, and operation. Many states developed such guidelines, but because of nonuniformity among states, the EPA developed what could be considered a minimum set of rules and regulations for sanitary landfills. Similar rules and regulations for other types of refuse processing system—in particular, incineration—are also the responsibility of the EPA. This is a dynamic process and the rules and regulations are generally in a state of flux as applied to new sources. Consequently, it is essential for the designer to remain current with the changing regulatory environment.

Congressional amendments to the RCRA in 1984 require that the EPA revise the Subtitle D criteria for facilities that may receive hazardous household waste (as most landfills do) or small-quantity-generator, a source of hazardous waste. The revisions must require groundwater monitoring to detect contamination, establish location standards for monitoring wells for new and existing facilities, and provide for corrective action. State programs must spell out performance standards to be met by leachate collection systems and landfill liners as well as requiring landfill operators to protect groundwater from the same list of hazardous constituents used in hazardous waste landfill regulations. New enforcement authority has been given to the federal government to ensure compliance with the regulations.

These 1984 amendments imposed significant new requirements for the design and operation of sanitary landfills. Although not as restrictive as the regulations for a hazardous waste site, the new regulations are much tighter than the regulations previously in effect. They require that the site must be designed to ensure no contamination of groundwater. Provision must be made for liners and leachate collection systems that will remove any leachate that is produced. Monitoring wells are required to ensure no groundwater contamination. The cost of refuse disposal by sanitary landfill will increase significantly.

The RCRA set the direction of much of the future legislation in this area. The focus was on hazardous waste. The Toxic Substance Control Act (TSCA) was passed in 1976 for the purpose of regulating the introduction of new toxic chemicals into the environment. This act was directed more toward the use of chemicals as pesticides, wood preservatives, and other uses that result in a wide environmental distribution. As such, it has little relationship to solid waste management except to further highlight the mood of the Congress at this time.

The discovery of the buried hazardous chemicals at the Love Canal Site in New York and the severe health problems that were attributed to this site simply added incentive for the Congress to tighten control even more. As the actual and projected costs for the cleanup of the Love Canal became known, Congress perceived a need to finance these costs from sources other than the federal treasury. In 1980, the Comprehensive Environmental Response, Compensation, and Liabilities Act (CERCLA) was passed. This legislation was designed to pass the cost of cleanup of the contaminated sites back to the producers of the waste. In cases where the origin of the hazardous waste can be determined, any site remediation costs can be recovered from the producer in proportion to the amount of waste contributed. Many companies were and still are recipients of substantial bills for site cleanup costs.

It was also clear that responsibility for some of the contaminated sites would never be assigned to any generator. In many cases there is no record of the source of the material, or the site has been abandoned and is in the public domain. In such cases a fund has been established, appropriately called "The Super Fund," that annually has several billion dollars available for orphaned site cleanup. These funds are provided by taxes on the chemical manufacturing industry.

The impact of CERCLA on conventional landfills is significant. If any known hazardous waste is disposed of by a landfill, it becomes the responsibility of the operator unless it is possible to trace the source of the waste. If a currently nonhazardous waste is accepted and this waste is added to the list of hazardous chemicals, the operator may be required to remove the waste from the site if it is found to be contaminating the environment. The liability associated even with small quantities of hazardous or potentially hazardous waste has had a major impact on the financial aspects of landfill disposal.

STATE RECYCLING LEGISLATION

As the difficulty in siting of sanitary landfills increased during the 1980s, means for reducing the flow of refuse into landfills were sought. Recycling of materials from the solid waste was one way to reduce the flow to the landfills. New Jersey was the leader in this area. In 1981, an Office of Recycling was established to encourage voluntary recycling. It was funded by a tax on tipping fees of $0.12 per cubic yard (approximately $0.36 per ton) which generated about $4.6 million per year. These funds were partially used for statewide educational programs on recycling, as well as providing grants and loans to communities in support of recycling activities. Forty-five percent of the surcharge was rebated to communities with active recycle programs, on a per-ton-recycled basis. In effect, the tax on landfills was used to subsidize recycling.

Although the voluntary program was successful, it was not diverting sufficient quantities of refuse. Consequently, in 1987, New Jersey passed a mandatory recycling law. It imposed an immediate ban on the disposal of leaves in a landfill. By March 1988, all counties had to recycle at least 15% of the previous year's total solid waste. By 1989, the total volume of waste recycled had to equal or exceed 25% of the urban solid waste stream. Partial funding of the program was obtained by a $1.50 per ton surcharge on refuse deposited in the state's landfills. One component of the refuse stream receiving special attention under these programs is yard and garden waste, especially leaves and grass clippings. This material accounts for the majority of the refuse that is recycled—or at least not placed in the landfill. Most of it is composted, and the compost is used for various purposes by the local agencies.

A number of states followed New Jersey's lead in the passage of various bills to encourage recycling. In 1986, Illinois passed the Solid Waste Management Act, imposing a fee of $0.20 per cubic yard of refuse deposited at a landfill. These funds were used for research and planning as well as innovative refuse management systems such as recycling. Public Act 85-1195, passed in 1988, changed the surcharge to $0.60 per cubic yard for the state and up to $0.45 per cubic yard for the municipality. In 1992, these rates are reversed, with $0.60 per cubic yard available to the municipality.

A companion bill, Public Act 85-1197, the Solid Waste Planning and Recycling Act, effective January 1, 1989, requires counties of more than 100,000 population and the city of Chicago to develop comprehensive solid waste management plans that emphasize recycling and alternatives to landfills by March 1, 1991. The plan must include a recycling program designed to result in 25% of the municipal refuse being recycled by the fifth year of the plan's adoption. This legislation also requires the elimination of all yard waste from landfills effective July 1, 1990.

MISCELLANEOUS LEGISLATION

State bottle bills that require as much $0.10 per container deposit for all beverage containers have had little or no effect on the quantity of solid waste generated. These laws have to some degree reduced the litter associated with the disposable container. That was the intent of the original legislation, but some have tried to use it to force a reduction in refuse generation.

At the local level, a host of ordinances are continually being passed in the hope that one can legislate away the refuse disposal problem. There is a reluctance for public officials to recognize that the demands of special interest groups for no landfills, incinerators, or other refuse disposal systems cannot be satisfied. It will not be possible to eliminate solid waste as long as society exists, especially with its consumer orientation.

Local ordinances are not likely to be effective and in some cases will simply shift business to adjacent municipalities. In 1988, Suffolk County in New York passed an ordinance prohibiting the use of certain plastics. After July 1, 1989, sale of food in packaging that is not biodegradable was prohibited. Banned items included foam "clamshells," foam meat trays, plastic grocery bags, and polystyrene and polyvinyl chloride containers. How successful this legislation will be remains to be seen.

STUDY QUESTIONS

1. What has been the historical relationship between public health and control of environmental pollution?
2. What is the common law doctrine of nuisance? How has it been used to control environmental pollution?
3. What are some of the major deficiencies related to the control of environmental pollution under the nuisance doctrine?
4. What is a statute law?
5. How does a legislative statute expedite control of environmental pollution?
6. What state agencies were created by the 1970 Illinois Environmental Protection Act, and what was the function of each agency?
7. How does a state requirement for a permit to construct and operate a pollution control facility facilitate control of environmental pollution?
8. What are the siting requirements specified under Section 39.2 of the Illinois Environmental Protection Act? How have these requirements affected the opening of new landfills?
9. What was the impact on solid waste management of the Solid Waste Disposal Act of 1965?
10. What was the most significant new issue introduced by the Resource Recovery Act of 1970?

11. What is the RCRA?
12. What is the objective of Subtitle C of the RCRA?
13. What is the objective of Subtitle D of the RCRA?
14. What is the potential impact of CERCLA on the operation of a sanitary landfill?
15. How has recycling legislation encouraged development of recycling programs at the local level?

CHAPTER 4

Sources and Characteristics of Urban Solid Wastes

The sources of urban solid waste have historically been consistent in that the refuse is generated from residential, commercial, and institutional establishments and from municipal services. Industrial solid waste is unique to a specific industry and is not normally considered as part of the urban refuse stream. If one includes industrial solid waste, there have been significant changes in the sources as new industries emerge and old industries are eliminated. It is not our intent in this chapter to discuss industrial solid waste because of the uniqueness of each waste stream.

Historically, the characteristics of refuse have not been consistent. There have been tremendous changes with time, and these changes are expected to continue. Refuse is a product of society and reflects the characteristics of the society: income level and economic health, consumer orientation, cultural persuasion, technological advancements, and social standards and aspirations. It is important to understand how refuse composition has changed with time so that one has an appreciation of potential future changes. The composition can be related to the various societal characteristics of the particular historical time.

HISTORICAL PERSPECTIVE

The composition of refuse in the United States at the turn of the century was indicative of the spartan life-style of the population. The refuse was truly rejected material. As shown in Table 4.1, combustion products (breeze, cinders, and ashes) accounted for over 60% of the reject stream. The balance was material of little value. This reflected two factors: the limited availability of consumer goods (i.e.,

Table 4.1 Refuse composition circa 1900

Constituent	Composition (%)
Breeze and cinders	50
Ash	12
Dirt and dust	20
Paper, straw, vegetable refuse	13
Miscellaneous (tins, 0.7%; metal, 0.2%; bottles, 1.5%)	5

throw-away items) and the economical need to use everything to the limit. Even the garbage was not wasted. It was fed to livestock, either at the residence or, in the cities, collected and used as food for swine. One finds these characteristics and practices today in many developing countries, where scavenging is an integral part of solid waste disposal, either at the source or at the disposal site.

Major changes occurred in the United States after the depression era of the 1930s and especially after the end of World War II. The economic prosperity of the country was greatly improved. Spin-off from technologies developed during the war provided numerous benefits to the consumer. Population density increases in the cities changed the life-style of the residents, all of which is reflected in the change in the refuse composition. The composition of the refuse from New York City clearly illustrates the change during this period of history (see Table 4.2). Of note is the large decrease in garbage and ashes and the large increase in paper and metal that occurred between 1939 and 1958.

CURRENT COMPOSITION OF URBAN SOLID WASTE

During the 1960s and 1970s numerous investigations were conducted in an attempt to determine the composition of urban refuse. When evaluating these studies, it appeared that each source of refuse was unique. This observation simply empha-

Table 4.2 Composition of New York City refuse

	Composition (%)	
Constituent	1939	1958
Garbage	17.0	4.8
Paper	21.9	56.5
Wood	2.6	0
Metal	6.8	14.8
Glass	5.5	5.7
Ashes	43.0	11.7
Miscellaneous	3.2	9.6

Table 4.3 Composition of typical urban refuse

	As-received composition (% wet wt)		
Constituent	1970	1984	2000
Paper and paper products	33.1	37.1	41.0
Garbage (food wastes)	11.5	8.1	6.8
Yard and garden waste	19.0	17.9	15.3
Metals	12.2	9.6	9.0
Glass	12.5	9.7	7.6
Plastic	2.7	7.2	9.8
Miscellaneous organic solids	8.3	8.4	8.4
Miscellaneous inorganic solids	1.7	1.9	2.1
Total (million tons)	110.2	133.0	158.8

Source: Waste Age, Oct. 1988, p. 46.

sized the variations among sources as well as seasonal variations in the composition of one source. It was clear that a "typical" solid waste probably did not exist. However, Table 4.3 presents the currently accepted composition of a typical urban solid waste as well as past and future changes in composition. This composition could be considered to represent an "average" for the refuse generated in the United States.

A different country, a different composition: The composition of refuse from a Canadian city and a U.S. city could be expected to be as similar as that between two U.S. cities. However, the refuse in other countries may be significantly different. Table 4.4 shows some of the variations that one may find among different industrialized countries. The data for all except Taipei represents the reported compositions in the late 1970s. The Taipei data were collected in 1982. Figure 4.1 shows the appearance of refuse from the United States [Figure 4.1(a)] and Spain [Figure 4.1(b)].

Table 4.4 Urban refuse composition of different countries

	Composition (%)			
Component	U.S.	U.K.	Madrid	Taipei
Paper	33	43	18	24
Putrescibles	31	17	50	29
Metal	12	9	4	5
Glass	12	9	3	11
Textiles	4	3	2	8
Plastics	3	5	4	12
Ash-fines	2	12	6	9
Miscellaneous	3	2	3	2
	100	100	100	100

(a) (b)

Figure 4.1 Physical appearance of refuse from the United States (a) and Spain (b).

The Madrid refuse has a low percentage of paper and glass. This is a result of the recycle ethic present in the Madrid population.[1] At this time the "throwaway" philosophy was not prevalent in Spain. Natural fibers are in short supply in Spain and much of Europe, so paper recycle is attractive. Also, reusable glass containers are used extensively for beverages. The high content of putrescible materials is a result of a cultural characteristic of marketing much of the food in fresh food markets. There is a preference for fresh fish, meat, and produce as compared to prepared foods.

The primary difference between the United States and the United Kingdom is in the ash component.[2] Solid fuels were used to a substantial degree for home heating. This fuel is gradually being replaced with gas and oil. The Taipei refuse has a relatively low paper content but a high plastic content.[3] In countries with a shortage of forests where the paper fibers are obtained from these forests or imported, plastic is cheaper than paper to use for packaging. Many of the other differences may be due to differences in the society or simply may be the inherent variability of these data.

The refuse in developing countries approximates the composition shown in Table 4.1 except for the composition of combustion products. This percentage would be low for tropical countries that do not use a solid fuel for space heating.

FACTORS CAUSING CHANGE IN SOLID WASTE COMPOSITION

There have been some significant events in the last 50 years that have had a major impact on the composition of urban refuse in the United States. These events were a result of either technological advances or legislative restrictions.

[1]ENADIMSA, Serrino 116, Madrid, Spain.

[2]Warren Springs Laboratory, Stevenage, U.K.

[3]Chung-Hsing Engineering, Inc. Taipei, Taiwan.

Technological Changes

Elimination of solid fuels for home heating. The ash content of refuse has decreased markedly since the end of World War II. To supply the large quantities of liquid fuel needed to power the machines of war, especially planes, tanks, and supply vehicles, the vast oil fields in the southwestern United States were developed. After the war, the production capacity of these fields was in excess of the peacetime needs for motor fuels, so the industry sought new markets for their products. In addition to the liquid fuel, vast quantities of methane (natural gas) were being produced. In many cases this gas was simply flared on site.

Recognizing the potential market for these fuels for space heating of all types of buildings, transportation systems, mainly pipelines, were developed to move the fuel to the population centers. The low price, the convenience of the automatic burners that were possible with both fuel oil and natural gas, and the lack of a residue resulted in a relatively rapid switch to these new fuels. This was particularly timely since there was a major construction boom in housing units for the millions of soldiers returning to civilian life and starting new family units. The loser in this market was the coal industry since it was the fuel being displaced. However, this competition kept the price of coal low. This encouraged the construction of electrical power generating stations with some of the electricity being used for space heating. In either case the ash residue was eliminated from the urban refuse stream.

The speed of this transformation is illustrated by the following numbers for New York State. These numbers represent the percentage of living units heated by coal. There was a more rapid transition in the urban areas during the decade of the 1950s, in part, because that was where most of the new home construction was occurring.

	1940	1950	1960
Urban	73.3%	39.1%	8.8%
Rural	81.0%	41.4%	17.4%
Statewide	74.6%	39.5%	9.9%

Similar examples can be found in other parts of the country. The trend has continued and today it is even difficult to buy coal for use in the home.

Packaging and marketing of food stuffs. This certainly is a major technological advance, encouraged by the general economic prosperity of this country and the marketing of consumer goods. This development affected many areas of the consumer market. Plastic film and metal foil were major contributors to the revolution in the food-packaging industry. The most common technique for food preservation in the 1940s and 1950s was canning in metal cans and glass jars. Prior to these changes, food was purchased either fresh or canned. Some dry food, such as flour, sugar, and salt, and some dried fruits and vegetables were available in paper packages.

Concomitant with the development of these new materials was the perfection of new processes for preservation of food (i.e., frozen foods, freeze-dried, and partially dehydrated). Various food items could now be prepared at a central plant and frozen for distribution to the consumer. The palatability of these foods was, in many cases, superior to the canned foods, and in some cases almost as good as fresh items. Freezing was a good technique for storing foods, so it was possible to have many foods all year that were generally not available as canned products, and therefore available only during certain periods of the year.

As this technology improved, the industry began marketing different types of prepared foods. Before long it was possible to buy a "TV dinner," essentially a complete meal that needed only to be thawed and heated in an oven. The development of the microwave oven made such meals even more convenient. Consequently, most of the food passing through the food markets has been processed to a greater or lesser extent. Portions of the food that would have become garbage in the household remains at the processing plant, where it becomes an industrial solid waste.

This technology affected the garbage component by eliminating all of the nonedible portion of the food. The pea pods, the carrot tops, the corn husks and cobs, and so on, remained at the processing plant. These materials were replaced with packaging materials such as paper, plastic film, and aluminum foil. The natural biodegradable food wastes were being replaced with a new synthetic material, plastic, and a metal product that was not degradable.

A majority of the food wastes that do reach the household are discharged into the sewer with a garbage grinder. This appliance is not present in all households, but most new residential units have them as part of the basic appliances. As a result of these technological advances, the decrease in the garbage content of solid waste indicated in Table 4.3 is understandable, as is the increase in plastic and paper. The garbage content will probably not decrease significantly in the future because of the relatively low percentage in the current refuse stream.

One example of the effect of new materials and consumer demands on the composition of refuse is the beverage container industry. Prior to 1950, essentially all of the beer and soft drinks were marketed in returnable glass bottles. During World War II, some of the beer was shipped overseas in metal cans. This was a special use of cans and it was not very popular in the domestic market. The development of the pop-top can was a breakthrough in gaining acceptance of cans for beverages. The can could be opened conveniently. Also, the nonreturnable bottle was being introduced at about the same time. There was definite competition between can and bottle manufacturers for this market.

The data in Table 4.5 show that the consumption of beverages increased from about 53 billion units in 1958 to near 80 billion in 1976. During this period, throwaway containers increased from 10 billion to almost 60 billion units, while returnable bottles remained essentially constant at about 2 billion units. From 1958 to 1976, the number of fillings per container decreased from 4.5 to 1.3. There has

Table 4.5 Effect of technology on beverage container use

	Containers used (million units)		
	1958	1965	1976
Nonreturnable			
Bottles	1,431	7,011	22,100
Cans	8,746	18,559	36,000
	10,177	25,570	58,100
Returnable bottles	1,628	2,499	1,660
Total containers	11,805	28,069	59,760
Total fillings	52,921	65,213	79,500
Containers/fillings	1/4.5	1/2.3	1/1.3

Source: Midwest Research Institute, *The Role of Packaging in Solid Waste Management—1966 to 1976* Publication SW-5c, Bureau of Solid Waste Management, Washington, DC, 1969.

been little change in this ratio in recent years due to the concern of a significant part of the population for resource conservation. This segment of the population continues to demand returnable containers, and the industry responds to this market as they do to any market.

In recent years, there has been a major decrease in disposable glass bottles and steel cans. Aluminum cans have assumed the dominant share of the market for disposable beverage containers, increasing from about 1% in 1964 to over 90% of the total number of containers in 1984. The aluminum can industry produced over 66 billion cans in 1985. Of course, it is a question of how long aluminum will retain this dominant position. Plastic containers have captured most of the market for large beverage containers (1 liter or greater), including milk, juices, and soft drinks. In 1982, approximately 21% of the soft-drink gallonage was sold in large plastic containers. If this trend continues, one may find aluminum and glass completely displaced from the beverage container market.

An additional complicating factor is the mandated recycling laws that have been passed in many states and will probably become nationwide in the near term. The ease of recycle may play a significant role in determining which material dominates. As discussed later, in 1988, about 50% of aluminum containers were being recycled. Plastic recycle was in the infant stage but being aggressively pursued by the industry.

Packaging and marketing of consumer goods. The marketing concept that resulted in large self-service stores required a new packaging philosophy. Individual items were selected from the display shelf by the shopper. Packaging was required to protect the item from damage or contamination by the customers. The item also needed pricing information for the checkout clerk, and depending on the size, may need to be packaged for security reasons. Consequently, a significant amount of packaging material, paper and plastic, as well as aluminum foil, is used for display of merchandise. When the purchase is completed, it is placed in one or more containers, adding more packaging material to be discarded.

Estimates of the quantities of packaging material contributing to the urban solid waste stream are given below.[4]

	1971	1980	1990
Metric tons	35.1	38.2	45.1
Percent of total	36.9	36.6	35.5

A significant increase in the quantity of discarded packaging material generated during each time period is apparent in the table. However, the percentage of the solid waste stream occupied by packaging materials remained essentially constant, or decreased slightly. During this time frame, there was almost a threefold increase in the amount of plastic in the waste stream. Since much of the plastic is used for packaging, it is reasonable to assume that the consumer activity increased substantially during this period. Lighter plastic material was substituted for the heavier paper products, which should have decreased the percentage by weight of the packaging material.

Communications (print) industry. This is a difficult area to quantify, but one that has had a significant impact on the quantity of solid waste. Newsprint is but one source of this paper. A vast number of mailings that bring the Madison Avenue message clog every mailbox. This advertising is heavily subsidized by the postal service through preferential mailing rates. The data in Table 4.6 reflect the expected quantities of paper used for different printing purposes.

Legislation: Air Pollution Control Laws

The legislation passed at the state and federal level for control of air pollution had a major impact on the composition as well as the quantities of solid waste generated in the urban areas. Few data were available on either the characteristics or the tonnage of refuse produced prior to the ban on burning. Therefore, one can only estimate the impact of this legislation. A typical refuse mix is about 25% ash and 25% water. If one were to assume that the combustion was 75% efficient in "burning" the 50% organic material and evaporating most of the water, only about 35 to 40% of the weight of the original refuse remained. This material would be about 65% inorganic (ash, glass, metals, ceramic goods, and dirt). This material has little value for recycle. It would not burn well in an incinerator and would only be suitable for landfill.

[4]*Waste Age*, Apr. 1980, p. 46.

Table 4.6 Paper used for printed communications

	Paper used (metric tons/year)		
	1971	1980	1990
Newspapers	6.62	6.47	9.11
Books, periodicals	4.43	5.25	6.96
Writing, other printing	2.40	3.52	5.71
Total paper	13.45	15.24	21.78
Percent of total refuse generated	14.1	14.6	17.1

Source: Waste Age, Apr. 1980, p. 46.

More significant is the increase in the quantity of material that has to be collected and hauled to the disposal site. After the "burn," the density of the refuse was probably about twice that of the unburned. With both a quantity reduction and density increase, the number of trips to the disposal site might be only one-fourth that required with unburned refuse. Therefore, the banning of open burning had a major impact on the number of vehicles required to collect the refuse and, consequently, the cost of the collection.

The apartment house incinerators in New York City that were causing major air pollution in the 1960s incinerated about 600,000 tons/year or about 2000 tons/day (6-day week) of refuse. Since these incinerators might be expected to destroy over 90% of the organic material, the density of the refuse in the collection vehicle would probably increase from 300 to 500 lb for unburned refuse to perhaps 1000 lb/yd^3 for the incinerator residue. If each vehicle could haul 30 yd^3 and make four trips per day to a disposal site, 10 trucks operating 6-day weeks could remove the residue from these apartments. Upon elimination of the incineration, the number of trucks required increases to about 75 to 80, depending on the density of the refuse in the truck. The cost of these air pollution control laws was not insignificant. Time is required to implement such changes due to the quantities of materials involved.

QUANTITIES OF URBAN SOLID WASTE GENERATED

The actual quantities of refuse produced nationally remains as elusive today as it was 20 years ago. Numerous estimates have been made and they have ranged from as little as 3 lb to as much as 10 lb per person per day. It is difficult to use numbers with such uncertainty. Therefore, it is always better to obtain actual tonnages for a specific project rather than trying to estimate the capacity requirements from such "average" per capita numbers.

Evaluation of the estimated per capita production rates provides some insight into the changing character of the refuse stream. It is interesting to see the projected increases in refuse production and the effect such things as recycling may have on future rates. It is also interesting to observe how consistent, or inconsistent, the projections have been over time. The 1968 National Survey of Community Solid Wastes Practices provided the first estimate of the national production rate at about 230 million tons per year. The per capita production was estimated at 5.2 lb per day. Subsequent studies conducted by various agencies, but funded by the U.S. EPA, produced different numbers.

One of the first updatings of the estimated quantities of refuse is presented in Table 4.7. The gross discards represent a product-by-product analysis of all major waste components. Using the industrial production figures, it is possible to obtain a better estimate of the amount of refuse a particular component will generate. For example, in the 1968 survey, the amount of paper in the solid waste stream significantly exceeded the production capacity of the paper mills. The total refuse production estimated with the latter technique was about half of the earlier estimates.

This evaluation also introduced the concept of resource recovery. With an allowance for recycling, the annual tonnage decreased to about 125 million for 1971. The substantial increase in refuse generation during the next 20 years, 133 to 225 million tons per year, was expected to be accompanied by a concomitant increase in the quantity recycled. As a result, the increase in the net refuse production is less and the per capita production increases only marginally. This projection was made prior to the rash of mandatory recycle legislation passed in the late 1980s. Therefore, the effect of recycling may be even more pronounced.

This report categorized the tonnage as to the general source. In 1971, the total estimated production was 125 million tons. This tonnage was allocated as follows: residential, 90 million tons; commercial, 23 million tons; and municipal, 12 million tons.

Refuse production was much greater in the United States than in other industrial countries. Data presented for the comparable time period (1971) show that the refuse generation rate in the United Kingdom was about 16.5 million tons per year, or about 1.5 lb per capita per day. Comparable data for Spain indicate a generation

Table 4.7 Estimates of solid wastes production

	1971	1980	1990
Gross discards			
10^6 tons/yr	133	175	225
lb/capita-day	3.52	4.28	5.00
Resource recovery			
10^6 tons/yr	8	19	58
lb/capita-yr	0.21	0.46	1.29
Net waste disposal			
10^6 tons/yr	125	156	167
lb/capita-yr	3.31	3.81	3.71

Source: Resource Recovery and Source Reduction, Second Annual Report to Congress, U.S. EPA Publication SW-122, Washington, DC, 1974.

Table 4.8 1980 Estimates of solid waste for disposal

	Waste (tons/yr)		
	1971	1980	1990
Rubbish	61.8	68.5	82.8
Garbage	20.6	22.1	24.3
Yard waste	22.5	27.6	33.0
	104.7	118.2	140.1

Source: International Research and Technology Corporation, *Forecasts of the Quantity and Composition of Solid Waste,* Final Report to the U.S. EPA, Washington, DC, 1980.

rate of 9.9 million tons per year, or 1.54 lb per capita per day. These countries did not have the same consumption level as that in the United States at this time and still are more conservative in the generation of refuse.

A report published in 1980 contains an analysis of a variety of waste materials. The categories comparable to the other studies are presented in Table 4.8. The total tonnage of what is generally considered urban refuse is a little less than that shown in Table 4.7. The tonnage figures in Table 4.3 represent the most recent estimates of the quantity of refuse generated. The more recent studies are indicating that the generation rate is less than initially estimated. Also, the two latter studies are relatively close in the predicted tonnage. This probably is a result of better estimating techniques and a better data base.

VARIATIONS IN THE QUANTITY AND CHARACTERISTICS OF SOLID WASTE

The discussion above identified the changes in refuse quantities and characteristics that are long term and more global in character. Additional variations are possible and must be considered when attempting to determine the refuse quantity and characteristics for a specific project. These variations can be categorized into three specific areas.

Seasonal Variations

Perhaps the most significant variation associated with seasonal factors is the amount of yard and garden waste. The onset of spring brings a sharp increase in the quantity of grass clippings from low-density residential areas. The quantities are highly dependent on the area of the community, specifically the yard area per living unit. Few data have been gathered on this specific source. As more states restrict the disposal of this material in the sanitary landfills, data on the quantities will become available because of the separate collection. In the fall, the grass clippings give way to leaves, which are also being excluded from landfills. Again, the quantities are determined by the number and types of trees present in the community. The burning of the leaves is prohibited in almost all metropolitan areas, so alternative means must be provided for collection and disposal.

Additional seasonal variations are less dramatic. During the Christmas season, there is a surge in consumer purchasing. This generally results in an increase in the quantity of packaging material discarded. After Christmas, the residue from the holiday can be significant, especially if one includes the millions of Christmas trees that are discarded.

Any refuse processing and disposal system must be designed to cope with the changing quantities of refuse as well as the changing characteristics. If disposal is by landfill, the primary question is one of capacity. With the restriction on the disposal of yard and garden refuse in the landfill, this problem is minimal. However, for an incineration process, the increase in the wet grass has a major impact on the energy content of the refuse and may require an auxiliary fuel to maintain the combustion temperatures.

The collection of refuse is also affected greatly by these variations. Collection routes are generally established so that the vehicle and crew can service it in about 7 to 8 hours. When significant increases in the quantity of refuse occur, additional vehicles and crews are needed, or overtime work is required. The added trips to the disposal site needed to handle the additional refuse can greatly increase the cost of operation. Special collection routes needed to collect the seasonal production of yard and garden waste require an added number of vehicles and crews for only part of the year. All of these variables significantly complicate the development of the collection, processing, and disposal system.

Regional Variations

There are obvious differences in the seasonal variation in different regions of the country. For example, Florida is a semitropical climate that does not exhibit the variability in yard and garden waste. In arid regions, the amount of yard and garden waste would be considerably less than would be found in regions with high rainfall. The effect of climate on this waste category is rather obvious. Moisture is another characteristic that may be influenced by regional climate. In areas with high precipitation, the moisture content of the refuse will be significantly higher than in arid regions. This would be a combination of more yard and garden waste as well as more precipitation to wet the refuse. In fact, the low humidity common in the arid regions may significantly reduce the moisture originally present in the refuse.

There may be other regional variations, but these are generally not major. The nature of the consuming public is rather uniform throughout the country. Certain marketing factors may affect some components. For example, if all beverages are marketed in disposable bottles rather than aluminum cans, the glass and aluminum content of the refuse would be substantially different. However, these regional mar-

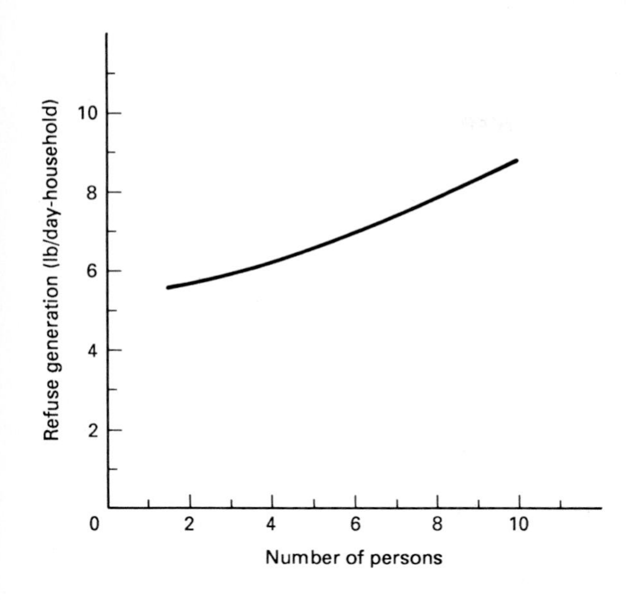

Figure 4.2 Refuse generation per household as a function of the number of residents. (From *Waste Age*, Apr. 1976, p. 35.)

kets are not as pronounced as they have been in the past. Legislation at the state level could have an impact. For example, a state that has a large plastic lobby may be encouraged to tax or ban all containers not made of recyclable plastic. Because of the monopolistic nature of such a law, it would probably not stand a court challenge. However, there is always the possibility that similar legislation will be passed that could have a significant impact on the composition of refuse.

Variations among Individual Households

The previous discussions did not address the variations that can be expected among households in the same community. According to the results of a study of refuse production from households of different sizes and characteristics, the primary factor influencing this variation is the number of persons per household.[5] Figure 4.2 presents the daily waste production per household as a function of the number of people. The total production increases from about 5.5 lb/day for two people to 8.8 lb/day for 10 residents. The increase is not in proportion to the increase in the number of residents.

This is not unexpected since some items are closely related to the individual consumer, while others are closely related to the household unit. The containers for food and beverages are more dependent on the number of people, while newspapers are more dependent on the unit. Also, there is an economy of scale for food and beverage items. A large family will probably purchase items in large containers.

[5] *Waste Age*, Apr. 1976, p. 29.

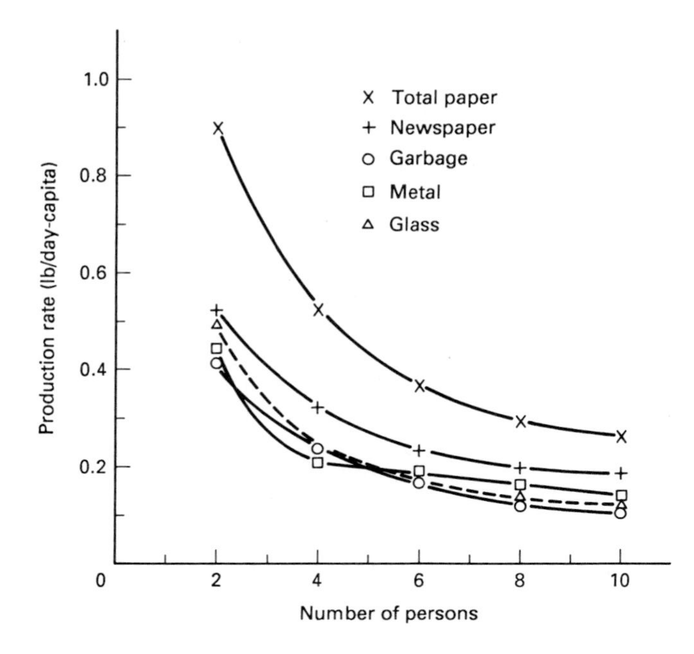

Figure 4.3 Refuse component generation per capita as a function of the number of residents. (From *Waste Age*, Apr. 1976, p. 39)

These relationships are illustrated in Figure 4.3, which shows the daily per capita production of different refuse components.

The curves show that the per capita production of most components are highly dependent on the number of persons until the number reaches four to five. All curves, except for paper and garbage, flatten to indicate no relationship with number of residents. The per capita production of paper decreases significantly with increasing size of unit. Again, one needs only one newspaper per household. This is true for many print items from bills to magazines.

The true impact of the results of this study is illustrated in Figure 4.4. The per capita production of refuse decreases from approximately 2.6 lb/day to about 0.9 lb/day when the number of persons increases from two to 10. This threefold decrease in per capita production will significantly reduce the quantity of refuse expected from housing units with large families. Except for extremely low income levels, these production rates appear to be independent of the income level of the family. Curves showing the per capita production of refuse as a function of dwelling size are almost horizontal. The number of family members has a much greater impact on refuse production.

PHYSICAL CHARACTERISTICS OF SOLID WASTE

The two most important physical characteristics of refuse are density and moisture content. The two are interrelated but have two totally different considerations in refuse management systems.

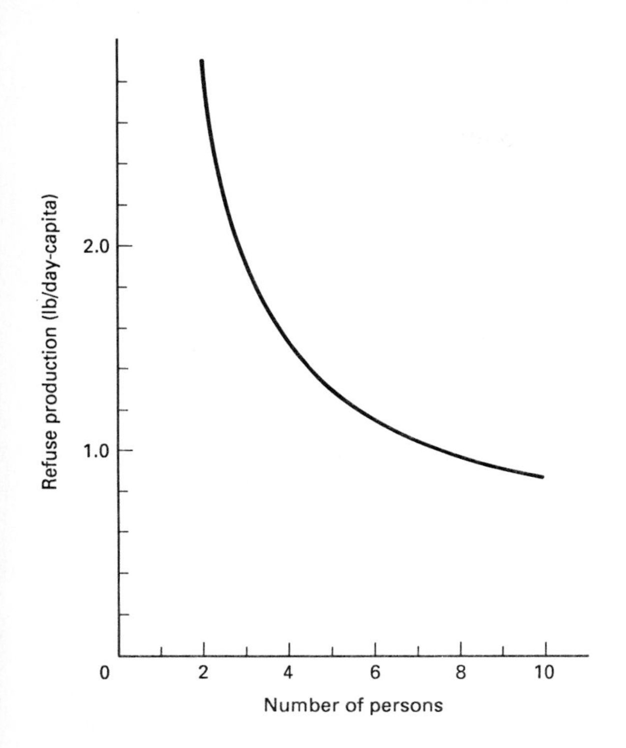

Figure 4.4 Daily per capita refuse production as a function of the number of persons per household. (From *Waste Age*, Apr. 1976, p. 38)

Density

Refuse as produced and deposited in the trash can has a density of perhaps 200 to 300 lb/yd^3. This low density is primarily a result of the shape of the material in the waste stream. A glass bottle will float on water even though the specific gravity of glass may be 2.5. The air trapped in the bottle reduces the overall specific gravity to less than 1. This is true for the milk cartons, the boxes, and a host of other objects present in the refuse. If the glass bottles are broken, the metal cans smashed, and the boxes collapsed, the density will increase. As the proportion of glass, ceramics, ashes, grit, and metals increase, so will the density. Therefore, the density of any refuse stream will be determined by the composition and by the degree of shape alteration (compaction).

Moisture also increases the density of refuse until the material becomes saturated with water. As the air in the voids is replaced with water, the density will increase. Additional water may actually decrease the density by displacing the solids, which have a greater density than water. To achieve high densities with a wet refuse, it is necessary to expel the excess water.

Reduction in volume is an important consideration in the management of refuse, both in the collection and transport and in the final disposal. For example, the volume (V_i) occupied by 1 ton of refuse at a density of 200 lb/yd^3 is 10 yd^3. This means that a truck with a 30-yd^3 capacity could only haul three tons of refuse.

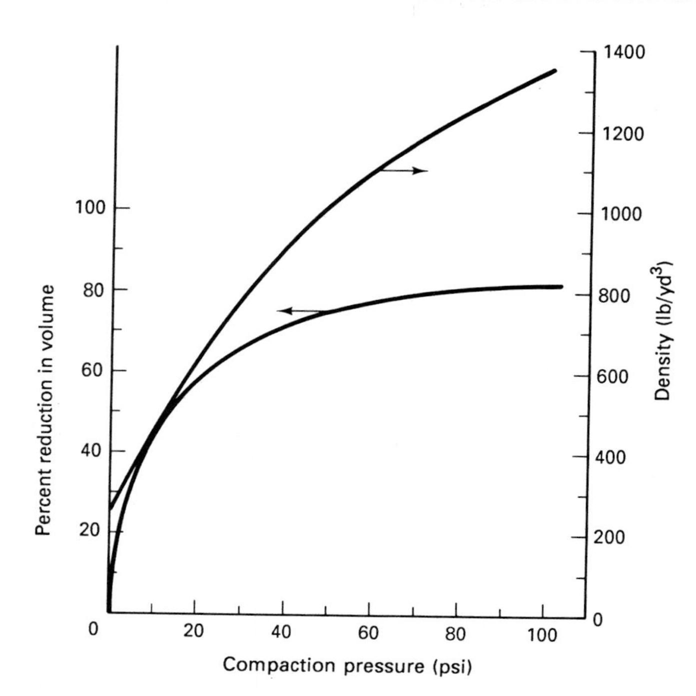

Figure 4.5 Effect of compressive forces on the percent volume reduction.

With a volume reduction of 50%, the density becomes 400 lb/yd^3. This same truck can haul 6 tons of refuse. If the density is increased to 1000 lb/yd^3, the volume (V_f) occupied by 1 ton is 2 yd^3. From equation (1), the PRV is 80%:

$$\text{percent reduction in volume (PRV)} = 100\left(1 - \frac{V_f}{V_i}\right) \tag{1}$$

Since the trucks are size limited rather than weight limited, the impact of such a reduction in volume on the cost of collection and hauling of refuse is obvious. Unfortunately, these volume reductions require a substantial compressive force.

Figure 4.5 presents data collected on the compaction of refuse. Internal pressures of 100 psi were required to achieve the 80+% volume reduction. It is possible to exert that level of pressure on the face of a hydraulic ram, but it is not practical to design a truck body to contain an internal pressure of 100 psi. Consider the load that would be placed on a flat surface with dimensions of 8 ft by 20 ft and a pressure of 100 psi, over 2,000,000 lb. It would require a very well supported truck body to contain such a load. Most collection vehicles strive to achieve a PRV of 50%. The density of the refuse is increased from approximately 250 lb/yd^3 to 500 lb/yd^3 by the normal compaction equipment on the collection vehicle.

High-pressure compaction has been investigated for increasing the density of refuse for long-distance transport. This compaction is accomplished with stationary balers that can exert great pressures on a ram that forces the refuse into a restricted

Table 4.9 High-pressure compaction of refuse

Ram pressure (psi)	Density (lb/yd^3) During stroke	After stroke
500	1620	1080
1000	2000	1380
1500	2210	1580
2500	2400	1600
3500	2490	1750

area. The refuse is strapped to prevent rebound. It is possible to increase the density significantly over the noncompacted refuse. Typical densities achieved with baling are shown in Table 4.9. It is necessary to restrain the refuse after the pressure is released. Otherwise, a rebound effect occurs and the density decreases. Operation at the very high pressures actually expelled water from the refuse. Water has a density of 1685 lb/yd^3. The only way to achieve densities as high as 2400 lb/yd^3 would be to eliminate the water. These data suggest that one approaches an area of diminishing return when the ram pressure reaches 1500 psi. It is very difficult to retain a density in excess of 1500 lb/yd^3 even in baled refuse.

The degree of compaction obtained when refuse is placed in a sanitary landfill is a critical operating parameter for the landfill. Volume is the determining factor for efficiency of landfill utilization. Efficient operation of the site requires placing the refuse and compacting it to an optimum density. Operational requirements for landfills are discussed in a later chapter. In the context of density considerations, the data in Table 4.10 are presented.

These data resulted from a full-scale test in which landfill cells were filled with refuse. A Caterpillar D-9 was used to spread and compact the refuse. Residential refuse was used and had the following dry weight composition; paper and paper products, 51%; grass and yard rubbish, 36%; inert materials, 11%; and garbage, 2%. The material was dumped into the cell and spread in layers at a depth of 1.0 to 1.5 ft. As each layer was placed, a track of the D-9 passed over it four times. The

Table 4.10 Refuse compaction in a sanitary landfill

Moisture Content (%)	Density (lb/yd^3) Before cover		After cover	
	Wet	Dry	Wet	Dry
23.6	1052	804	1209	924
35.4	1094	706	1186	765
53.2	1199	561	1318	605

Source: Public Works, May 1969, p. 111.

volume of the cell and the wet and dry weight placed in the cell were measured in the field. To determine the effect of moisture on the compaction process, the moisture content of the refuse was adjusted prior to placement.

Several observations can be made from these data. First, the effect of increasing the moisture content of the refuse is detrimental. The wet density increase is small, and the dry density decreases significantly at the higher moisture levels. In effect, the landfill is being filled with water rather than solids. Second, the soil cover plays an important role in containing the refuse. In this case, 2 ft of soil was placed on the top of the cell. The imposed load from the soil (about 300 lb/ft^2) prevented the refuse from rebounding, resulting in a measurable increase in both wet and dry density. Finally, there is a limit to the increase in density that can be achieved. The maximum density based on the dry solids was slightly more than 900 lb/yd^3. With this careful operation, the maximum density of as-received refuse was only 1300 lb/yd^3. One can use a conservative estimate of in-place density for refuse in a sanitary landfill of about 1000 lb/yd^3 "as received." If the density of the refuse is increased from about 250 lb/yd^3 at the source to 1000 lb/yd^3 in the landfill, the PRV is approximately 75%.

Moisture Content

The moisture content of refuse is defined as the ratio of the weight of water to the weight of the wet refuse. The following equation can be used to calculate the percent moisture of a sample:

$$\% \text{ moisture} = \frac{\text{wet weight} - \text{dry weight}}{\text{wet weight}} \tag{2}$$

A known quantity of wet sample is dried in a drying oven to a constant weight. The source of this moisture is either from the components of the refuse (i.e., garbage and yard wastes) or from precipitation. A typical range for moisture content is 20 to 40%. The lower represents refuse from an arid region, while the higher would be for a region with high precipitation. These are only average, and values greater than 40% are not uncommon.

The moisture in the refuse affects the weight of the material handled. Refuse with a dry weight density of 250 lb/yd^3 is capable of absorbing a considerable weight of water before additional volume is required. The water simply fills the void spaces. If this refuse has a moisture content of 25%, the wet density increases to 333 lb/yd^3. At 50% moisture, the wet density is 500 lb/yd^3. Water simply adds weight to the solid waste, and the cost to collect and transport increases. It is desirable to eliminate extraneous water from contacting the dry refuse.

Moisture is very important in refuse processing that involves thermal systems. The water must be evaporated before a thermal process can proceed. The energy penalty can be a significant factor in the economic success of these processes. The heat loss associated with the evaporation of water and the heat associated with the

Table 4.11 Enthalpy of water and water vapor

	Enthalpy (Btu/lb)	
Temperature (°F)	Liquid	Vapor
32	0	1076
77	47	1096
212	180	1150
600	617	1165

Source: R. H. Perry and C. H. Chilton, Eds., *Chemical Engineers Handbook,* 5th ed., McGraw-Hill, New York, 1973, (3) p. 110.

elevated temperature of the water vapor is significant when the moisture content is greater than 15 to 20%. In Table 4.11, the data on the enthalpy value of liquid water and water vapor at various temperatures show how much energy can be associated with the latent heat of water.

The latent heat of water is temperature dependent, but since the evaporation in a thermal process generally occurs at temperatures below 200°F, the heat of vaporization is approximately 1050 Btu/lb. The heat capacity of liquid water (the change in enthalpy at constant pressure C_p) is 1 Btu/lb-°F, while water vapor has a C_p of approximately 0.5 Btu/lb-°F. Consequently, the mass of water evaporated and heated to some elevated temperature is a major determinant in the energy balance for a thermal process.

CHEMICAL CHARACTERISTICS OF SOLID WASTE

There are several ways to characterize the chemical composition of refuse, including the general classes of chemical compounds, elemental analysis (ultimate analysis), or proximate analysis used for combustion system evaluation.

General Classes of Chemical Compounds

Classification provides some insight to the type of material contributing to solid waste. Knowledge of the general classes of compounds and their characteristics enables one to have a better understanding of the behavior of refuse.

Lipids. These compounds are commonly called fats, oils, and grease. Refuse will contain approximately 8 to 10% lipids on a dry weight basis. The primary sources of lipids are garbage, cooking oils, and fats. Because of the chemical structure of lipids, they have a high energy value, about 16,000 to 17,000 Btu/lb. A refuse with a high lipid content is a good candidate for an energy recovery process. Lipids also become fluid at temperatures slightly above ambient. This can add to the liquid content of the refuse and cause a change in the physical properties of the refuse due to wetting of the paper and paper products. Although lipids are biodegradable, they have a low solubility in water, which substantially reduces the rate of biodegradation.

Carbohydrates. Carbohydrates, hydrated carbon, have the general formula $(CH_2O)_x$. This category includes a variety of sugars and polymers of sugars, such as starch and cellulose, which are both polymers of glucose. One additional factor may be included to subdivide these compounds [i.e., the resistance of the polymer to hydrolysis (breaking into the single monomer of sugar)]. If one uses this classification, carbohydrates include only the sugars and starch. The starch polymers hydrolyze easily to glucose and the sugars are soluble in water and readily biodegraded. They also attract vermin since they are a good food source for flies and rats. The sources of carbohydrates are primarily garbage and yard wastes. Sugars account for 4 to 6%, and starch accounts for 8 to 12% of the dry weight of solid wastes.

Crude fibers. This category includes the natural fibers, which are resistant to degradation. The major polymers are cellulose and lignin. Cellulose is a very large polymer of glucose. Lignin is composed of a number of monomers with benzene rings being the primary monomer. Cellulose and lignin occur together in many fibers and result in material that is highly resistant to biodegradation. This is the source of the humus material found in nature. Natural fibers found in paper products, garbage, and yard waste are the major source of these polymers. Cellulose may account for 25 to 30% of the dry weight of refuse, while lignin may be 8 to 10%.

Proteins. Proteins are nitrogenous compounds that consist of an organic acid (R-COOH) with a substituted amine (NH_2) group. Garbage and yard wastes are a source of proteins, which comprise about 5 to 10% of the dry solids in refuse. Proteins provide a valuable source of nutrients for the biodegradation of solid waste. The refuse is deficient in nitrogen, and rapid biodegradation requires additional nitrogen for the microbial growth. Also, partial decomposition of proteins can result in the production of amines, which have very intense odors. Common names for some of these amines are "putrescine" and "cadaverine" and are indicative of the association they carry.

Synthetic organic materials. The remaining organic material is composed of synthetic compounds that are classified as plastic. The textiles made from natural fibers, cotton and wool, would be classified as crude fibers, as would leather. Synthetic materials have recently become a significant component of solid waste, accounting for 5 to 7%. The resistance to natural degradation makes these materials of special concern, especially when placed in an environment where natural processes are expected to destroy the material.

The regulatory pressure on the plastics industry has spurred the development of polymers that are either photodegradable or biodegradable. Several photodegradable polymers are currently used in special cases. An ethylene–carbon monoxide copolymer has been developed for the manufacture of plastic six-pack carriers. When exposed to ultraviolet light, the polymers disintegrate. Other photodegradable plastics combine copolymers of ethylene–propylene with ketone-containing monomers to increase the presence of a light-sensitive group that will interact with ultraviolet light. Of course, these plastics must be exposed to light before they degrade.

Biodegradable plastic contains a starch-based additive along with the normal synthetic monomers. The polymers that form contain a certain fraction of natural starch polymers that is biodegradable. As the microorganisms attack the starch, the polymer structure of the plastic deteriorates. With certain modifications, these synthetic monomers themselves can be biodegraded.

There is considerable resistance to use of degradable plastics. One of the primary attributes of plastic is its stability under many different environmental conditions. Degradable plastic would be defeating the purpose of the plastic. The presence of these plastics in the waste stream may complicate the recycling effort. It will be difficult to separate the degradable and nondegradable polymers. If plastic recycling is fully developed, much of the incentive for degradable plastics will be eliminated. It is probable that they will not become a significant component of the refuse stream.

Another factor of interest is the high energy content of the plastic, which is in the range of 23,000 Btu/lb. It is an excellent fuel, and an increase in plastic content will increase the potential for energy recovery from solid waste. A negative factor associated with the combustion of plastic is that some of the plastic is polyvinyl chloride, a source of chlorine, which has been associated with the formation of dioxin, and a source of acid gas, which increases corrosion in the combustion system.

Noncombustibles. The balance of the material is considered to be noncombustible: glass and ceramics, metals, dust and dirt, and ashes. This category may be in the range of 12 to 25% of the dry solids. It is essentially the residue that remains after combustion.

Ultimate Analysis

The ultimate analysis is an elemental analysis that determines the percentage of each element present in the refuse. This analysis is necessary for conducting any mass balance calculations for a chemical or thermal process. The five primary elements and their percentage in the refuse are shown in Table 4.12. Carbon, hydrogen, and oxygen constitute the majority of the mass of material in refuse. The low sulfur content is important when considering a combustion process. There is little concern with air pollution from sulfur oxides. The ash fraction contains the residual

Table 4.12 Ultimate analysis of solid waste

Element	Range (% dry wt)
Carbon	25–35
Hydrogen	2.5–6
Oxygen	15–30
Nitrogen	0.25–1.2
Sulfur	0.02–0.12
Ash	12–30

from combustion of the organic material as well as the original inorganic components in the refuse. Because of its composition, this ash can have significant environmental consequences. Where does the ash go—up the stack or to a landfill? The glass and ceramic materials and the metal containers account for most of the ash fraction. If an efficient resource recovery system is used to process the refuse prior to combustion, the magnitude of the ash fraction can be reduced to less than 10% of the ash from a conventional incineration system.

Table 4.13 shows some typical values for the toxic metal content of the ash residue from combustion of raw refuse and of refuse-derived fuel (RDF). The bulk inert materials, such as cans and bottles, are excluded from the fine ash produced by an incinerator. If the weight of these materials are also considered as part of the ash stream, the weight percent of the metals in Table 4.13 for incineration would be reduced by a factor of 3. The percentage of heavy metals in the ash from RDF combustion is higher. Many of these heavy metals are present in the organic components (plastic, leather, rubber, and paper) of the refuse. However, remember that much less ash is produced from the combustion of RDF.

The heavy metals present in the ash pose a significant disposal problem. The mobility of these metals may cause the ash to be declared a hazardous waste, requiring a special ash landfill. The U.S. EPA has specified a standard procedure for determining the leachability of these metals. These ash fractions also contain sig-

Table 4.13. Metal content of ash from refuse combustion

	Fine ash (lb/ton)	
Metal	Incineration	RDF
Cadmium	0.186	0.614
Chromium	1.396	3.574
Mercury	0.024	0.102
Nickel	1.162	1.022
Lead	14.418	21.464
Tin	2.094	1.022
Zinc	23.256	51.104

Source: Waste Age, Sept. 1978, p. 51.

Table 4.14. Proximate analysis of solid waste

	Solid waste (wt %)	
	Range	Typical
Moisture	15–40	20
Volatile matter	40–60	53
Fixed carbon	5–12	7
Ash	15–30	20

nificant quantities of other metals, such as iron, manganese, calcium, magnesium, and sodium. These metals are not toxic and do not present the same problem for disposal.

Proximate Analysis

The proximate analysis shown in Table 4.14 is of importance in the evaluation of the combustion properties of a fuel. There are four principal categories: moisture, ash, volatile matter, and fixed carbon. Moisture adds weight to the fuel without having any heating value, and the evaporation of water will reduce the heat release from the fuel. Therefore, the higher the moisture content, the lower the value of the material as a fuel. Ash has a similar effect. It adds weight without generating any heat during combustion. The heat loss with the ash is small, due to the low heat capacity of the material (about 0.2 Btu/lb-°F) and the exit temperature of the ash from the furnace.

The volatile matter and the fixed carbon define the combustion characteristics of the fuel. The volatile matter represents the portion of the fuel that is converted to gases when the temperature increases. This gasification will occur even before combustion is initiated. The gases are transported to the secondary combustion chamber, where rapid combustion of a fuel gas occurs. The chemical reaction between the fuel and oxygen occurs in a molecular dispersion. Heat release is rapid and combustion is complete in a very short time. Fixed carbon represents the carbon that remains on the furnace grates essentially as charcoal. Combustion occurs on the surface of the solids and the combustion rate is controlled by the surface area of the fuel. Consequently, a fuel with a high percentage of fixed carbon will require a longer retention time on the furnace grates to achieve complete combustion than will a fuel with a low percentage of fixed carbon.

High Heating–Low Heating Value

A knowledge of the heating value of any material is critical for the evaluation of its potential for use as a fuel in a combustion system. This parameter is a function of the composition of the refuse, the percentage of materials that have higher Btu values, such as plastics and lipids. The Btu value is determined experimentally using the bomb calorimeter test. This test measures the heat release at a constant temperature of 25°C (77°F) from the combustion of a dry sample. This is the standard reference temperature for all heat balance calculations. The following chemical equation represents the reaction occurring during combustion of a fuel, say cellulose:

$$(C_6H_{10}O_5)_n + 6nO_2 \rightarrow 6nCO_2 + 5nH_2O \tag{3}$$

One product of this reaction is the water that results from the oxidation of hydrogen. At the standard test temperature, this combustion water remains in the liquid state. This condition produces the maximum heat release and is defined as the *high heating value* (HHV). This is always the value obtained from the bomb calorimeter test. In equation (3), 162 lb (1 lb-mol) of cellulose will produce 90 lb (5 lb-mole) of water during combustion. For every pound of cellulose burned, 0.56 lb of combustion water are produced. The quantity of combustion water depends on the hydrogen content of the fuel.

However, during combustion, the temperature of the combustion gases remain above 212°F until discharged to the atmosphere. Consequently, the water is always in the vapor state when in the combustion system. The heating value associated with this condition is termed the *low heating value* (LHV). The following equation shows the relationship between HHV and LHV:

$$\text{LHV} = \text{HHV} - (\Delta H_v \times 9H) \tag{4}$$

ΔH_v is the heat of vaporization of water, and H is the hydrogen content. Each pound of hydrogen will produce 9 lb of water (18/2).

This water is only that resulting from the combustion process. If the fuel has moisture associated with it, this free water must also be evaporated. The energy required may be substantial and may result in a very inefficient combustion process. Three factors must be remembered when evaluating refuse as a potential fuel.

1. Only dry organic material yields energy.
2. The ash content of the fuel reduces the proportion of dry organic material per pound of fuel and retains some heat when removed from the furnace.
3. The moisture contained as free water in the refuse reduces the amount of dry organic material per pound of fuel and requires a significant amount of energy for evaporation.

The heating value of refuse on an as-received basis can be greatly reduced as the ash and moisture content increases. "As received" is a term used to indicate that the refuse has received no processing but is in the same condition as when it was discharged from the collection vehicle. This topic will be discussed again in greater detail when considering heat balances on combustion systems.

STUDY QUESTIONS

1. What are the six major components in present-day solid waste? How are these components projected to change by the year 2000?
2. What factors are responsible for the variations in refuse composition in different countries?
3. What significant technological changes in the past 50 years have influenced the composition of refuse?
4. What effects have environmental legislation during the 1960s and 1970s had on the quantities and characteristics of refuse?
5. How is the per capita production of refuse changed during the past 20 years, and what has been the effect of resource recovery on this production rate?
6. What seasonal and regional variations in refuse composition can be expected?
7. What is the significance of density in solid waste management systems?
8. What is the role of water in refuse-processing systems?
9. What is ''ultimate analysis''? What is a typical elemental composition of refuse?
10. What toxic metals are commonly found in ash from refuse combustion systems?
11. What is the meaning of ''proximate analysis''? What is a typical proximate analysis of refuse?
12. If the high heating value of methane (CH_4) is 382,000 Btu/lb-mol, what is the low heating value?
13. What is the significance of free moisture and combustion water in the operation of a thermal refuse processing system?

CHAPTER 5

Recycling

As discussed in Chapter 3, the thrust of much of the state and local legislative action in the mid-1980s was directed toward recycling of various components of solid waste. A dictionary definition of recycle is rather straightforward, "to put through a cycle again or through a new cycle." The returnable glass beverage bottle was a classical example of recycle. Prior to the advent of disposable glass, plastic, and aluminum containers, most beverages were sold in returnable glass bottles. The bottles were used again and again, with only the broken bottles being discarded. In the present-day "disposable" society, such a practice would not be considered recycling because the container is never discarded. The commonly accepted definition of solid waste recycling would be to utilize one or more components in such a way that they are not deposited in a sanitary landfill. With this definition, the uses discussed below may be available for a refuse component.

USES FOR RECYCLED MATERIALS

Replace Virgin Material

Historically, it has been cost-effective to manufacture consumer goods from virgin materials rather than reusing the existing material. There are many reasons for this practice, but a primary reason was the known quality of the raw material. The composition of the virgin material was consistent, having little or no contamination. Continued reuse of the material may change its intrinsic properties or may increase the level of foreign materials that interfere with the product quality. Quality control was an original justification for use of virgin materials and probably will continue to be the limitation on the percentage of material that can be recycled.

Today, paper is recycled for the manufacture of new paper products. The various paper products recovered from solid waste can be repulped and made into new product (i.e., newsprint is pulped and made into new newsprint, etc.). The proportion of recycled paper blended with virgin fibers will depend on the quality of the recycled material, in particular the length of the paper fiber. The more the fibers are processed, the shorter they become, soon reaching a size that is no longer acceptable in the final product.

Glass has recycle value as a replacement for the silica (sand) that is used to manufacture glass. Again, quality control is a problem, especially with colored glass. As we discuss later, the value of recycled glass is low because the cost of virgin material is low. Much of the cost is associated with the energy required for melting the glass and forming a new product. In this case the recycled container rather than the container material has the real value for resource conservation. Aluminum, especially cans, has a high recycle value. The cost of the virgin material is high because of the energy required to produce the metal from the ore. The 33.1 billion cans recycled in 1985 resulted in an energy savings of 9.5 billion kilowatts. As long as the aluminum is free from contamination, which is true for cans, it can be re-formed economically. A major problem with the recycle of ferrous metals is the contamination with other metals used to make special steels. Copper contamination, in particular, imposes a limitation on the amount of steel that can be recycled.

Raw Material for New Product/Use

Some refuse components have been used as a raw material for the production of other products. Recycled paper has a large market for use in manufacturing building materials, such as roofing felt, insulation, and wallboard. It has also been used to manufacture special containers such as egg cartons. It could see a bigger market in this area, especially if the legislative bans on the use of plastic foams expand. Recycled plastic has been used to produce fiberfill, an insulating material used in lightweight cold-weather clothing and gear. It has also been used to manufacture structural plastics for playground and park equipment. As the quantity of recycled plastics increases, it will continue to provide a raw material for a low price. There will be many new uses found.

All of the organic material present in the solid waste stream contains significant quantities of energy. Incineration of solid waste with energy recovery is currently used in many large communities. The material is burned in a steam boiler dedicated to refuse combustion, or burned along with another fuel in a central power generation station. As the separation of paper, plastics, and yard and garden waste at the source increases, the value of the solid waste as a fuel decreases. This is a case where the recycling and resource recovery schemes are incompatible.

Table 5.1. Major commodities recycled in the United States, 1984

	Generated (10^6 tons)	Recycled (10^6 tons)	Percent
Paper and paperboard	62.3	12.9	21
Glass	13.9	1.0	7
Metals	13.7	0.9	7
Plastics	9.7	0.1	1
Rubber and leather	3.4	0.1	3
Others	45.1	0	0

Source: Characterization of Municipal Solid Waste in the United States, 1960 to 2000, Prepared for U.S. Environmental Protection Agency, Franklin Associates, Washington, DC, 1984.

EXTENT OF RECYCLING

The data in Table 5.1 show the extent of recycling in 1984. Paper is the major commodity recycled, with glass and metals a distant second. Certain metals have a much higher recycle rate than the general category. Plastics, rubber, and leather make a small contribution to the recycled stream. Plastics are receiving considerably more attention as the proportion of plastics in the solid waste increases. The decade of the 1990s should see a significant increase in the amount of plastics recycled.

Paper and Paper Products Recycling

Paper has the longest history for recycling of any component of refuse, except for perhaps ferrous metals. It was not uncommon to save paper and ferrous metals for collection by civic groups such as the Boy Scouts. This was occurring prior to World War II and increased significantly during this war. The demand for recycled paper has continued. Over the years, 20 to 25% of the paper produced has been from recycled paper fibers. In 1986, about 17.8 million tons of waste paper were used in the United States. Two hundred paper mills processed only waste paper, and another 150 mills used waste paper for part of their production. One limitation on the amount of waste paper that can be used in any year is the capacity of these paper mills. One can not build such a mill in 6 months. The investment is large and the corporation will need to be certain that the supply of waste paper will continue at a competitive price. Consequently, the growth in the capacity for use of recycled paper will be gradual, and until the mill capacity catches up with the supply of waste paper, there will be an oversupply and price instability.

The price of waste paper is very volatile and regional. In 1987, the price paid to municipal recycling systems ranged from around $5 to over $80 per ton. The extreme fluctuation in price is reflected in an article from the *Wall Street Journal* (Jan. 25, 1989, p. B4), which reported: "Municipalities that just two months ago were receiving as much as $25 per ton for their newspapers from brokers now are

Table 5.2. Projection of paper production and recycling

	Millions of tons				
Year	Production	Recoverable	Actual	Fuel	Landfilled
1984	77	70	21	2	48
1990	86	76	27	5	45
2000	110	95	34	9	52

Source: Waste Age, Sept. 1986, p. 53.

having to pay these brokers $5 to $35 a ton just to haul the old papers away.'' One factor that is blamed for the extreme fluctuations in the value of waste paper is the export market. The excess paper that is diverted to this market can keep the supply tight, but the export market is not the dominant market for recycled paper. In 1987, 24 million tons (28.6% of the paper produced) were recycled. Domestic consumption was 19.6 million tons and 4.4 million tons (18.3% of the total recycled) were exported.

Table 5.2 presents some projections as to the future potential for this activity. These data project that about 90% of the paper produced could be recovered, as shown in the third column. A more realistic percentage might be 85%. It is estimated that some 15% of the paper is removed from the waste stream as permanent products, such as building materials, books and ledgers, and sanitary paper. The actual recovered paper is considerably less, for the reasons discussed earlier. The column marked fuel represents the paper that is burned for power production. If these production projections are accurate, it is possible that the amount of paper being hauled to the landfill will have increased by 2000. There are, of course, many factors that impact these numbers. They are nothing more than educated guesses using historical data and expected expansion of the mill capacities.

It is interesting to know the types of paper and paper products that are recycled. In 1987, corrugated containers, including the clippings from the fabrication of these containers, accounted for 11.2 million tons (47%) of the recycled paper products. Newsprint provided 4.4 million tons (18%); mixed papers, 2.8 million tons (12%); and high-grade papers from print shops, 5.5 million tons (23%). The major sources of recycled paper have been and continue to be the industrial and commercial paper and paper product users. The data in Table 5.3 show how much recycled material is used in each category. Clearly, paperboard leads in total tons recycled, but construction paper and others (45%) and tissue paper (43%) uses a larger percentage of recycled fiber to manufacture new product. Paperboard uses about 34% of recycled material for the manufacture of new containers. High-quality paper used in printing and writing paper uses a very low percentage of recycled fibers, only about 5%.

The historical sources for recycled paper and paper products developed for a reason. The large producers of these waste papers were prime candidates for recycle because the quantities of waste paper generated were large and the cost for

Table 5.3. Paper recycling in the United States, 1986

	Millions of tons	
	Total production	Waste paper used
Newsprint	5.6	1.4
Printing, writing, etc.	19.6	1.0
Industrial, etc.	5.1	0.3
Tissue	5.1	2.2
Paperboard	35.4	12.0
Construction paper	2.0	0.9

Source: Paper Recycling and Its Role in Solid Waste Management, American Paper Institute, New York, 1987.

disposal was significant. Therefore, there was an economic reason for recycling the paper; it was more cost-effective than hauling the paper to a disposal site. Economic considerations continue to drive paper recycle. Legislative dictates may increase the quantity of paper available for recycle, but there will be strong resistance on the part of waste paper processors to make the large capital investments necessary to increase their plant capacity. State legislatures are very fickle and they could eliminate the source of funds for recycle subsidy just as rapidly as they made them available.

Aluminum Recycling

The recycle of aluminum, especially aluminum cans, has been a major success. There was a resistance on the part of the industry to become involved in reuse of the aluminum. However, the social and economic pressure convinced them that it was in the industry's best interest to use the recycled material. They have recognized the marketing advantage of recyclable aluminum beverage containers. The major aluminum companies use this in their advertising to induce the use of more aluminum containers. Recyclability and convenience of the aluminum can has increased its market share in the number of beverage containers from about 1% in 1964 to over 90% in 1984.

Table 5.4 presents data on the quantity of aluminum recycled. In 1975, about 25% of the cans were recycled. This percentage remained relatively constant until about 1980. During this period, a number of states passed laws requiring deposits on all containers. The mandatory deposits of from $0.05 to $0.10 per container provided an additional incentive for recycle.

In recent years the value of recycled aluminum cans has increased such that this alone is incentive for recycle. Contracts for the purchase of aluminum cans for as much as $1400 per ton were reported in 1989. Most new refuse processing systems include an aluminum recovery process for the express purpose of recovery of these cans prior to disposal. There are about 25 cans per pound, so the value of a

Table 5.4. Aluminum can recycling

Year	Millions of pounds	Billions of cans	Percent recovered
1975	180	4.1	26.9
1976	212	4.9	24.9
1977	280	6.6	26.4
1978	340	8.0	27.4
1979	360	8.5	25.7
1980	609	14.8	37.3
1981	1017	24.9	33.2
1982	1124	28.3	55.5
1983	1144	29.4	52.9
1984	1226	31.9	52.8
1985	1250	33.1	51.0

Source: Waste Age, Mar. 1987, p. 120.

single can will be almost \$0.03. This value encourages the pickup of cans discarded as litter. There are many ''street people'' that make a special effort to collect these discarded cans and take them for recycle. Such automated facilities as the Golden Goat buy-back centers make it very easy to recycle the beverage containers (see Figure 5.1). In fact, the value of the mixed aluminum scrap at \$500 to \$700 per ton is sufficient that reports of the theft of aluminum highway signs and aluminum siding from buildings have been in the news media. There are very active legal recycle programs for mixed aluminum scrap and aluminum cans in all communities. These programs are generally self-sufficient and, in some municipal programs, provide an income to subsidize the other recycling activities.

The future of aluminum as a beverage container will depend upon the inroads made by the plastic industry. At present, the larger beverage containers—soft drinks, juices, and milk—are marketed in plastic containers. The ability of the aluminum industry to retain its market share will depend, in part, on the image it projects regarding recycling. In 1987, the plastics industry instituted a major

Figure 5.1 Golden Goat franchised automated buy-back unit.

research and public relations program for the purpose of developing an image of plastic as a material that is easily recycled.

Glass Recycling

Glassmakers have been returning glass broken during manufacturing to the glass furnace since the advent of glassmaking. This cullet, mixed at a ratio of 15% with raw materials, is used for new product manufacture. When the supply of broken glass from the manufacturing plant was inadequate to meet this ratio, some plants would initiate a buy-back program from the public in the vicinity of the plant. As long as the recycled glass is of the same color, it can be used without additional refining. General public collection of glass for recycle was initiated in 1970 with a total of about 24,700 tons collected during that year. This quantity has increased to about 1 million tons, approximately 4.5 billion containers, in 1986. The Glass Packaging Institute estimates that 25% of the glass bottles are made from recycled glass.[1] At present, there are 85 container manufacturing plants that buy reclaimed glass from thousands of reclamation centers.

It is necessary to separate glass according to colors and run it through a crusher to break the containers and remove the metal caps and the labels. This reduces the volume of the glass and substantially reduces shipping costs, which are a significant cost factor in glass recycle. The weight of the glass restricts the shipping distance since the price paid for the recycled glass may not cover long-distance shipping costs. Many glass plants are located throughout the country, but they may produce only certain colors or quality of glass. It is not always possible economically to market all colors of glass from a given recycling center. A careful market analysis must be conducted before glass recycling is undertaken. Prices in 1988 were about $25 per ton for mixed glass and $38 to $45 per ton for color-sorted glass.

Glass has been losing its share of the beverage container market. In the 1950s, over 90% of the beverages was packaged in glass containers. By 1975, glass containers had only 40% of the beer market. The glass container share of the soft-drink, juice, and milk market was equally affected. Aluminum and plastic containers were assuming a larger share of these markets. Glass is still used extensively for many food items, but plastic has been slowly displacing some of this market.

Ferrous Metal (Can) Recycling

Historically, there was a market for tin cans. The cans were passed through a detinning process for recovery of tin and ferrous metal. Other cans were, and still are, shipped to copper mines for use in the recovery of copper from low-grade ores. The detinning operations were considerably reduced during the 1970s and 1980s because of the increased cost of the process and a reduced supply of "tin" cans. The location of the copper mines in the western part of the country has eliminated use of

[1] *World Waste*, Aug. 1988, p. 40.

cans from east of the Mississippi River due to the high transportation costs. The concern with metal contamination of the steel product has substantially reduced the market for recycled steel cans. The current recycling activities have increased the availability of tin-coated steel cans. Consequently, the detinning operations may increase. There are some changes in the cans that may present problems, especially the plastic coatings used on the can interior. The quantity of tin on the can is reduced accordingly. Even under the best scenario, ferrous can recycling is marginal.

However, ferrous metal is easily recovered by magnetic separation. In 1982, some 3.5 billion cans were recovered from resource recovery facilities. Most of these cans are processed through a scrap dealer and included with a variety of other ferrous scrap metal. ''White goods,'' which are the appliances and other enameled fixtures discarded from households, are also processed by these dealers. The price for cans may range from $0 to $60 per ton, depending on many site-specific conditions. Much of the scrap metal is exported to countries that have metal processing plants designed for a high percentage of scrap metal. The separation of ferrous metals is convenient and inexpensive. The removal of these metals will reduce the maintenance cost of the subsequent processes, so most refuse processing plants recover these metals even if they have to give them away.

Plastic Recycling

Plastic has only recently become a significant portion of the refuse stream. Prior to 1970, it was not listed as a component of solid waste. The data for 1970 show plastic to be between 2 and 3% of the wet weight of the refuse. The use of plastic for containers, in food packaging, and in packaging in general has increased this percentage to about 7% in 1990. In 1987, 14.7 billion pounds of plastic were used in packaging. This quantity is expected to increase to 34 billion pounds by 2002.[2] With the emphasis on recycle, the industry has recently launched an intensive public relations effort and research program to improve the image of plastic as an easily recycled material.

There are two types of plastics that are currently receiving the major attention for recycle: polyethylene terephthalate (PET) and high-density polyethylene (HDPE). PET is the plastic in the clear bottles used for beverage containers, and HDPE is used as the base cap on PET bottles and for white or milky containers. PET is used for large soft-drink containers and in 1985 accounted for 21% of the soft-drink volume. For that year, PET production was about 450 million pounds and about 110 million pounds were recycled. HDPE production in this year was 1.8 billion pounds, but very little was recycled.[3]

The process for manufacturing of these plastics makes it difficult to use recycled plastic to produce the original product. Some in-plant waste plastic may be

[2]*Chemical Engineering*, Nov. 23, 1987, p. 22.

[3]*Waste Age*, Jan. 1985, p. 50.

Table 5.5 Estimated plastic wastes

Year	Waste (billion pounds)	
	Manufacturing	Post-consumer
1984	2.5	30
1990	3.5	35
1995	4.5	45

Source: Waste Age, July 1987, p. 55.

used, but most of the recycled plastic is used to manufacture other plastic products. PET can be used as the starting material for a number of other products and the potential market for recycled PET has been estimated at 1.3 billion pounds.[4] Fiberfill, unsaturated polyester, strapping, and other extruded products are possible uses. Fiberfill has an estimated market potential of 250 million pounds per year.

Unsaturated polyesters are made by chemically modifying the clean PET. When mixed with glass fibers, it can be used for a variety of components ranging from boat hulls to no-rust exterior panels on automobiles. Marble dust can also be mixed with the liquid unsaturated polyesters and allowed to solidify to produce materials for sinks, countertops, and tubs. A number of other uses for PET are being developed, so it appears that there will be a readily available market for this plastic.

The primary market for the HDPE plastic is in the production of extrusion products such as flowerpots, toys, plastic lumber, and car components. There were approximately 1.8 billion pounds of HDPE used in 1985, with little recycled from the solid waste stream. Most of the recycled plastic was from in-plant waste. The effort being invested by the industry will lead to a variety of new uses for the various plastic products in the solid waste stream. The quantities of plastic wastes expected are given in Table 5.5.

ECONOMICS OF RECYCLING

The economic justification for recycling as it is being practiced in many municipalities today is very dubious. There are three basic sources of revenue that can meet the cost associated with recycling. These income streams must be adequate if the venture is to be successful.

Avoided Cost

If the material is not recycled, it must be deposited at a disposal site, either a refuse processing plant or a sanitary landfill. There is a "tipping fee" that must be paid for each ton or cubic yard of material delivered. Since the recycled material is not

[4] *Waste Age*, Jan. 1985, p. 50.

brought for disposal, this fee is not paid. Also, the transportation cost may be reduced by not having to haul this material to a distant landfill site. This avoided cost can be significant. Historically, tipping fees at sanitary landfills were about $5 to $10 per ton. As a result of the more restrictive controls placed on sanitary landfills, these prices are in excess of $20 per ton at the landfill site. These tighter controls and the resistance of communities to the location of sites close to urban areas may add transportation costs of $20 to $80 per ton to the tipping fee at the landfill. Comparable costs are also incurred at the mass-burn (incineration) plants.

Consequently, there may be a built-in savings for the producer of the refuse of $40 to $100 per ton if the recycle can be effected. The large producers of recyclable materials, such as supermarkets, department stores, and printing plants recognized these savings many years ago. The quantities were sufficient to warrant special collection and haul to the recycle center even though the disposal costs were considerably less than today. As the disposal costs escalated, these recycling activities became more cost-effective. These sources of recycled material are the foundation of the recycling industry. They will continue to generate these wastes and because it is cost-effective, will continue to recycle. This is not necessarily true for the many municipal recycling systems. When the public recognizes the cost of these municipal recycle systems, the support could disappear.

The avoided cost will be significant only if the quantity of material recycled is significant. The recycling of aluminum can reduce the weight of refuse for disposal by a maximum of less than 1% if all of the aluminum is recycled. Because aluminum is such a small fraction of the refuse stream, avoided cost will have little effect on the economics of aluminum recycle. Conversely, paper and paper products are a major component of refuse, especially from certain generators. If 50% of the paper in urban solid waste is recycled, the avoided cost will be significant, and perhaps even more important, the life of the landfill will be extended significantly. The tipping fee is the same for the aluminum as the paper. The difference in tonnage diverted is important because of the economies of scale associated with the recycle operation.

Value of Recycled Material

The second income stream from recycling is the intrinsic value of the recycled material. Some of these values have been discussed previously, but a range of values for the main recyclable material is presented in Table 5.6. With the exception of aluminum and plastic, the values of the other recyclable materials are marginal, and under certain market conditions, may be negative. The avoided cost may be the only economic benefit from recycling these low-value materials. Conversely, aluminum is the other extreme. Its value is sufficiently high that mechanical systems have been developed solely to recover aluminum from the solid waste stream. Because of this high recycle value, more and more aluminum is not discarded. It is reclaimed before it becomes a solid waste. If the solid waste contains 1% aluminum,

Table 5.6 Value of recycled materials

Material	Range ($/ton)
Paper	−25–50
Ferrous metal	0–60
Aluminum	600–1400
Plastic	
PET	160–180
HDPE	60–200
Glass	10–100

Source: Recycling Times, Aug. 15, 1989.

a recovery system that is 90% efficient would provide a revenue stream of $9 per ton of refuse processed if the aluminum is marketed for $1000 per ton. Careful evaluation of the quantity of material present in the refuse stream is essential when conducting an economic evaluation of recycling.

State Government Subsidy

The laws passed by states that require recycling frequently include a tax on the tonnage of refuse landfilled in the state. This generates a revenue stream that can be used to subsidize the cost of recycling. This revenue can be significant. For example, Illinois has a charge of $1.05 per cubic yard landfilled. The approximately 25 million cubic yards landfilled annually will generate about $26 million. In 1992, the municipalities will receive 60% of this money to use for support of various solid waste disposal activities.

Recycling Costs

The cost of recycling varies considerably from community to community. The materials collected and the type of collection are big factors in defining these costs. Source separation with separate collection has been most common. The costs reported have varied from negative (income earned from recycling) to over $100 per ton. Figure 5.2 shows a typical curbside container used for recycle. There are a variety of types of containers used for setting out recycled materials. Color coding may be used to assist in the collection. Multibin vehicles such as the one shown in Figure 5.3 are used to collect the recycled materials.

As the recycling systems become better planned and operated, these costs will become more uniform and may be more attractive. For example, in 1982, the city of Woodbury, New Jersey, collected 177 lb of paper and 74 lb of glass per person.[5]

[5]F. P. Mulvey, *The Economics of Recycling Municipal Waste: Background Analysis and Policy Approaches for the State and Local Governments,* New York Legislative Commission on Solid Waste Management, Albany, NY, 1986.

Figure 5.2 Curbside containers used for setout of recycled materials to be collected by curbside collection.

The collection cost was $61 per ton, but these costs were offset by the town's ability to reduce the frequency of regular refuse collection from twice to once weekly. When this savings and the value of the recovered material were considered, the reported net cost of the recycling program was $13 per ton of material recycled. This was considerably less than the landfill cost.

Economic success depends on participation by a large percentage of the population, so that a large quantity of recycled materials can be generated. When separate collection is used, it can only be cost-effective if it collects enough tonnage to use the collection vehicles efficiently, and enough to reduce the collection costs of the regular solid waste. When the avoided cost can be attributed to both collection and disposal, the cost-effectiveness of recycling can improve significantly.

Figure 5.3 Multibin vehicle for the curbside collection of recycled materials.

STUDY QUESTIONS

1. What uses can be made of recycled materials?
2. Does recycling have a long history, or is it a recent innovation (last 20 years)? What are some examples?
3. What are the technical factors that limit the percentage of a material that can be recycled?
4. How significant is the recycle of major constituents of solid waste?
5. Why has corrugated paperboard enjoyed such a success in recycling?
6. What factors were responsible for initiating the recycle of aluminum cans?
7. How has the plastic industry attempted to use the recycling issue to improve the market for plastic products?
8. What economic factors are used to support recycling?

CHAPTER 6

Storage, Collection, and Transport

Refuse is generated at the source on a somewhat continuous basis. However, collection occurs on an intermittent basis—once or twice a week or perhaps daily, depending on the quantity generated at a specific site. Therefore, it is necessary to provide on-site storage for the solid waste until it is collected. The storage and collection are separate operations but must be closely coordinated. The type, size, and location of containers are very important factors in determining the most efficient collection system.

ON-SITE STORAGE

Single-Family Residential Sources

The single-family home located in suburbia presents the most expensive unit cost scenario for refuse collection. The quantity of refuse generated is low and it is spread over a large area. The common storage mechanism is one or more ''garbage'' cans with a capacity of 20 to 30 gallons each. These cans are manually transferred to the collection vehicle. Roll-out containers are becoming more popular in some areas of the country. These containers have a capacity of about 50 gallons and are on wheels. On the day of collection, the resident rolls the container to the street and it is mechanically loaded onto the collection vehicle. Figure 6.1 shows a typical garbage can and a single-family roll-out container.

The size and number of containers as well as the location of the containers are regulated by ordinance for a municipal agency and by contract for a private collection

Figure 6.1 Typical on-site storage containers for single-family residence.

company. Collection efficiency and cost dictate the number and location of the containers. For collection, the refuse must be at curbside. When alley collection is possible, the containers are at the back of the property. However, only older neighborhoods have alleys, and then the alley width may be insufficient to allow passage of the modern collection vehicle.

Usually, the refuse is not stored at curbside because of the aesthetics of having solid waste setting on the street. It must be transported to the curb on the day of collection. This may be done by a setout program in which the residents carry the containers to the curb, or a special crew member of the agency goes through the neighborhood for this purpose. In either case, it is the responsibility of the resident to reclaim the cans. This may not always happen promptly and the cans or lids may create a nuisance on a windy day.

An alternative to can setout is the use of a ''shoulder barrel'' (see Figure 6.3). A plastic container with about a 50-gallon capacity is carried by a member of the collection crew to the can location. The cans are emptied into the shoulder barrel and the refuse is transported to the collection vehicle. This has several obvious advantages.

1. The collection crew does not lose time carrying the cans from storage to the collection vehicle.
2. Can setout is not required, which eliminates the inconvenience to the resident or the cost of added crew members for can setout.
3. The cans remain in a more secure location and are not as likely to be a nuisance.

Shoulder barrels have become so popular that they are generally used no matter where the cans are stored. It simplifies the collection if the cans are not transported to the vehicle, even if for a short distance.

The shoulder barrel imposes a definite limit on the amount of refuse that can be handled by one person. If the 50-gallon barrel is filled with refuse at a density of 250 lb/yd^3, it will weigh 60 to 70 lb. That would be equal to about two typical cans full of refuse. A much heavier load would require two trips to the collection vehicle. It is common for the city ordinance to regulate the weight and number of containers so that this delay does not occur. A private agency contract will also have a limit. If exceeded, an additional charge will be assessed to the resident to cover the added cost due to the extra time required for the second trip.

Any oversize solid waste such as furniture and appliances that are generated at a single-family residence require special handling. These objects are too heavy for a crew member to carry to the collection vehicle. The typical vehicle is not designed to accept these oversize items. Consequently, a special collection is required. Many municipal systems provide this service when requested. This may also be true for a private collector, but there may be an additional charge. There are a variety of procedures for providing this special collection, but all are costly.

Multifamily Residential/Commercial/Institutional Sources

These sources generate larger quantities of refuse. Historically, multifamily housing units had one or two cans for each unit. However, transfer of these containers to the collection vehicle was very time consuming. It is not cost-effective to use individual cans for storage. Also, little care was exhibited by the users and these storage areas were frequently overflowing with refuse, attracting vermin. To reduce costs and litter and to improve the aesthetics of these storage areas, large containers are rapidly replacing the individual containers.

The Dempster Dumpster was one of the early container units on the market. This container was manufactured by the Dempster Corporation, but the name *dumpster* is frequently used to identify all containers. Load Lugger is another name used to identify these containers. These containers are manufactured by a number of companies in a variety of sizes, ranging from 0.5 to 40 cubic yards. Sheet steel is used in the construction, so these containers have a significant empty weight. The smaller containers may be mounted on rollers to facilitate movement from the storage location to the collection vehicle. The larger containers are too heavy to move manually and require a location that will permit access by the collection vehicle.

The refuse in containers that have a capacity of 5 yd^3 or less will be transferred to the collection vehicle by a mechanical loading system. The container is lifted by the mechanism and the refuse is dumped into the truck body, where it is compacted (see Figure 6.2). Containers between 5 and 10 yd^3 may be dumped into the truck, or they may be hauled directly to the processing plant or disposal site. Containers greater than 10 yd^3 are generally too large for the contents to be transferred to the collection vehicle. They are loaded onto the truck chassis and the container becomes the truck body. After unloading, the container is returned to the on-site storage location for additional use.

Figure 6.2 Typical container and collection used for on-site storage for significant generators of refuse.

The density of the loose refuse in these containers is about 200 to 250 lb/yd^3. In locations where a substantial quantity of refuse is produced, it is common to combine these containers with a compaction unit, which may increase the density by a factor of 2. The refuse accumulates in a hopper that feeds a hydraulic ram. When the hopper is full, or at a preset time, the ram is activated and the refuse is shoved into the container. The density of the refuse is increased by the action of the ram. In the transport of solid waste, volume is the limiting factor, not weight. A 40-yd^3 container with refuse at a density of 500 lb/yd^3 will have a net load of only 20,000 lb (10 tons). Even if one includes the weight of the truck and the container, this is well below the maximum allowable highway gross weight of 80,000 lb.

TECHNIQUES FOR SOLID WASTE COLLECTION

Collection Procedures for Single-Family Residences

The manner in which the refuse is transferred from the storage containers to the collection vehicle depends on several factors. Two key considerations are the number of members in the collection crew and the type of collection vehicle. A survey of municipal collection agencies conducted in the 1960s found that one municipality had nine members in their collection crews. That was a bit excessive, but the same survey found that most crew sizes were equal to or less than three. A number of collection agencies have found that a very cost-efficient collection can be accomplished with only one crew member, who both drives and loads the truck. The single crew member is particularly efficient when containers are located at curbside.

There are essentially two types of vehicles to consider, the side load and the rear load. The side-loading truck is particularly suited to the single-member crew. The hopper into which the refuse containers are emptied is immediately behind the cab of the truck. The driver can simply stop the truck next to the refuse containers and step out of the cab, pick up the containers, and dump them into the truck. These trucks are dual drive, with operating controls on both sides of the cab. The driver can operate the vehicle as either left-hand or right-hand drive, whichever is most convenient to the refuse location. This combination saves substantial time since the driver neither has to walk to the rear of the truck to empty the containers, nor around the truck to get to the controls when the containers are located on the right side of the vehicle. This type of collection is also very efficient when collecting along busy streets that require collection from only one side at a time. Figure 6.3 illustrates the two types of vehicles most commonly used.

There are numerous ways in which the collection crews can function. Figure 6.4 illustrates two techniques, method A for a two-person crew and method B for a three-person crew. In both cases, refuse is collected from both sides of the street. In method A, the truck stops between the residences to be serviced, four in total. The driver (D) goes forward to collect residence 2', returns to the truck to empty the shoulder barrel, then services residence 1'. The second crew member is servicing the residences on the other side of the street. The driver then returns to the cab and drives forward to the next pair of residences. This type of scheme works for both curbside and backyard collection.

Method B involves a three-member crew and a total of six residences are serviced for each stop. As the vehicle moves forward, the two loaders (L_1 and L_2) disembark to collect residences 1 and 1'. They bring the refuse to the vehicle, which has stopped adjacent to residences 2 and 2'. The driver leaves the cab to service residence 3'. During this time the loaders are emptying their barrels and servicing residences 2 and 2'. The driver empties the shoulder barrel and moves to residence 3. One of the loaders becomes the driver and moves the vehicle forward, picking up the driver, who serviced residence 3. This cycle is repeated and one of the other members of the crew becomes the driver.

One can envision a number of other ways in which the crew members can be directed to service the residences. The objective of specifying these procedures is the development of the most efficient collection system. Time is the critical consideration and the procedures are designed to minimize the time required to carry out the steps involved in transferring the refuse from the on-site storage containers to the collection vehicle. There have been numerous time–motion analyses conducted on refuse collection to improve the design of equipment and to develop efficient procedures for the collection crews.

Methods-Time-Measurements (MTM) Techniques

The task of transferring refuse from an on-site storage container to a collection vehicle may appear to be a simple step. Just pick up the can, dump it into the

Figure 6.3 Side- and rear-loading compactor vehicles

shoulder barrel, and carry the barrel back to the collection vehicle. If done only once, it would not matter how you did it. However, it takes the refuse from about 200 residences to fill a 30-yd^3 truck. A 10-second reduction in the time required to service a residence translates into 2000 seconds per truck (about a half-hour). If an agency has 100 trucks collecting two loads per day with a three-person crew, the daily reduction in labor cost would be $4500 assuming a labor cost of $15 per hour. The reduction in the operating cost of the vehicle must also be included, so the magnitude of the savings becomes substantial.

The steps involved in the collection of refuse are many. Assuming that a shoulder barrel is used and the driver also collects, the first question is: What does

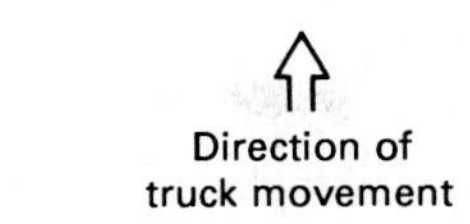

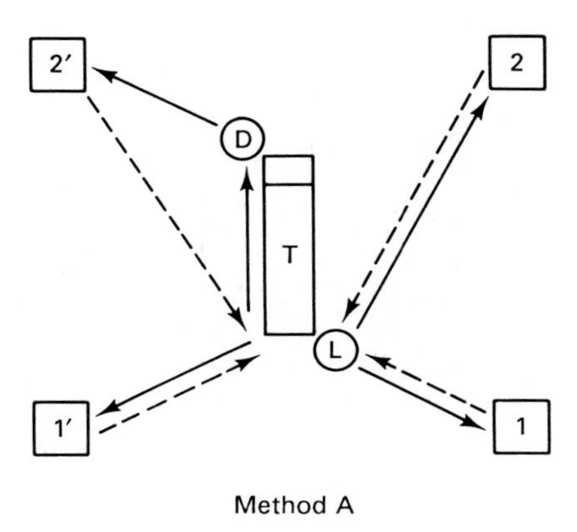

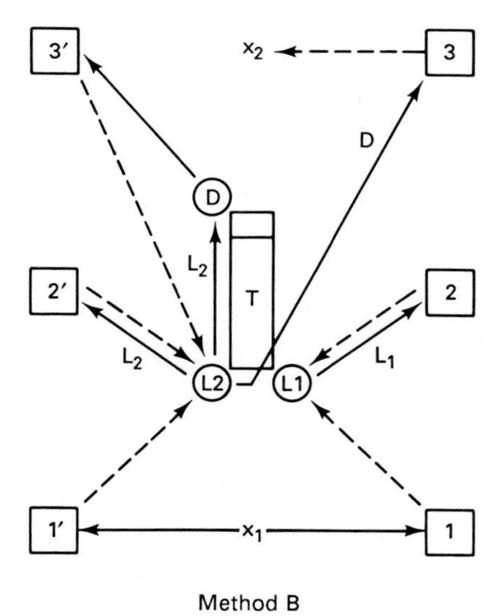

Figure 6.4 Schematic of possible procedures for collection of refuse.

the driver do with the shoulder barrel when operating the vehicle? Is it left at the back of the truck, or is there a place for it right outside the cab? It takes several seconds to walk to the back of the truck to retrieve the barrel. As indicated in Figure 6.4, there are several paths that the collectors and the driver can take in a multimember crew. What is the most efficient procedure? One can identify many other actions that are part of the collection process that affect the time required to service a residence.

Efficiency experts in the industrial engineering area have developed techniques to measure the time required to perform a specific task. This technique involves monitoring the actions of a crew that is collecting refuse. The time required for each move is measured. A "standard time" for the task is determined. This is the actual time required to complete the task and excludes all time delays not related to the task (i.e., fatigue, personal delays such as lighting a cigarette, equipment malfunctions, accidents, etc.). The standard times measured for each procedure can be compared and the most efficient procedure can be identified.

Table 6.1 presents the results of such an analysis. A shoulder barrel was used except when the cans were located in the alley. Only two cans were permitted at a residence. A comparison of can location and crew size was made. One can see from the person-minutes per stop that the alley location requires the least time for collection. There are other factors that may influence collection from an alley location. First, alleys exist only in the older sections of most communities. Subdivisions developed after about 1950 tended not to allocate land for alleys. Even if an alley exists, it may not be practical to operate the large vehicles normally used in refuse collection. If a small vehicle is used, this must be considered in the context of the overall cost of collection. This is discussed in more detail later in the chapter.

The curbside location offers the second-best location, requiring only 0.20 to 0.30 person-minute more per stop than the alley location. The crew size is an im-

Table 6.1 Standard times for refuse collection

Crew size	Can location	Standard time[a]	Services per stop	Person-minutes per service-stop[b]
3	Backyard	2.35	6	1.33
2	Backyard	2.38	4	1.33
1	Backyard	2.40	2	1.35
1	Alley	0.86	2	0.58
2	Alley	0.49	2	0.79
3	Alley	0.35	2	0.97
3	Curbside	2.06	6	1.18
2	Curbside	2.11	4	1.12
1	Curbside	1.58	2	0.88

[a] Minutes per collection stop for both sides of the street.

[b] Includes travel time between collection stops on the route.

Source: A Study of Solid Waste Collection Systems, U.S. Public Health Service Publication 1892 (SW-96), Washington, DC, 1969.

Table 6.2 Results of a two-week field study of specific crews

Agency	Crew size	Cans per service	Items per service	Time (min/service) Collect	Time (min/service) Travel	Refuse per service	Person minutes per ton
A	1	2.45	3.40	0.68	0.15	77.1	26.3
B	2	2.65	3.55	0.59	0.17	81.2	43.0
C	3	2.70	4.01	0.58	0.17	73.2	63.5
D	1	1.79	2.60	0.57		56.9	37.6
X	1	2.74	4.04	0.99		88.1	33.8
Y	1	2.07	2.87	0.59		60.5	39.0

Source: A Study of Solid Waste Collection Systems, U.S. Public Health Service Publication 1892 (SW-96), Washington, DC, 1969.

portant factor in determining the person-minutes per service stop for both alley and curbside locations. About 0.3 person-minute can be saved per service stop by reducing the crew size from three to one. However, the time required for collection from the backyard is not sensitive to crew size. The time required for backyard collection is substantially greater for all crew sizes.

A field survey of collection crews in southern California was conducted to evaluate the time required to service residences. The results of this study are presented in Table 6.2. There were four municipal collection agencies (A, B, C, D) and two private agencies (X, Y). Crew sizes and the number of cans per residence are shown in the table. The number of items per service include the cans as well as any other type of container or bundle. The time required to service the residences and the weight of refuse collected at each residence permits computation of the labor required per ton of refuse collected. This can be translated to collection costs by multiplying the person-minutes/ton by the cost of labor and adding the equipment costs. A very clear message is the lower cost associated with the single-member crew.

Crew size and the number of containers as well as container location are significant factors in determining the cost of refuse collection. Table 6.3 compares the collection time required for increasing numbers of containers and for different crew sizes. A larger number of containers will significantly increase the time required at any stop regardless of the crew size. The larger crew will reduce the time required to service a residence, but not in proportion to the increase in crew size. A three-member crew does not collect the refuse three times as fast as the single-member crew.

The times previously discussed are all standard times. They do not include any time loss due to nonproductive activities. During the period of a day, there are many events that are considered nonproductive. These events can be classified as unpredictable or predictable. These unpredictable delays fit the following categories: personal actions, traffic congestion, equipment failure, and accidents. Personal actions include a variety of activities of a personal nature, such as lighting a cigarette, telling a joke, or adjusting clothes. Fatigue also reduces the efficiency of the crew. A loader that loads a 20-yd^3 truck with refuse in an 8-hour period has lifted and carried about 10,000 lb.

Table 6.3 Collection time as a function of crew size and number of containers per service stop

Crew size	Time (min/service stop) for number of containers per service stop				
	1	2	3	4	5
1	0.278	0.456	0.620	0.793	1.161
2	0.223	0.383	0.558	0.705	0.863
3	0.214	0.311	0.454	0.558	0.716

Source: *A Study of Solid Waste Collection Systems*, U.S. Public Health Service Publication 1892 (SW-96), Washington, DC, 1969

These delays are difficult to control. They depend on the morale of the crew and the level of training and supervision. Traffic congestion is difficult to predict. However, careful planning of vehicle routes can avoid travel on streets with high traffic density during rush-hour periods.

Equipment failure can be minimized through a good maintenance program but is still unpredictable. A well-trained crew will minimize equipment failure by knowing how to operate the equipment. Accidents are also related to crew training. Traffic accidents are less controllable, but accidents associated with the equipment are definitely related to crew training. In recent years, considerable attention has been given to training to reduce this time loss as well as the cost of worker compensation associated with job-related accidents.

Predictable nonproductive time is related to specific activities that are a necessary part of the collection system. These times can be factored into the cost of collection. These activities include the travel time required to get the collection vehicle and crew from a storage yard or central base at the beginning of the day to the collection route and to return from the tip site at the end of the day. Also included is the time allocated for lunch and breaks, time for dispatch to another route when finished with one route, route retracing, and relief time. Route retracing refers to the need for traveling over streets that have already been collected. Because of the configuration of certain routes, it is not possible to avoid retracing some of the streets. When a collection crew has filled a truck and there is not sufficient time left in the day to return to the collection area, collect a reasonable partial load, return to the tip site, and unload, this may be the end of the day for this crew. This is designated as the relief time. These times will be utilized in estimating collection times.

Collection Procedures for Large Producers

Collection of refuse from single-family residences is labor intensive, while collection of refuse from large producers requires the proper mechanical systems. As described previously, the refuse is stored in a metal container with capacities ranging from 0.5 to 30 or more cubic yards. The weight of even the smaller containers precludes manual loading. The refuse in the containers is either transferred to the collection vehicle, or the container is attached to the truck chassis and hauled to the tip site. The collection vehicle and the containers must be compatible. As was illustrated in Figure 6.2, smaller containers are engaged with mechanical arms that lift the container and transfer the contents to the collection vehicle. The empty container is replaced and the vehicle moves on to the next container. This sequence is repeated until the vehicle is filled, at which time it will go to the tip site to be unloaded. The contents of the larger containers are not transferred to the collection vehicle, but the containers themselves are loaded onto the truck chassis and hauled to the tip site. The container is emptied and returned to the original site to be refilled.

DETERMINATION OF COLLECTION TIME/COSTS

A rather simple equation can be developed to calculate the time and, consequently, the cost required to collect refuse. This equation can be applied to either single-family residences or to large producers. A minor modification is required to reflect the various steps involved. The following four time factors must be considered in this formulation.

Pickup Time

Pickup time is the time required to load the collection vehicle. It is determined by the following factors: the mean quantity of refuse per stop, the vehicle capacity (volume), and the density of the refuse in the vehicle, which determine how many service stops are required to fill the vehicle. This number can be combined with the time required to service a particular stop and the time required to travel between stops. V and D are characteristics of the collection vehicle, and t_s and Q are characteristics of the collection area. This time element is defined as

$$\text{Pickup time} = \frac{VDt_s}{Q} \tag{1}$$

where V = volume of the collection vehicle, yd^3
D = density of refuse in the collection vehicle, lb/yd^3
t_s = mean pickup time per stop, min/stop
Q = pounds of refuse per stop, lb/stop

Therefore, if the truck size, refuse density in the truck, the mean quantity of refuse per service stop, and the mean time required to service each residence or stop are known, the mean time required to fill the truck can be determined.

The truck size is a very important variable in this equation. Proper sizing of the collection vehicle will be necessary to obtain a cost-effective collection system.

In this part of the equation, truck volume is the only significant variable. The density is usually fixed at about 500 lb/yd^3, due to the limits on internal pressure in the truck body. The mean quantity per stop can be controlled to some extent by ordinance or contract. However, it is necessary to collect all of the refuse generated or it will be discarded in some other manner. The time per stop is a function of the area—the distance between houses, the container location, the crew size, and other factors, as discussed previously. In designing a collection system, it would be possible to select a crew size. This would be a variable that would change the time per stop.

Haul Time

The location of the tip site, be it a transfer station, an incinerator, or a sanitary landfill, relative to the collection route becomes a very important time element. When the truck is full, it must travel to the tip site and unload. If one workday is required to fill the truck, after unloading it will go to the storage yard to wait for the next day's operation. However, if only part of the day is required to collect the first load, the truck will return to the collection area and collect another full or partial load. The actual haul time for multiple loads is complicated by the destination of the truck after unloading. The time required for haul is given as

$$\text{haul time} = (2n - 1)b \tag{2}$$

where n is the number of trips from the collection route to the tip site and b is the travel time from the collection route to the tip site (one-way travel time only). The $2n$ accounts for the fact that for more than one load per day, the vehicle must return to the collection area after each load except the last.

On-Site Time

When the collection vehicle reaches the tip site, a time element must be included that allows for the time required to unload the vehicle. There are two components to this element, the actual unloading time and the waiting time. The unloading time will depend on the traffic pattern on the tip floor and the design of the vehicle. When the truck enters the tip area, it must position itself properly to dump the refuse in an area that accommodates further handling of the refuse. At the landfill, the refuse must be dumped at the working face of the fill so that the equipment can place it properly. At a transfer station, the refuse may be dumped into a pit or on a floor, where a front-end loader loads it onto the transfer vehicle. Several minutes may be required to position the vehicle, especially in a poorly designed tip area. The vehicles do not handle like a small car. Considerable space for the long turning radius is necessary for easy maneuvering.

The compaction mechanism must be securely attached to the body of the truck; otherwise, it would not be able to exert the required pressure during compaction. In a rear-loading vehicle, this mechanism must be disengaged. This frequently requires the driver to exit the cab to unlock fasteners that release the mechanism. The mechanism can be raised hydraulically and the front of the truck bed can be elevated and the refuse dumped from the truck. The time required to complete all of these tasks may total 10 minutes or more. A side-loading vehicle has the compaction mechanism located in the front of the truck bed. In many cases the rear end of these truck bodies can be opened hydraulically from the cab and the compacting ram used to push the refuse out the rear of the truck body. The time required is still about the same, due to the time required for the ram to travel to the rear of the bed and back.

A major problem at all tip areas is scheduling the arrival of the collection vehicles. This is especially acute for residential collection. All of the trucks will tend to complete the first load at about the same time. The difference in the travel time from the route to the tip area is the only time factor that will prevent all the trucks from arriving at the tip site at the same time. Collection from commercial areas and apartment buildings will tend to be better spaced over the operating day. The queue that forms at the tip site can consume considerable time. Careful attention must be given to the traffic flow at the tip site.

The on-site time can be designated as s and is usually a fixed value for each trip to the tip site. For multiple trips, the on-site time becomes ns.

Off-Route Time

The off-route time refers to the nonproductive time, the time that does not result in the collection or transport of refuse. The predictable and nonpredictable times discussed above fall into this category.

1. Travel time from storage yard to collection route at the beginning of the day
2. Travel time from tip site to storage yard at the end of the day
3. Lunch, relief time, dispatch time, and route retracing
4. Miscellaneous time loss, such as traffic congestion, accidents, and equipment failure

Estimating these time factors requires a through knowledge of the area served. The location of the storage yard relative to the collection routes and the tip area have a significant impact on this time element. Route retracing is an inefficient use of the collection vehicle and can be minimized by careful design of the collection route. As discussed later, selection of the right-size collection vehicle can reduce such time elements as relief time and dispatch time. Off-route time is generally fixed for a given collection area and can be designated as w.

Collection and Disposal Time

The time required for the collection and disposal of one load (either full or partial) during the working day can be computed as

$$L_1 = \text{pickup time} + \text{haul time} + \text{on-site time} + \text{off-route time}$$

$$= \frac{VDT_s}{Q} + b + s + w \qquad (3)$$

In this case the truck comes from the storage yard, collects the refuse needed to fill it, drives to the tip area and unloads, and then returns to the storage yard. Essentially, this involves making only one collection trip during the day. If a full working day is not required to collect one load, what happens to the vehicle after it has been unloaded? This is determined by the restrictions placed on the allowable overtime. Assume that the workday is 8 hours (480 minutes) and that the maximum overtime allowed on the average is 30 minutes. The following considerations will determine what happens to the vehicle after the first load.

1. If $L_1 = 480$ minutes, it is the end of the day.

This is the case where a full day is required to complete the task of collecting a load and unloading at the tip site and returning to the storage yard.

2. If $L_1 < 480$ minutes and $L_1 + 2b + s \geq 510$ minutes, it is the end of the day.

There may be more time left in the day to collect refuse, but after the first load is unloaded, the round-trip time from the tip area to the collection area plus the unloading time exceeds the allowable overtime. Therefore, the crew has completed the workday and the time short of the 480 minutes is considered relief time.

3. If $L_1 > 510$ minutes, it is the end of the day with a partial load. The magnitude of the partial load, a, is calculated by the equation

$$510 = \frac{aVDt_s}{Q} + b + s + w \qquad (4)$$

4. If $L_1 + 2b + s < 510$ minutes and $L_1 < 480$ minutes, the truck returns for one or more additional loads.

The collection crew and vehicle will make a total of n trips, where n is an integer. The total number of trips made will depend on how many full and partial loads can be collected in the time allotted. The minimum allowable partial load will also be a factor. It is impractical simply to collect from a couple of residences. The constraint on the size of the minimum partial load may be 25% of the truck capacity. The time required to collect the additional loads can be calculated as

$$L_n = (n + a - 1)\frac{VDt_s}{Q} + (2n - 1)b + ns + w \quad (5)$$

This equation has the constraint that $L_n \leq 510$ minutes and $a \geq 0.25$. If $a < 0.25$, only full loads will be collected. The actual number of loads, full and partial, is given by

$$N = n + a - 1 \quad (6)$$

where N is the number of loads and n is the number of trips. The parameter a represents the portion of a load obtained during the last trip, and must always have a value. If the last trip had a full load, the value of a would be 1.

Cost Considerations Associated with Collection Systems

The time elements discussed above have associated cost factors that may vary from one element to the other. The three cost categories are: labor, vehicle operation and maintenance, and vehicle capital amortization.

Pickup costs. In this operating mode, the labor cost for the full crew is applicable. All members will be needed to operate the process as designed. The normal rate of vehicle amortization will apply. Amortization is calculated as an annual cost, but this cost can be expressed as an hourly rate by dividing the annual cost by the number of hours that the vehicle can be used annually. Theoretically, the vehicle is available for use 24 hours per day, but the hourly cost must be based on the hours that personnel are available to operate it. Therefore, for the typical 8-hour shift, 5-day week, the number of hours annually would be 2080.

The capital costs are significant. These vehicles can cost from $75,000 for the smaller units to over $150,000 for the large units. The expected life of these vehicles may be only 5 years, but with care they may last longer. For cost purposes, the 5-year period is used. A $100,000 unit at a 10% interest rate and a 5-year life will have an annual cost of $26,380, or $12.68 per hour. In addition to this capital cost, one needs to add the operation and maintenance cost. Depending on the maintenance program, this cost can equal the annual capital cost. Total costs of $20 to $75 per hour are not uncommon for collection vehicles.

Haul costs. Once the collection vehicle is filled, it travels to the tip area to unload. The full collection crew may ride to the tip site, or they may be transferred to another empty vehicle and continue collection. In this case only the labor costs for the driver would be incurred. These cost savings are significant if a large crew is employed. At a labor cost of $15 per hour, the savings would be $30 per hour of haul time by using only the driver for the trip to the tip site, as contrasted to a crew size of three. When the haul time is short, the savings may not justify the more complex routing required to transfer the loaders to another vehicle. The amortization and operating and maintenance costs for the collection vehicle apply to these time elements the same as for the pickup time. Operating costs may be a little higher, due to the higher fuel consumption for higher operating speeds. The on-site time (unloading time at the tip site) would have the same cost considerations as the haul costs.

Off-route time. The time associated with the off-route activities carries the same costs as the pickup time except for the relief time. The collection crew has to be transported from the base location to the collection area and returned at the end of the day.

Relief time. The relief time occurs when there is not sufficient time in the day to collect additional refuse and unload it within the time constraints. The collection vehicle is returned to the storage yard before the 480-minute day is complete. The crew will be paid for a full day and the charge for the amortization of the vehicle will apply. However, since the vehicle is sitting on the lot, there will be no charge for vehicle operation and maintenance.

Analysis of the time required for the various time elements associated with refuse collection and the costs associated with these elements will allow the computation of the cost of refuse collection. Sensitivity analysis can be conducted to determine the effect of vehicle size, crew size, haul distance, and a number of other factors on the collection costs. The most cost-effective collection system can be selected. Table 6.4 presents the results of such a calculation. The specific values for the parameters in equation (5) are as follows: $D = 500$ lb/yd^3, $Q = 75$ lb/service stop, $t_s = 0.63$ min/service stop, crew size = 3, $b = 20$ min, $s = 10$ min, and $w = 60$ min.

The first part of Table 6.4 shows the time required to collect the indicated number of loads. The number of loads that can be collected in 480 minutes can be determined from equation (5). The value of n equals the number of full loads that can be collected in 480 minutes or less plus one. Inserting this value of n and the values for the other parameters into equation (5) and equating it to 480 minutes will determine a, the size of the partial load that can be collected. If a is less than 0.25, the partial load is not considered. The tons per day are computed from the number of loads and the weight per load. Labor costs are based on three crew members for 8 hours at \$15 per hour (this includes the typical fringe benefits). In this example, no overtime is allowed. The final figure is the cost in dollars per ton. This identifies the most cost-effective vehicle. There is not a major difference among the different-sized vehicles. The 15- and 25-yd^3 vehicles have essentially the same cost, with the 20-yd^3 only slightly more costly. Therefore, other factors may dictate the size of vehicle to select.

Table 6.4 Effect of truck size on collection costs

	Truck volume (yd^3)				
	10	15	20	25	30
Time required to collect indicated number of loads					
1 load	132	153	174	195	216
2 loads	224	266	308	350	392
3 loads	316	379	442	505	568
4 loads	408	492	576		
5 loads	500				
Loads/480 min	4.5	3.8	3.0	2.8	2.3
Tons/day	11.3	14.2	15.0	17.3	17.3
Labor (at $15/hr)	$360	$360	$360	$360	$360
Operating time (hr)	8	8	7.37	8	8
Relief time (hr)			0.63		
Vehicle cost					
Per hour	$20	$26	$34/17	$42	$52
Per day	$160	$208	(250.6 + 10.7) = $261.3	$336	$416
Total cost					
Per day	$520	$568	$621.3	$696	$776
Per ton	$46.0	$40	$41.4	$40.2	$44.9

The lack of sensitivity of collection costs to vehicle size in Table 6.4 is a result of the relatively short haul time. The tip area would have to be very close to the collection area to have such a short haul time. Table 6.5 shows the effect of a haul time, *b*, of 40 minutes on the most cost-effective vehicle size. The longer haul time increases the cost of collection with all sizes of vehicles, but the 15- and

Table 6.5 Effect of truck size on collection costs

	Truck volume (yd^3)				
	10	15	20	25	30
Loads/480 min	3.0	3.0	2.25	2.0	2.0
Tons/day	7.5	11.25	11.3	12.5	15.0
Labor (at $15/hr)	$360	$360	$360	$360	$360
Operating time (hr)	6.93	8	8	6.83	7.53
Relief time (hr)	1.07	0	0	1.17	0.47
Vehicle cost					
Per hour	$20/10	$ 26	$ 34	$42/21	$52/26
Per day	$149.30	$208	$272	$311.43	$403.78
Total cost					
Per day	$509.30	$568	$632	$671.43	$763.78
Per ton	$ 67.91	$ 50.49	$ 55.93	$ 53.71	$ 50.92

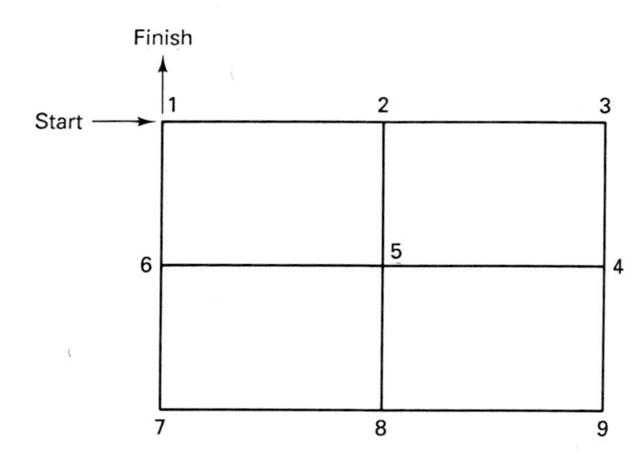

Figure 6.5 Simplified solid waste collection route.

30-yd^3 vehicles are the most cost-effective. That is primarily because full loads are collected with little or no relief time. As the haul distance increases, the vehicle that can obtain a full load and make the least number of trips to the tip area will generally be the most cost-effective.

COLLECTION VEHICLE ROUTING

Figure 6.5 illustrates a very simple collection district with only four loops. The most cost-effective collection procedure is to have the collection vehicle travel over each street only once. A quick examination of this figure clearly shows that it is not possible to eliminate street retracing. The best that can be achieved is to minimize this retracing. It would be possible by trial and error to find a way for the vehicle to collect in this district and minimize the retracing. It would be much more difficult if there were multiple districts that contained hundreds of loops. This issue of vehicle routing has been addressed by numerous researchers in the past two decades. There are three publications that make a complete and understandable presentation of a technique for routing refuse collection vehicles.[1]

The overall collection vehicle routing consists of three parts:

1. The division of the collection area into smaller districts, each of which generates one load of refuse
2. Determination of the vehicle path or tour from its entry into the district until its departure
3. The aggregation of the districts and their associated tours into a full day's workload for the vehicle and crew, which is called a route

[1]J. C. Liebman and J. Male, *Journal of the Environmental Engineering Division, ASCE,* June 1975, p. 399; D. Krabbe, *Solid Waste Management,* July 1979, p. 68; J. Reindl, *Solid Waste Management,* July 1979, p. 72.

Each tour will include a trip to the tip area to unload the filled vehicle. The location of the district boundaries relative to the location of the tip area will have a significant impact on the costs of the trips to and from the tip area.

A procedure based on a network representation of the collection district has been developed by Liebman and Male.[2] This procedure is summarized in the following discussion. Streets over which collection must take place are the links of the network, and their intersections are the nodes. If collection occurs on both sides of the street at the same time and the street is not one-way, it is represented by a single undirected link. If the collection must take place separately on each side of the street (as may be the case for a major four-lane street), it is represented by two links, each with an assigned direction that is the direction the collection vehicle must travel.

There are three cases to consider:

1. Networks that are totally undirected: all are two-way and collection takes place on both sides simultaneously
2. Networks that are entirely directed: either all streets are one-way or collection takes place separately on each side of the street
3. Networks that are a mixture of directed and undirected links

Associated with each link is a cost, and the objective of routing is to obtain the minimum cost. The cost of travel over any link is treated as invariant. It is true, however, that traffic conditions as related to the time of day may affect the actual cost, but it is not practical to consider such a factor systematically. There is sufficient flexibility in constructing the final tour to allow for avoiding thoroughfares during rush-hour periods. The cost of actual collection is fixed with respect to routing. Collection must take place over every required link, and rerouting will change neither the time required nor the distance traveled while collecting. The only variable under the control of the router is the cost of "dead heading": the cost of traveling over or retracing streets that have been collected, that will be collected in a future pass, or that do not need collection. If a tour can be found that does not have street retracing, that tour clearly has the minimum cost. If that tour does not exist, the tour that has the minimum amount of retracing is the tour with the minimum cost.

Definition of a Unicursal Network/Euler Tour

What conditions are necessary in a network to have a tour that does not have any street retracing? This answer was formulated over two centuries ago, in 1736, by Euler.[3] He was concerned with routing a parade across the seven bridges on the

[2] J. C. Liebman and J. Male, *Journal of the Environmental Engineering Division, ASCE*, July 1979, p. 399.

[3] *Scientific American*, Vol. 189, 1953, pp. 66–70.

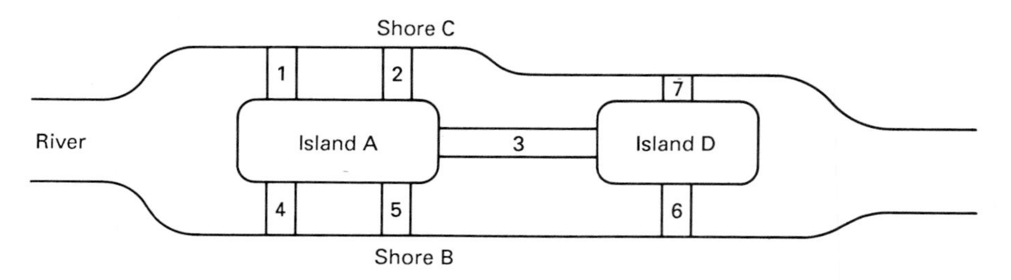

Figure 6.6 The seven bridges of Königsberg.

River Pregel in Königsberg, a town in eastern Prussia (see Figure 6.6). The object was to find a tour that started on one bank, traversed each bridge exactly once, and returned to the staring point. Euler generalized the problem by deriving two conditions that are necessary and sufficient in any undirected network for there to be a tour that travels over every link exactly once and returns to the starting point: (1) the network must be connected so that there is a way from every node to every other node, and (2) the number of links touching every node (called the degree of the node) must be even.

The necessity of this condition is made intuitively clear by considering the entrances and exits at each node. Each time you enter a node (except at the starting and ending node), you must exit from it. You must travel a pair of links touching the node. If any node is of odd degree, you will ultimately enter it and find no remaining untraveled link on which to exit. The same reasoning applies to the starting and ending nodes, except that the links are used in the opposite direction. Thus it is clear that there can be no tour without street retracing on any network that does not have all nodes of an even degree. It is also true that a tour can be constructed without retracing any streets on any network that does have all nodes of even degree. A network that is connected and has all nodes of even degree (and therefore has an Euler tour) is called a unicursal network.

Euler's condition for unicursality has been extended to directed networks and mixed networks. An Euler tour can be found in a completely directed network if and only if the network is connected and the number of links directed toward the node equals the number of links directed away from the node. Thus it is also easy to determine by inspection if an Euler tour is possible on a completely directed network.

In a mixed network, consisting of some directed and some undirected links, the requirements for existence of an Euler tour are much more complex. The network must, of course, be connected, and there must be an even number of links touching every node. In addition, the number of undirected links touching any node must be at least as great as the difference in number between incoming and outgoing directed links. Intuitively, this requirement must hold so that the undirected links may be used in the appropriate direction to make up the difference between incoming and outgoing directed links and provide an equal number of entrances and exits from the node. It is possible to satisfy these requirements at every node and still have some portion of the network that has more entrances than exits, thus making an Euler tour impossible. To insure the existence of an Euler tour, the condition that the number of undirected links be at least as great as the difference in number between the inward and outward directed links must be satisfied, not only at every node, but also at every group of nodes.

Construction of a Unicursal Network

The effects of the unicursality condition are more important than simply permitting the identification of a unicursal (an unusual condition) collection district in which no street retracing is required. They suggest a procedure for developing a minimum-cost tour in a nonunicursal network. One can start to develop a tour simply by tracing the path of the collection vehicle through the network of streets, retracing streets wherever necessary. The resulting network must be unicursal with new links inserted for the streets that are retraced because a tour exists that travels over every link in this network exactly once. Instead of simply setting out to trace a tour in a nonunicursal network, retracing streets wherever necessary, it is possible to make the network unicursal by adding links. The added links must correspond to existing streets and they represent retracing of these streets. When the network has been made unicursal, a tour must exist with street retracing exactly equal to the added links.

In an undirected network, the unicursality conditions are simple and violations are easy to identify. If the network of required collections streets is not connected, additional links representing optional streets must be added to make the network connected. It is also easy to identify nodes that are of odd degree, but their conversion to even degree is slightly more complicated. Clearly, any node of odd degree must have at least one added link incident upon it. However, that added link is incident at its opposite end upon another node. If this node was also originally of odd degree, the added link makes both nodes even. However, if this node was of even degree, this added link makes it odd. Addition of another link incident upon this node will convert it back to even degree, but may make another node odd.

The ultimate conclusion is inescapable; if progress is to be made toward unicursality, each added link must connect two nodes of odd degree, or be part of a continuous chain of added links that begins and ends at odd nodes. Figure 6.7 illustrates the technique for making the network unicursal. The odd nodes are identified (2, 4, 6, 8) and links (shown by the double lines) are added to make them even. By adding a link between nodes 2 and 5, node 2 becomes even but node 5 becomes odd. As shown in Figure 6.8, extending a link from node 5 to node 8 makes not only node 5 even, but also node 8. The same can be applied to nodes 4 and 6; a link is added between node 4 and 5 and between node 5 and 6. The network is now unicursal.

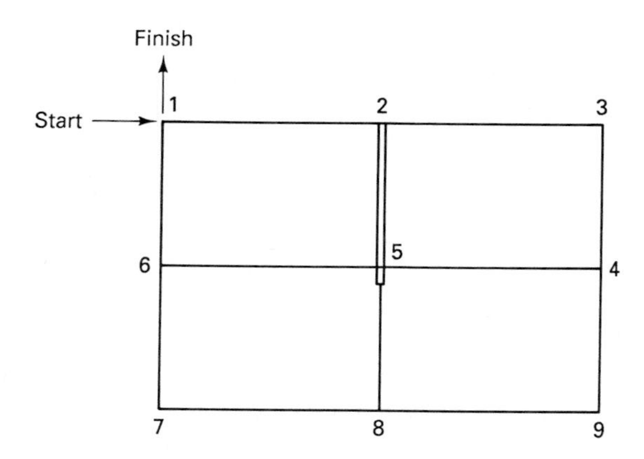

Figure 6.7 Building a unicursal network.

In a completely directed network the procedure for obtaining unicursality is similar. If the original network is not unicursal, there will be some nodes with more incoming than outgoing links, and other nodes with more outgoing than incoming links. Unicursality requires that each node be balanced, so it is necessary to provide the appropriate number of outgoing links at each node of the first group and the appropriate number of additional incoming links at each node of the second group. Nodes that were originally balanced require no additional links; if an incoming link is added to such a node, a balancing outgoing link must be provided, and vice versa. Thus to make a directed network unicursal, one must add links in directed chains leading from nodes with an excess of incoming links to nodes with an excess of outgoing links. In a mixed network there is no simple method for obtaining unicursality. It is possible to treat the network as undirected and add links to make every node even, and then add additional undirected links to ensure that the number of undirected links at each node is at least as great as the difference between the number of incoming and outgoing links. It remains necessary, however, to ensure that this condition is met for all subsets of nodes as well as individual nodes. This is rather difficult but can usually be done by inspection.

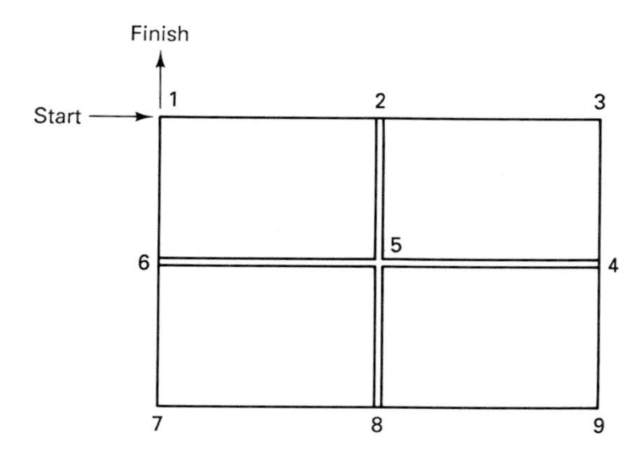

Figure 6.8 Completed unicursal network.

Minimizing Street Retracing

The assignment of links required to convert the odd nodes to even nodes will not necessarily produce the minimum distance of retraced streets. It is possible, but difficult, to select the right links so that the sum of their distances is minimum. This is especially true for a tour having a large number of added links. To solve a similar problem, Kwan[4] observed that any unicursal network can be decomposed into a set of closed loops, in which each link appears in exactly one loop. In any given loop, or cycle, some of the links may be required collection links, and the others may be retraced streets, the added links. All nodes are of even degree, since the cycle is part of a unicursal network. In any such cycle, if all retraced streets (added links) are removed, and the previously *nonretraced* streets now become *retraced* streets, the even-degree property at every node is maintained, and thus the network remains unicursal. However, such an exchange is advantageous only if the cost of the links removed is greater than the cost of the links added. Kwan proposed that a unicursal network be inspected for cycles in which such exchanges reduce the length (cost) of retraced streets and proved mathematically that when no more such exchanges are possible in the network, the minimum cost for retracing of the streets has been achieved. Kwan's procedure is not practical for guaranteeing minimum cost because the number of cycles that must be inspected is very large in any size network. It is, however, quite valuable as a method of seeking improvement in a unicursal network by inspection.

Construction of an Euler Tour

As soon as all of the links needed to make the network unicursal are added and Kwan's procedure is applied, an Euler tour can be constructed. It is usually possible to construct this tour by inspection. Care is required to be sure that all the links, including those added at the odd nodes, are traversed once and only once. Figure 6.9 illustrates the direction of travel for this simple route. The numbers indicate the order in which the streets are collected. The dashed lines represent retracing. If U-turns are permitted, which is generally not true, it would be possible to construct a different tour.

There are a number of road conditions and traffic constraints that can influence the final tour. Many intersections prohibit left turns, and U-turns are prohibited at most intersections. The grade of the street may have the effect of fixing the direction of travel for the collection vehicle. It is not desirable to have frequent start and stops on a steep up-grade. Travel downhill is much preferred. Each network will have variations that will require a little foresight in development of the tour.

[4]M.-K. Kwan. *Chinese Mathematics*, Vol. 1, 1962, pp. 207–218.

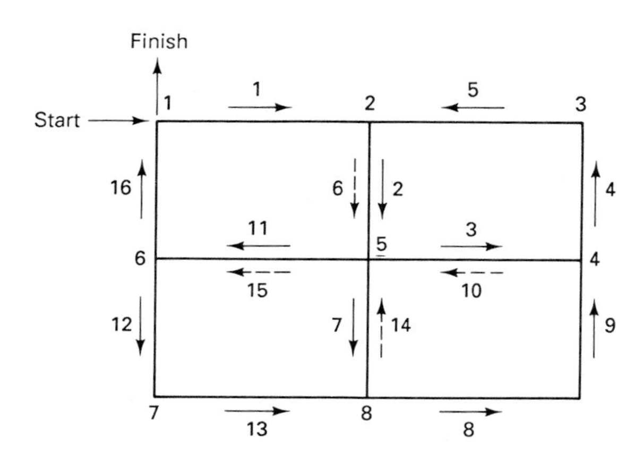

Figure 6.9 Completed Eular tour of a unicursal network.

LONG DISTANCE TRANSPORT/TRANSFER STATION

As discussed earlier, the haul time to the tip site has a significant impact on the cost of refuse collection. It is desirable to use the collection vehicle for collection and minimize the time it is used to transport the refuse collected. In cases where the disposal site (sanitary landfill, for example) is at a significant distance, or requires significant travel time to reach, it may be more economical to use other means of transport. The economic comparison is made on the unit cost associated with using the collection vehicle as the haul vehicle versus the cost of construction and operation of a transfer station and the cost of the transfer vehicle. One can speculate that the ton-mile cost of haul with the collection vehicle is much greater than the ton-mile cost of the transport vehicle, or the distance must be very great to warrant the investment in the transfer station.

Cost-Effectiveness of Transfer Stations

Figure 6.10 shows the cost of operating a collection vehicle as a transport vehicle. A detailed cost breakdown developed by Schaper[5] in 1982 dollars was used to construct this curve. It is for a 20-yd^3 vehicle with a two-person crew. The total collection vehicle haul cost is obtained by adding the fixed and variable cost. This curve shows the margin available for the transfer operation if the vehicle travels at the indicated speed. Note the higher operating costs at the higher speeds, but also remember that the distance traveled is greater, so the cost per ton-mile will be less. A 45-minute haul time will add approximately $16 per ton to the cost of collection and disposal if the vehicle travels at about 35 mph.

The cost of the transfer operation has two components: a fixed cost resulting from the construction cost for the station and the capital cost associated with the equipment and transfer vehicles, and an operating cost related to the tons processed. The fixed cost is determined by the cost of the money used to finance the

[5] L. Schaper, *Waste Age*, Dec. 1982, p. 28.

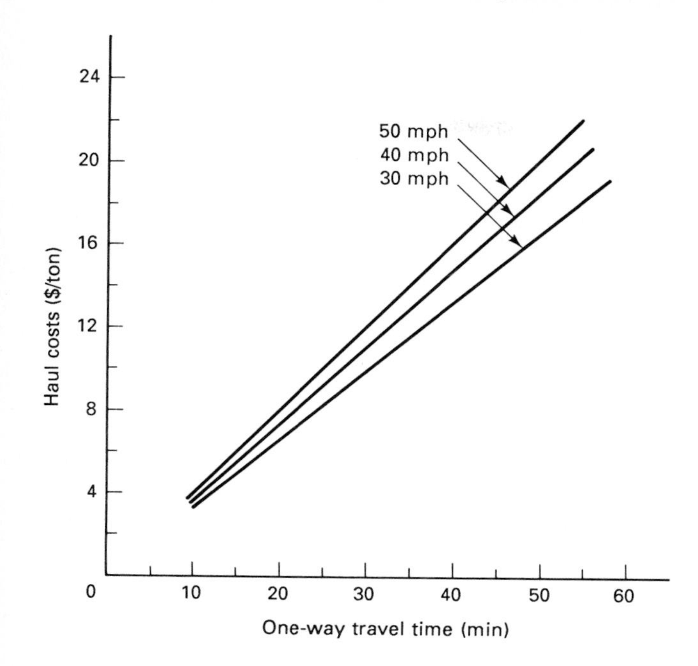

Figure 6.10 Collection vehicle haul cost. (From L. Schaper, *Waste Age*, Dec. 1982, p. 29.)

facility and does not depend on the tons processed. The lower the number of tons processed, the higher the cost in dollars per ton. In making economic comparisons, the capital cost for the transfer station and transfer vehicles are separated. The capital cost of the transfer vehicles is included in the vehicle cost. Figure 6.11 shows the range of capital costs expected for transfer stations. The large cost range results from the complexity of the facility. It may be nothing more than a tip floor and a couple of truck bays for loading the transfer vehicle with a front-end loader, or it may contain a processing system that compacts and containerizes the refuse for rail or barge shipment.

The important cost figure is presented in Figure 6.12. The annualized capital cost plus the operation and maintenance cost for the transfer station determine the first costs that are incurred simply by transferring the refuse from one mode of transportation to another. The annualized capital cost is based on a 12% interest rate. Obviously, the annual capital cost is very sensitive to interest rates. Added to this first cost is the cost of the transfer vehicle, which is a function of the distance the refuse is hauled. The first cost decreases with the size of the facility, due to economies of scale. The larger the capacity, the lower the cost as long as the capacity is being used.

Figure 6.13 presents the cost of transfer vehicles. The total transfer haul cost is the sum of the vehicle fixed cost per ton plus the variable cost, which depends on the travel time to the disposal site. The total cost of the transfer system is the sum of the cost of the transfer station (Figure 6.12) and the transfer haul cost (Figure 6.13). This total cost is compared with the cost of operating the collection vehicle (Figure 6.10) to determine the most cost-effective means to transport the refuse to the disposal site.

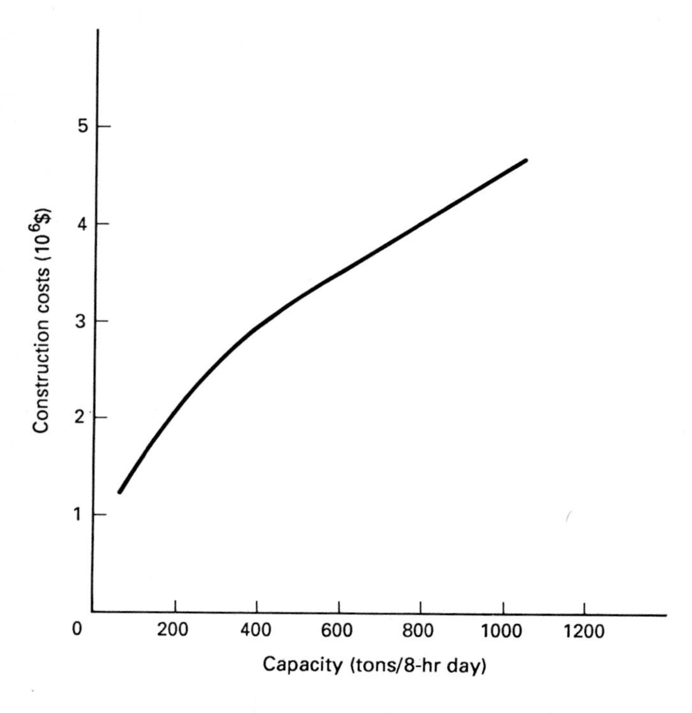

Figure 6.11 Transfer station construction costs. (From L. Schaper, *Waste Age*, Dec. 1982, p. 29.)

Transfer Systems and Equipment Utilized in the Industry

Highway transport is the primary mode of transporting refuse to distant disposal sites. This is not necessarily the most economical for large quantities of refuse, but may be all that is available in many urban areas. Aluminum and steel trailers with

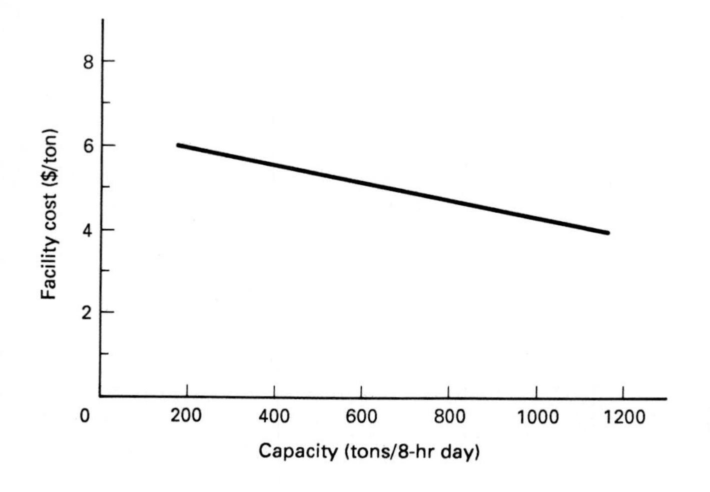

Figure 6.12 Construction and operation cost of transfer station. (From L. Schaper, *Waste Age*, Dec. 1982, p. 29.)

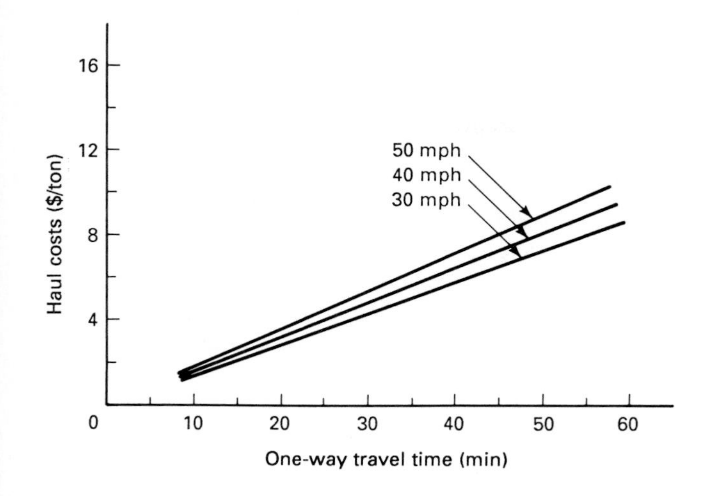

Figure 6.13 Transfer vehicle cost. (From L. Schaper, *Waste Age*, Dec. 1982, p. 29.)

capacities up to 96 yd^3 are used for hauling the refuse. Volume is the limitation for the amount of material that can be transported by one trailer. Densities are in the range of 500 to 600 lb/yd^3. Even at 96 yd^3, the weight of the refuse is only between 50,000 and 60,000 lb. Including the vehicle weight, this is still less than the 80,000-lb weight limit for primary highways. There are, however, other factors that may limit the load these trailers carry. Access to the landfill may be over unimproved roads that will not support heavy loads. Moving these heavy vehicles on the landfill site is difficult in wet weather. Therefore, not all vehicles will be sized for maximum capacity.

Offloading of the trailer at the site varies with the type of equipment. Many trailers have a cable winch that pulls a panel in the front of the trailer to the rear, forcing the refuse out the rear end. In other cases, the cable may be attached to a landfill tractor and the refuse pulled out of the trailer. Other systems involve large dump platforms that lift the entire truck into the air, with the refuse falling out the rear of the trailer. Dump trailers are not generally used because of the length of the trailer. When the bed is raised into the air for dumping, the raised trailer bed is very unstable, and moving forward to make room for the refuse discharging from the trailer could result in damage to the trailer.

Some transfer operations simply load baled refuse on flatbed trucks. After the refuse passes through the baler, the bale is picked up by a forklift tractor and placed on the trailer. After the trailer is loaded, it is covered with a tarpaulin to keep the refuse completely contained. At the disposal site, the baled refuse is unloaded by a forklift tractor and stacked in the landfill. These systems have been designated as ''bale fills.'' They appear to be economically attractive for long-haul distances where good roadway conditions exist. With the density of the bales approaching 1000 lb/yd^3, it is possible to exceed the maximum highway weight limit.

Baling of refuse has also been used to improve the load-carrying capacity of other modes of transportation, such as rail and barge. These transport systems can carry large weights, so they are always volume limited with a low-density material. The bales may weigh 4 to 5 tons, which facilitates transferring the refuse at the transfer station as well as the disposal site. With rail or barge haul, the refuse must be transferred to another vehicle at the landfill site to move the refuse to the working face of the fill. Also, containers are used to facilitate loading and offloading rail and barge transport systems. The refuse can be compacted when placed into the containers to a density of perhaps 700 to 800 lb/yd^3. The refuse is isolated from the environment during transport. Special cranes are used to load and unload the transporter. At the disposal site, the containers are transferred to trucks for haul to the working face of the landfill. Special unloaders dump the containers and place them back on the truck for return to the transporter, which in turn takes them back to the original transfer station for a new cycle.

The selection of a specific transfer system is very site dependent. The available transportation links between the source (municipality) and the disposal site (sanitary landfill), the distance (time) required to make the trip, and the quantity of refuse being transported will be the primary factors influencing the selection of a system. The trade literature describes a number of unique systems in operation.[6]

FACILITIES LOCATION AND WASTE ALLOCATION

When establishing a network of collection, transfer, and disposal operations, it is desirable to determine the cost-effective location for each component of the network. It is, indeed, a rare occasion when there is a choice in the site selection. There are so many external constraints on the location of refuse processing and disposal facilities that the basic question is "Where can I find a site?" not "Where can I find a site that has a cost-effective location?" It therefore becomes pointless to address the facilities location question. It is, however, desirable to establish the most cost-effective allocation of refuse once the sites have been identified.

Consider the locations of the sanitary landfill sites (L), an incineration site (I), and transfer stations (T) in Figure 6.14. How does one determine the amount of refuse that goes from each transfer station to each processing or disposal site? The primary objective of the distribution of the refuse among the various disposal options (i.e., the sanitary landfill and the incinerator) is the development of the least-cost disposal system. The costs considered are the cost of each disposal option, which may not be the same even for each sanitary landfill, and the cost of hauling the refuse from each transfer station to each disposal site. This cost function is described by

$$\begin{aligned} &X_{11}(H_{11} + D_1) + X_{12}(H_{12} + D_2) + X_{13}(H_{13} + D_3) \\ &+ \cdot \cdot \cdot \cdot + X_{mn}(H_{mn} + D_n) \end{aligned} \tag{7}$$

[6]*World Wastes,* Mar. 1984, p. 20; Mar. 1985, p. 18; Nov. 1985, p. 26.

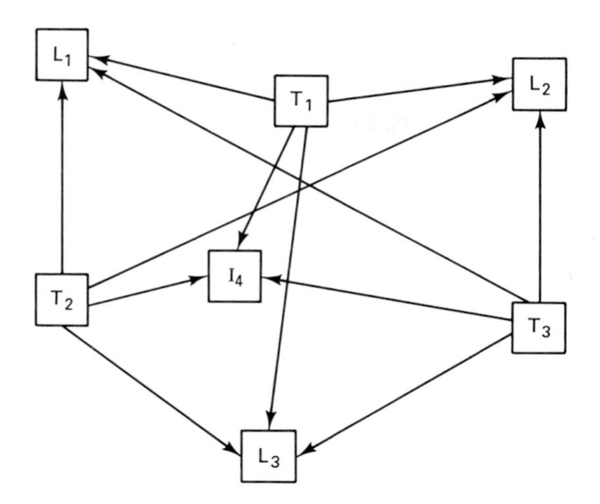

Figure 6.14 Allocation of refuse among disposal sites.

where X_{ij} = tons of refuse hauled from transfer station i to disposal site j
H_{ij} = cost of haul of refuse from transfer station i to disposal site j
D_j = cost of disposal at site j

Thus

$$\text{minimize cost} = \sum_{i=1}^{m} \sum_{j=1}^{n} X_{ij}(H_{ij} + D_j) \tag{8}$$

subject to the following constraints:

Total refuse available. There is only so much refuse available at any of the transfer stations, R_i. It is impossible to manufacture refuse in the transfer stations. Therefore, the total amount of refuse shipped out of a transfer station must not exceed the amount available. Conversely, all of the refuse available at a transfer station must be shipped to a disposal site.

$$\sum_{j=1}^{n} X_{ij} = R_i \tag{9}$$

Total capacity at the disposal site. Each site has a defined maximum capacity that cannot be exceeded, Q_j. The facilities simply cannot handle additional refuse. This constraint is defined by

$$\sum_{i=1}^{m} X_{ij} \leq Q_j \tag{10}$$

Additional site capacity constraints may also apply. For example, operation of the incinerator (disposal site 4) at partial capacity increases the unit cost. Therefore, an incinerator may require that the refuse flow to it equals its capacity. This is defined by

$$\sum_{i=1}^{n} X_{i4} = Q_4 \tag{11}$$

Direction of refuse flow. It is not practical to consider hauling refuse from one disposal site back to a transfer station and then to another disposal site. It is not likely that such a scenario will occur, but to ensure that it does not, the constraint shown in equation (12) is applied:

$$X_{ij} \geq 0 \tag{12}$$

This constraint controls the direction of refuse flow.

While these equations are rather simple to solve with some of the available linear programming software, the cost functions are difficult to define with accuracy. Therefore, such an exercise will be only as valuable as the data provided in the formulation of the cost functions for the haul and disposal costs.

STUDY QUESTIONS

1. How does the location of the storage containers for single-family residences affect the collection costs?
2. What is a "shoulder barrel"? How does use of a shoulder barrel affect collection costs, aesthetics of refuse collection, and quantity of refuse that can be produced per service stop?
3. How does collection from single-family residence service stops differ from that of multifamily/commercial service stops?
4. What is "standard time"? How is the methods-time-measurements technique used to improve the efficiency of refuse collection?
5. What is the effect of crew size on the standard time required for each service stop? How does container location affect the standard time?
6. What time factors contribute to nonproductive time and are not considered as part of standard time?
7. What time factors comprise the collection time and cost?
8. Find the nearly optimum size of collection vehicle for use under conditions (a) and (b) below. What observations can you make on the effect of haul time to the disposal site on the optimum size of the collection vehicle? *Given:* Compacted refuse density, 500 lb/yd^3; production rate, 90 lb/service stop;

mean time per service stop, 1.0 min; crew size, 3; labor rate, $12/hr; off-route time, 60 min; disposal time, 10 min; maximum overtime, 30 min at $18/hr; $a = 0.25$; vehicle capacity/costs:

Capacity (yd^3)	Operating cost ($/hr)	Relief cost ($/hr)
10	28	18
15	36	23
20	45	26
25	52	30
30	60	37
35	70	45

Condition (a): Haul time to disposal site is 15 minutes.
Condition (b): Haul time to disposal site is 60 minutes.

9. What are even- and odd-degree nodes?
10. What is a "unicursal network"? How is one constructed from a group of streets in a collection route?
11. What is Kwan's rule for minimizing the retracing of streets?
12. What are the cost components of a refuse transfer system? Under what conditions does a transfer station result in lower refuse collection and disposal costs?
13. What transportation systems have been used for the long-distance transport of refuse?
14. What characteristic of refuse has a significant impact on the cost of long-distant transport?
15. What are the constraints imposed on the distribution of the refuse collected from an urban area among the various operating disposal sites?

CHAPTER 7

Engineering Economics

METHODS TO FINANCE SOLID WASTE MANAGEMENT SYSTEMS

The funds necessary to finance the construction of publicly or privately owned systems must be obtained from the commercial lending market. This is accomplished by the use of bonds which are sold by bonding institutions at an interest rate that reflects the current market conditions and the security used to back the bonds. A brief discussion of the possible financing mechanisms is necessary to understand the impact of the bonding method on the cost of the facilities. The types of bonds available to public and private agencies are different.

General Obligation Bonds: Municipal Corporations Only

General obligation (GO) bonds can be used by any municipal corporation or agency that has a defined property tax base. This tax base is pledged as a guarantee that the bonds will be retired as agreed. A tax rate based on the assessed valuation of the property is levied annually. Because the tax base is pledged to support GO bonds, they generally have the lowest interest rate, as they are considered the safest possible investment. The property tax is collected by the county in which the municipal agency is located and transferred to this agency for making the required annual payment on the bonds. The actual tax rate levied can change as the assessed value of the property changes.

The revenue generated from the operation of the facility is not required to retire these bonds. The annual bond payments are available from the property tax collected each year. Historically, this has led to abuse of this funding mechanism. Systems were built that were not technically sound, and since the revenue for paying of the bonds was guaranteed by the property tax, little effort was made to improve the technical operation. In fact, many of these systems were abandoned, leaving the municipal agency (property tax payers) saddled with the bill.

The security of this investment is protected by the states, which limit the amount of general obligation debt that a municipal corporation can accumulate. It is easy to sell GO bonds because of the security associated with them. The experience with the near bankruptcy of New York City in the 1970s did considerable damage to the public perception of these bonds. However, the federal government provided assistance that prevented New York from defaulting on any bond payments. The new controls that now exist have recaptured the public's confidence.

Revenue Bonds: Public and Private Financing

Revenue bonds are becoming the preferred method of financing solid waste management systems. The constraint on the percentage of assessed valution that can be used for GO bonds has reserved their use for projects that do not have a defined revenue stream, such as storm drainage, municipal buildings, and matching funds for street construction. A sanitary landfill, incinerator, or collection system can have a revenue from the tip fee or the service charge assessed the producers of the solid waste. The revenue bonds are a little more risky in that they rely on the income from the operation of the system for the funds to retire the bonds. Consequently, these bonds cannot be sold unless the economic analysis shows that the income stream is adequate to meet the expenses and bond payments. Binding contracts will probably be required to guarantee the source of funds (i.e., a tip fee). This is where the need for flow control of the refuse is most acute. Also, the projections of the refuse tonnage processed is a critical number in determining the economic validity of the project. Is the projected tonnage available, and if so, can the system process it at the required rate? This is frequently a deficiency in the cost projections. If with a nominal capacity of 1000 tons/day, only 800 tons/day are processed, the income stream is only 80% of the projected value for 1000 tons/day.

Revenue bonds can be used to finance both public and private projects. The total capital required for public agencies to finance projects is significantly less than that required for private companies. There are many reasons for this differential, one of which is the tax status of municipal bonds and municipal agencies. Municipal corporations do not pay any federal income tax or property tax. Also, the interest rate is generally lower because of the tax-exempt nature of the income from municipal bonds. The income from bonds used to finance privately owned systems are not exempt from federal income tax, so the interest rates are higher. Also, there are other financing requirements that affect the amount of money that must be borrowed. These requirements are discussed in detail later.

Innovative Financing Methods

To assist municipalities in attracting new business, a number of mechanisms have been developed to provide a financial incentive to encourage building industrial facilities. Industrial development bonding authority is one mechanism that has been given to units of local government. These bonds are similar to other municipal bonds in that the interest is exempt from federal income tax. Therefore, the interest rates are lower than the current market rate. These bonds are issued by the municipality to finance a particular privately owned project. The revenue from the project is to be used to retire the bonds. Again, because of excessive use of this bonding mechanism, most states have placed a limit on the amount of these bonds that a municipality may issue.

The creation of a tax-increment financing (TIF) district by a municipality is a mechanism to postpone property tax payment on the improvements associated with the construction of a facility. The intent of a TIF district is to encourage the redevelopment of blighted areas of a community. In addition to the stable low property tax rate, the municipality may improve the infrastructure of the area as an incentive for the private development.

Other mechanisms may be developed in the future to assist private developers in reducing the cost of a facility. However, a bond payment plan must exist that will provide the necessary funds to meet the interest payments and the face value of the bonds when they are due. A revenue stream must be generated to meet this payment plan.

COST COMPONENTS

To consider alternative systems for solid waste management, it is necessary to compare them on the same economic basis. Costs are divided between fixed costs (capital costs) and reoccurring costs (operation and maintenance).

Capital Costs

Capital costs are those up-front costs associated with the purchase and installation of the system being constructed. There are several costs that are considered capital costs and are included in the total value of the bond issue.

On-site construction costs. The construction costs are associated with on-site and off-site activities. The on-site, or battery limits costs represent those costs that are common for the system no matter where it is constructed. These costs are less site dependent and more predictable. The following cost categories are included:

Equipment. All systems will require various pieces of equipment to make the system complete. These items are priced by the manufacturer and generally do not include installation costs. For initial cost estimates, it is possible to obtain estimates of the installation cost from the equipment supplier. These installation costs are not sufficiently accurate to use for the economic evaluation needed for financing purposes.

Foundations. Foundations are required for all significant pieces of refuse-processing equipment as well as buildings and other structures. The cost of these foundations can be determined based on specified loading parameters. They are site dependent to some degree and require knowledge of the soil and foundation characteristics of the site.

Buildings. The construction costs of the buildings required to house the system is another cost category to include in construction cost. This cost is very closely tied to the type of process system and the quantity of solid waste generated. Buildings can be used for a number of purposes and are likely to be a significant component of the cost.

Utilities. The utility system for the site must be developed to distribute the electricity, steam, and water and to provide the necessary storm drainage and sewers and perhaps a treatment system for the sanitary sewage and any process wastewater. The utility needs are system dependent and can be a significant construction cost.

Site Development Costs. The site development costs refer to the provisions of on-site roads and other pavement, landscaping, and the like. All of these costs relate to the actual site.

Off-site construction costs. Providing the necessary roadways or railroads and utilities (electricity, sewers, and water) to a facility may be a significant cost item. The availability of adequate electrical power may be a determining factor in the facility location. A system with several thousand horsepower operating will need access to a major power supply. These cost categories are very site specific and may be deciding factors in site selection for certain systems. These costs are part of the infrastructure development costs that the municipality may be willing to assume.

Land acquisition costs. Land requirements vary with the system. A sanitary landfill may require several hundred acres. Conversely, a transfer station may need only 2 or 3 acres. Land costs are both system and site dependent.

Professional services. Several professional services are required to assemble a project of any size. The development of the necessary design drawings and specifications and the construction supervision will require a substantial effort from a consulting engineering firm. The cost of these services is negotiated but may range between 7 and 10% of the project cost, on-site and off-site, for the preparation of the design documents. Construction supervision may require a similar expenditure.

Additional professional services may be required that relate to contract negotiations, fiscal management, and the general administration of design and development phase. Funds have to be made available to cover these costs as they occur. Bills are submitted for work completed and are expected to be paid. There is no income stream yet, so it is necessary to borrow the funds to cover these costs.

Financing costs. This is a cost category that is frequently missed in a casual economic evaluation. This category includes the cost of selling the bonds as well as the funds needed to meet the financial commitments until a revenue stream is developed. These costs are covered by actually borrowing money by issuing more bonds so that these payments can be made before the revenue stream starts and for money that is to be placed on reserve for future bond payments in case of a break in the revenue stream.

Bond Issue Costs. There are brokerage houses that specialize in raising capital by selling bonds for a client. The magnitude of the fee for this service will vary with the size of the issue, but it is approximately 4% for a normal-sized bond issue.

Bond Payment Reserve. A requirement that is imposed on the borrower of the money is the establishment of a reserve that can be used to make the necessary bond payments in case of process failure or plant shutdown. The magnitude of this reserve is a reflection of the technical maturity of the processing system. If there is some uncertainty about the ability to keep the system on line, the required reserve will be much higher than if the system has a long history of successful operation. A typical reserve may be equal to 1 or 2 years of bond payments. Of course, the higher the reserve, the higher the financing costs. Since this money is a reserve, it is placed at interest and actually earns money until it is used to make the last payments to retire the bonds.

Bond Payment During Construction. The time required to complete the construction of a processing system may be one or more years. During this period, there will not be any income stream to meet the bond payment requirements. The bonds may be retired at a uniform annual rate that starts during the first year after the bonds were issued. In other payment schedules, principal payments may not be required for years, but interest payments are required during the first year. If the financing arrangements require payments starting at once, the funds to make these payments must be borrowed.

Startup costs. Once construction is complete, a period of time is required before the facility becomes operational. Much of the equipment will require adjustment while operating, some equipment may be installed incorrectly, operators must be trained, and so on. This will prevent the facility from operating at capacity and the expected income stream will not exist during this period. The full labor costs and most of the operating costs will be incurred during this time. It may take from 1 to as many as 6 months, depending on the complexity of the project, before the facility becomes operational. The allowable startup time is generally in the contract, and if longer periods are required than specified, it may be at the expense of the contractor or the equipment supplier. Funds are made available to cover these costs by borrowing additional money.

Contingency. A contingency is always provided for in any economic evaluation. This is a measure of how much confidence one has in the data used to generate the design and compute the economics. The larger the contingency, the less confidence exhibited in the system data base and performance. Contingencies of 10% are common in many designs.

Cost Indexing

Frequently, the cost for the equipment or the installation cost may be obtained from a project that was bid in previous years. Other sources of cost information may be cost curves available from company files. In general, these costs are all referenced to a specific year and have a cost index value associated with them. The cost index is designed to correct the cost information for inflation. A number of indices exist and each is somewhat specific. The following are three that may be useful in the environmental field.

Engineering-News Record. There are two indices: one that covers heavy construction—concrete and earth moving, and one that covers building construction. These indices are published weekly in *Engineering-News Record.*

U.S. Environmental Protection Agency. The U.S. EPA has developed a cost index for small and large wastewater treatment plants. Until 1989 these indices were published annually in the Washington Notebook section of a monthly issue of the *Journal of the Water Pollution Control Federation.*

Chemical Engineering Plant Cost Index. *Chemical Engineering* magazine publishes an index that shows how much the chemical plant construction costs have escalated with time. This index may be particularly useful for refuse-processing systems because many of these systems have a lot of processing equipment.

The chemical plant cost indices listed in Table 7.1 illustrate the change that can occur in the costs of facilities. The cost of chemical plants during the period 1957–1959 is used to set the reference index of 100. The ratio of these indices will permit the extrapolation of historical cost data to present costs. The changes can be significant. For example, if the cost of a plant in 1980 was $10,000,000, the same plant would cost $11,380,000 in 1981, a 13.8% increase. This was a period of high inflation. Conversely, a $10,000,000 plant built in 1984 would cost only $9,845,000 in 1986. These indices clearly reflect the national pattern for economic growth and

Table 7.1 Chemical Plant Cost Indices

Year	Plant cost index	Percent change
1980	261	
1981	297	13.8
1982	314	5.7
1983	317	1.0
1984	323	1.9
1985	325	0.6
1986	318	−2.2
1987	324	1.3
1988	343	5.9
1989	355	3.5
1990	357	0.6

demand for materials and equipment. A negative index will occur only when the manufacturers have an oversupply of product and are selling at a discount, if not below cost.

Operating and Maintenance Costs

Operation and maintenance costs are variable costs that depend on the way in which the facility operates. Many of the costs are in direct proportion to the quantity of material processed. For example, the electricity required will depend on how many hours the facility operates and the quantity of refuse processed. Labor may be less dependent on the quantity processed. The labor contracts will generally require payment for an 8-hour shift even if the desired through put is achieved in 6 hours. Maintenance costs will also be dependent on the hours operated and the tons processed. It is necessary to be specific in identifying these costs; what determines the actual cost for each category?

Labor. The operating and maintenance personnel requirements are specific for each facility. The skill levels and trades required must be determined before the labor costs can be specified. There are two cost categories for labor: base wage rate and fringe benefits. The base rate represents the hourly wage, which is based on job skill requirements and experience. Certain fringe benefits are applied uniformly to all employees (e.g., health insurance, dental insurance, clothing allowance, etc.). The cost is independent of the wage rate. Other fringe benefits are tied to the wage rate (e.g., paid vacation and holidays, retirement, etc.). However, for cost estimating, it is common to apply a percentage to the base wage to determine the magnitude of the fringe benefits. A typical number might be 35%, but this percentage could be much higher for certain employee contracts.

Utilities. There is always a requirement for utilities at any refuse management facility, if only for the personal needs of the employees. These costs are relatively insignificant and generally not evaluated. However, the utility costs for a processing system may be significant, especially energy costs. Careful evaluation of these costs is important because they can have a significant impact on the economic feasibility of a system.

Supplies. This category includes the various operating and maintenance supplies required in the facility. A variety of operating supplies are included, ranging from process chemicals to fuel for steam generation. Refuse management systems have a limited demand for these supplies, except for fuel for the collection and transport vehicles and the wheeled equipment used on site. The maintenance supplies can be significant for certain processing systems and for the collection and haul vehicles. The economic comparison can be only as good as the estimates for these costs. Some are very significant, whereas others are insignificant, but all are highly system dependent.

Administrative Costs

Personnel. All facilities must have an administrative structure, including a facilities manager, business manager, legal staff, and a support staff. The normal personnel costs apply to this group, both the base wage rate and the fringe benefits. The size of this group will be system dependent.

Insurance. Insurance has not been considered to be a significant cost factor in the economic health of a system. In addition to the normal insurance requirements, such as worker's compensation, some processing and disposal systems may have a very large potential liability. Some agencies choose to discount the need for this type of insurance, while some of the large companies are self-insured. The future requirements in this area are likely to become more restrictive. It is possible that some of the new regulations for disposal sites will require the operator to carry insurance against any damage claims, environmental or otherwise.

Legal and fiscal. These costs are associated with the normal operation of the facility. The need for legal assistance will depend on the characteristics of the facility and how much resistance exists from the community. Fiscal costs are associated with the required audits and contract accountability. This can be a significant activity when a private company has a contract that has a profit-sharing clause for the host community. Record keeping must be detailed and complete.

Taxes

This category applies only to privately owned facilities. Municipalities are not required to pay either property or income tax. Under normal circumstances, private companies are required to pay both. Mechanisms are available to municipalities to provide property tax relief as an inducement to encourage private development of facilities. However, these inducements are not very common in this industry. They have been used to make some of the large mass-burn facilities more economically attractive as an investment for private money.

ECONOMIC COMPARISONS: PUBLICLY OWNED SYSTEMS

Publicly owned systems do not have profit motivation as a goal. The issue is not the return on investment. The municipal agency has a problem that needs a solution, so the objective is to find a system that has the lowest cost. Because fixed and variable costs are involved in determination of the economic comparisons, some procedure is necessary to express these costs on a common base. There are a number of ways to achieve this goal, but two common procedures are used in the environmental field, annual costs and present worth.

Annual Costs

The determination of annual costs provides more information for making cost comparisons because it is possible to reduce the annual costs to a common parameter such as dollars per ton. Operating and maintenance costs are readily available as annual costs since they accrue with the operating time. The first cost (capital costs) can be converted into annual costs by using the appropriate compound interest formula. The following equation gives the capital recovery factor (CRF) for repaying a loan over a period of n years at an interest rate of i per annum.

$$\text{CRF} = \frac{i\,(1 + i)^n}{(1 + i)^n - 1} \tag{1}$$

For example, the CRF for a 20-year loan at 10% interest is 0.11746. A \$10,000,000 loan would require an annual payment of \$1,174,600 for 20 years. The CRF can be computed from equation (1) or can be obtained from any number of handbooks.

Since the operation and maintenance costs are available on an annual basis, the same basis as that used for the annual capital costs, these numbers can be added to obtain the total annual costs associated with the facility. Knowing the capacity of the facility, the tons processed per year can be computed. Dividing the total annual costs by this tonnage will yield the cost per ton processed.

Present Worth

When the federal government initiated funding of the construction grants program for wastewater treatment, it chose to use a different basis for cost comparison, the present worth. Present worth is the amount of money that would have to be invested at i interest to generate a fixed amount of money every year for n years. It is the inverse of the capital recovery factor, and the present worth factor (PWF) given by the following equation is the inverse of CRF:

$$\text{PWF} = \frac{(1 + i)^n - 1}{i\,(1 + i)^n} \qquad (2)$$

In fact, equation (2) is the inverse of equation (1). Multiplying the annual costs associated with the operation and maintenance by PWF determines the present worth of these annual costs. This present worth value is comparable to a capital investment and therefore can be added to the capital cost of the facility. The resulting total cost, present worth value, can be used for economic comparisons of the systems.

ECONOMIC EVALUATION FOR PRIVATE FINANCING

The following example is an economic evaluation of a facility that is planned to process 400 tons/day on a 5 day/week basis or 2000 tons/week. This is a privately financed facility in which the developer has taken an equity position (invested some of their own funds). Construction and startup of the plant are assumed to begin in 1988 and take 2 years to complete, and the facility is scheduled to operate for 20 years. A bond is floated in 1988 with payments to begin immediately. During construction, only interest payments are made. Once the plant is operational, both principal and interest payments are made. A bond payment reserve (1.5 years) is set aside in 1988 and kept until the last year of operation (2009), at which time it is recovered. Annual interest is earned on that reserve. Construction costs are paid during the course of the construction period, and short-term interest is earned on the unspent money raised by the bonds.

The definitions used to determine the necessary bond issue are presented below.

Total Capital Cost (TCAP). This includes the cost of land, equipment, installation, design, construction supervision, startup, capital contingencies, furnishing, market development, mobile equipment, and so on.

Equity. This is the amount of cash the developer has invested in the plant.

Bond Payment. The semiannual bond payment is determined by using the capital recovery factor (CRF) for an annual interest rate of 8% and 20 years (since the bond payments are semiannual, $i = 0.04$ and $n = 40$ payments):

Bond Payment Reserve (BPR). A reserve of three semiannual bond payments (1.5 years) is set aside at the beginning of the project. That reserve will earn interest both during construction, shown as interest earned on reserve (RINT), and during the life of the plant. That interest is estimated as 8% annually. The reserve is recovered at the end of the bond term (2009).

Bond Payment During Construction (BPDC). Two years of bond interest must be paid during construction. Those payments must be made with borrowed money. The money is to be received at the beginning of 1988, while the four payments are made at the end of 6-month intervals. The money is invested and earns short-term interest [bond payment interest (BPINT)] until these semiannual payments are made.

Interest on Construction Funds (IDC). During the construction period, payments are made according to completed work. The estimate assumes that the required money is borrowed at the start of construction and that it is placed into a short-term investment until needed. While the interest received will depend on the payment schedule, it is estimated here as 5.3% of the total capital cost.

Bond Issue Costs (BIC). This is the fee paid to an investment firm for the sale of the bond issue. Generally, it is some percentage of the bond issue. In this example, 4% was used.

Bond Issue (BOND). The bond issue is the amount of money that must be borrowed and is determined as follows;

$$\text{BOND} = \text{TCAP} - \text{EQUITY} + \text{BIC} + \text{BPR} + \text{BPDC} - \text{RINT} - \text{BPINT} - \text{IDC} \quad (3)$$

Other pertinent equations are:

$$\text{EQUITY} = \text{FEQ} \times (\text{TCAP} + \text{BIC} + \text{BPR} + \text{BPDC} - \text{RINT} - \text{BPINT} - \text{IDC}) \quad (4)$$

$$\text{BIC} = 0.04 \times \text{BOND} \quad (5)$$

$$\text{BPR} = \text{CRF} \times 3 \times \text{BOND} \quad (6)$$

$$\text{BPDC} = i \times 4 \times \text{BOND} \quad (7)$$

$$\text{RINT} = (\text{BOND} \times \text{CRF} \times 3) \times \text{INTR} \times 2 \quad (8)$$

$$\text{BPINT} = (\text{BOND} \times i) \times 2.5 \times \text{INTR} \quad (9)$$

$$\text{IDC} = 0.053 \times \text{TCAP} \quad (10)$$

where FEQ is the fraction of equity that is put forth and INTR is interest on the short-term investments (8% per annum).

Once the plant is in operation, the operating expenses and bond payments must be offset by its revenues. The expenses that are included here are those for personnel, power, water, equipment replacement, maintenance materials and supplies, landfill disposal, and general supervision and administration (GS&A) expenses. These expenses are inflated each year at 5%.

Table 7.2 1990 expenses and revenues

Expenses		Revenues	
Personnel	$1,533,200	Methane	$ 1,578,500
Power	1,688,000	Carbon dioxide	1,351,000
Water	2,920	Tipping fee	5,564,000
Equipment replacement	181,700	Sewage sludge	952,000
Maintenance (M&S)	990,800	Interest income	353,000
Landfill disposal	235,040	Metals	275,000
GS&A expenses	788,375		
Mobile equipment	288,500		
Contingency	209,000		
Total expenses	$5,937,535	Total revenues	$10,073,500

To illustrate an economic evaluation, a process for the anaerobic fermentation of refuse to generate methane has been selected (see Chapter 10). This facility receives revenues from the sale of methane, carbon dioxide, and recovered metals, from the tipping fee for the refuse, from the fee for the sewage sludge disposal, and from the interest earned on the bond payment reserve. The revenues and expenses for the first year of operation for this example are presented in Table 7.2. The estimate of the capital requirements and the cash flow for the facility are presented in Tables 7.3 and 7.4 for an equity contribution of 25%.

The cash flow presented in Table 7.4 shows the revenues exclusive of that generated by the tipping fee. The tipping fee is determined based on the difference between the bond payment and the operating expenses and other revenues plus the required pretax income. The income from the bond reserve and the expense of the bond payment are constant over the life of the project (with the exception of the return of the 1.5-year bond payment reserve at the end of the project). Expenses and revenues other than methane are assumed to increase with inflation (5%). Methane revenues are inflated according to the Gas Research Institute's (an industry group) gas price projections. The methane revenues increase because of inflation as well as resource depletion.

Table 7.3 Capital requirements

Total construction costs	$ 28,862,000
Equity	7,034,000
Bond capital requirements	21,828,000
Bond issue costs	1,179,000
Bond payment reserve	4,467,000
Bond payment during construction	4,716,000
Interest earned on reserve	(715,000)
Interest on bond payment	(472,000)
Interest on construction funds	(1,530,000)
Bond issue	$ 29,474,000

Table 7.4 Cash flow

Year	Other revenue ($1000)	Bond reserve ($1000)	Operating expenses ($1000)	Bond payment ($1000)
1990	4,157	358	5,938	2985
1994	5,383	358	7,218	2985
1999	7,289	358	9,211	2985
2004	9,962	358	11,757	2985
2009	14,014	4771	15,005	2985

The tipping fee required to cover the difference between the expenses and other revenues presented in Table 7.4 may be determined quite readily. For example, in 1990, the tipping fee would have to generate $4,408,000 ($4,157,000 + 358,000 − 5,938,000 − 2,985,000). Since this facility is expected to process 104,000 tons/year (2000 tons/week), the tipping fee must be $42.38 per ton. This is the break-even tip fee; no profit is included. This number does not change substantially until the last year, when the bond reserve is treated as income. The additional tipping fee required to provide a profitable operation depends on the measure of acceptable return. Three measures are investigated here:

1. At least 10% first-year pretax income ratio
2. At least 25 to 30% average pretax income ratio (PTIR)
3. At least 25 to 30% average annual pretax return on invested equity (ROI).

The PTIR, also known as the profit margin, is calculated by dividing the pretax profits by the gross revenues. As the equity increases, the expenses for the bond payments decrease, which then increases the profit for a fixed tipping fee. The ROI is determined by dividing the pretax profits by the amount of equity invested.

The objective of this exercise is to find the equity contribution yielding the minimum tipping fee that satisfies all three profit measures presented above. Figure 7.1 is constructed by calculating the income and expenses for each equity position and applying the appropriate profitability objective. Each line plots the 1990 tipping fees that are required to satisfy the foregoing constraints for the range of equity that might be put forward. The tipping fee required to satisfy a given pretax income ratio decreases with increasing equity because the annual cost of the bond payment decreases. The tipping fee required to give the desired return on investment increases with increasing equity because the initial investment, and thus the required return, is larger.

The feasible regions shown in Figure 7.1 are equity/tipping fee combinations that satisfy all three constraints. Region 1 satisfies the condition of 25% ROI and 10% first-year PTIR. Region 2 satisfies the constraints of 30% ROI and 30% PTIR. Notice that the requirement for a 10% first-year PTIR exceeds the requirement for a 25% average PTIR. Since this objective is binding, the tipping fee will be higher than that required for 25% average PTIR and ROI. Putting forth 12 to 15% equity

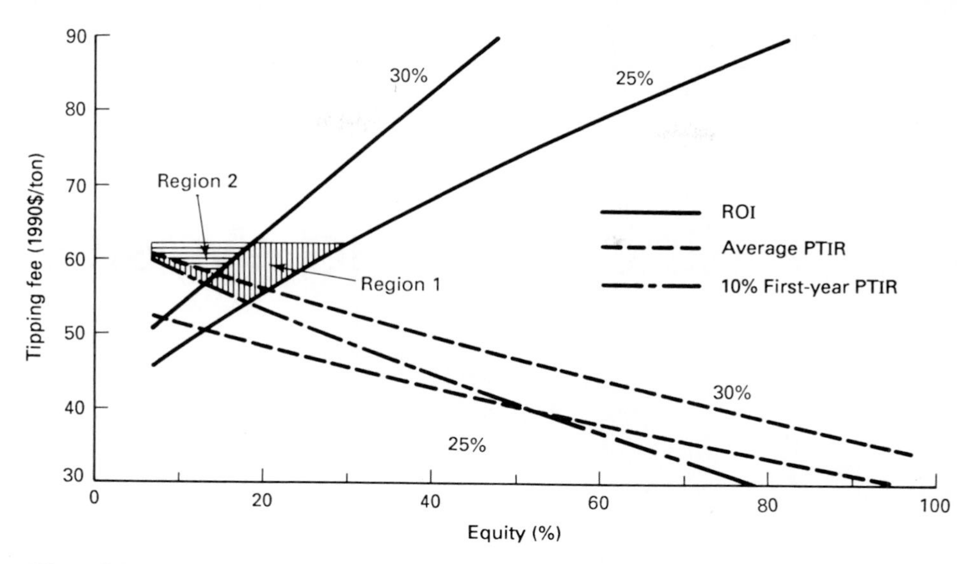

Figure 7.1 Economic evaluation for private financing.

gives the lowest tipping fees for the 30% PTIR and ROI requirements. The tipping fees would be about $57 per ton. The first-year PTIR constraint requires a tip fee of about $52 per ton and an equity position of about 18% if the 25% ROI is used.

STUDY QUESTIONS

1. What are the advantages and disadvantages of GO bonds and of revenue bonds?
2. What is the source of revenue for each type?
3. What cost elements are considered to be capital costs?
4. Distinguish between capital and construction costs.
5. What is the value of cost indexing?
6. What costs are considered to be operation and maintenance costs?
7. Compute the cost of a publicly owned processing system in dollars per ton given the following information.

Plant capacity, 100,000 tons/yr	Bond issue, $10,000,000
Interest rate, 10%	Economic life, 20 years
Operating costs, $800,000/yr	Maintenance costs, $350,000/yr

8. Calculate the present worth for the facility in Question 7.
9. Distinguish between fixed and variable costs.
10. How does present worth compare fixed and variable costs?
11. How does the actual bond issue differ from the capital costs for private industry financing?

CHAPTER 8

Processing of Solid Wastes

Historically, refuse has been considered as a waste material and hauled to the dump for disposal, or in the urban areas to an incinerator for partial combustion. With recognition of the magnitude of the solid waste management problem and the impact that improper management was having on the environment, alternative methods for management of the refuse were sought. These methods involve a variety of processing systems that are used to upgrade the quality of the refuse, as is the case with refuse-derived fuel (RDF), or to recover specific streams from the refuse, such as aluminum and plastic. In this chapter we present the various processes for preparation of the refuse stream for use as RDF and/or for the recovery of useful materials.

RECEIVING AREA

After the collection vehicles have been filled, the refuse is transported to a tip area for additional processing or final disposal. The facility receiving the collection trucks must be designed to facilitate the movement of the trucks in and out as well as unloading of the refuse. The flow of traffic becomes a critical issue in the design of the receiving area, whether it is a processing plant or a sanitary landfill. Of the following components, only the weigh station would be required at a sanitary landfill.

Weigh Station

This station consists of a platform scales of sufficient size to accommodate the largest vehicle expected to use the facility. The vehicle has an identification number that is fed to the computer as the truck drives on to the scale. This number identifies the owner of the vehicle and the empty weight. The gross weight is recorded and the net weight computed. This information can then be used to bill the owner of the truck for the tons of refuse received. This tip fee is a major source of revenue for the facility.

These stations can be operated manually by having the data inputted by a keyboard. This requires an operator in the scale house. There may be advantages to having more control of the access to the facility by having this checkpoint. When access control is not a problem, the facility can be completely automated. The vehicles have magnetic cards that are inserted into a card reader. The information is collected and computed automatically. The degree of sophistication of the weigh station must be balanced against the cost of the system and other needs that may be served by having personnel on duty at this station.

The weigh station serves other purposes by providing data on the rate at which refuse is processed by the facility. The input tonnage is important for making cost calculations for the operation of the plant. In the case of a sanitary landfill, the accumulated tonnage received indicates the rate at which the capacity of the fill is being used. The weigh station also provides a data base for determining the total refuse production of the area served by the facility. Because the vehicles are coded, it is possible to track the rate of solid waste production on the routes served by the specific vehicle. This will assist in revising the routes as the refuse production changes because of changing land uses.

Receiving Area/Tip Floor

The sizing of the tip floor must consider the number of trucks that will be unloading in a given period. If the facility receives only 100 tons/day, that is equivalent to about 20 loads. Distributing this number over 480 minutes results in one load every 24 minutes. Therefore, if the unloading time is 10 to 15 minutes, space for only one truck at a time is necessary. Of course, the truck arrival is never uniformly spread over the day. In this case it might be appropriate to have space for three or four trucks to unload at the same time. If 12 ft is allowed for each unloading area, a space of 35 to 50 ft would be sufficient. Figure 8.1 is a photograph of a receiving area that uses a portion of the tip floor for short-term storage of the refuse (slab storage).

As the facility becomes larger, the number of trucks increases in proportion. A figure of 1000 tons/day would be the equivalent of about 200 loads. Distributing these loads over a 480-minute period would result in one truck ever 2.4 minutes. A 10- to 15-minute unloading time would require five or six spaces for unloading. Again, the trucks tend to arrive in bunches. When they start the route at the same time, they will generally finish at approximately the same time. The only factor that will cause a distribution of the truck arrival will be the time differential between the route and the tip area. Planning of collection routes can assist in distributing the arrival times over the workday. A particularly difficult time is the end of the

Figure 8.1 Collection vehicle unloading on a tipping floor.

day. If the day ends at the same time for all collection crews, all vehicles will arrive at the tip floor at the same time. This will result in a long queue and a considerable loss of time. It is not practical to provide space for perhaps 100 trucks to unload at the same time. Therefore, some thought should be given to a staggered starting time. Still, the number of spaces required would be 10 to 15, or a distance of 120 to 180 ft. There will be many other factors that may limit the space available for truck unloading.

Temporary Storage

All refuse-processing facilities are faced with the problem of matching the rate at which the refuse is received with the rate at which it is processed. As discussed above, the solid waste does not arrive uniformly over the operating day. The S-shaped curve in Figure 8.2 illustrates how 500 tons of refuse might accumulate over an 8-hour operating period. This mass accumulation curve shows periods of low receipts and periods of high receipts. If the refuse is being fed into a processing line at a constant rate, there are times when not enough refuse is being received to meet this demand rate, or times when too much refuse is being received. A ''surge'' or storage facility must be available to accommodate these differences. Curve 1 shows the demand rate for a processing system processing the 500 tons in an 8-hour period. It is assumed that the operating day for the processing line starts at 8:00 A.M., the same time as the collection day. A total of 135 tons of storage would be required for this operating condition. It is clear that the amount of storage could be reduced if the processing line was not started until a sufficient quantity of refuse was received. Shifting the demand curve 1 hour later could reduce the storage requirements significantly.

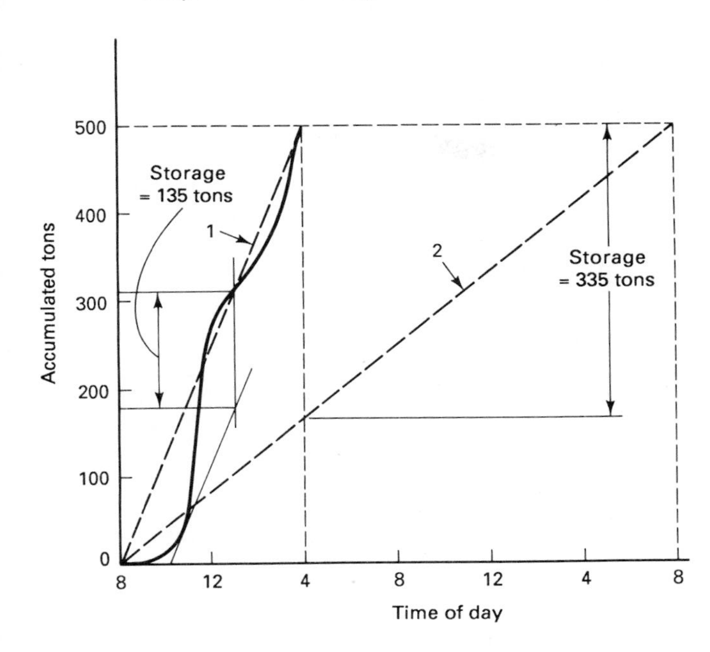

Figure 8.2 Accumulated mass flow rate of incoming refuse.

The required storage is complicated by the period of operation for the facility. If the refuse is processed 24 hours/day, there must be enough refuse in storage at the end of the collection day to supply the facility until the trucks arrive to unload the next day. This is illustrated by curve 2 in Figure 8.2, which shows the demand rate for processing the 500 tons in a 24-hour period. Storage capacity for 335 tons would have to be provided for this operating condition. If the refuse is incinerated, the storage must be sufficient to feed the incinerator from the end of the collection day on Friday until sufficient refuse arrives on Monday. This would require a storage capacity for over 64 hours of operation of the facility.

Pit storage. A common method of storage for the large mass-burn (incineration) plants is the use of a large pit, as shown in Figure 8.3. The tip floor may be 25 to 30 ft above the bottom of the pit and the trucks back up to the edge of the pit and dump the refuse into it. The refuse is reclaimed with an overhead crane using a "claw." The refuse is loaded into a feed hopper and conveyed to the furnace. The crane operator can monitor the quality of the refuse in the pit and select a balanced feed to the furnace. It is possible to alternate, for example, wet and dry refuse to obtain a more even heat load on the furnace. Operator control of the feed stream is a very important consideration that tends to discourage total mechanization.

Alternative techniques for reclaiming refuse from storage include "live bottom" pits. The bottom of the pit is a heavy pan conveyor that moves the refuse to one end of the pit, where it is transferred to another conveyor for feed to the furnace. This system has had limited success with raw refuse because of bridging of

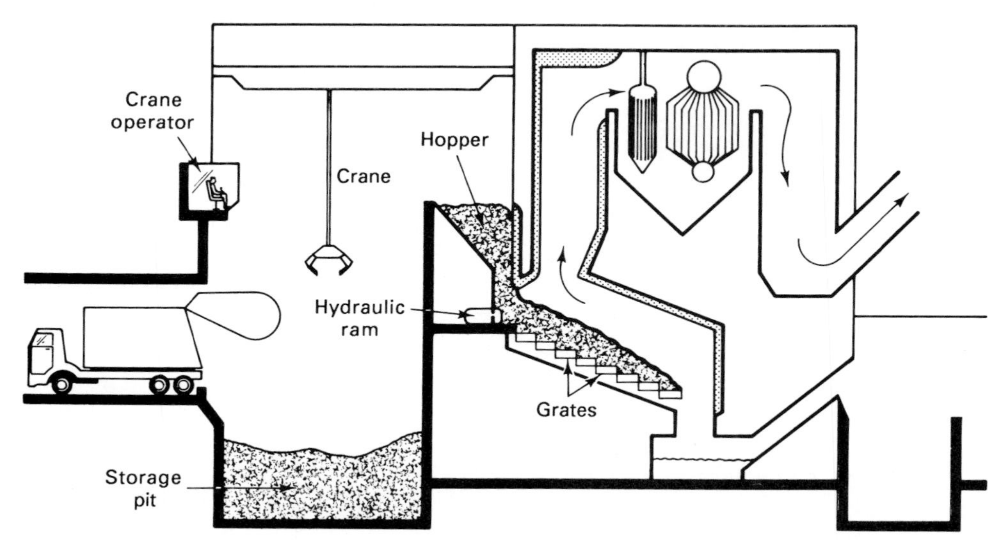

Figure 8.3 Schematic diagram of a typical incineration system using pit storage with retrieval of refuse by overhead crane.

the refuse over the conveyor. There are very few of these units in operation except on shredded refuse.

The required storage capacity is based on volume requirements. The tonnage that must be stored as determined above must be converted to volume. The density of raw refuse after it is discharged from the collection vehicle will be approximately 300 to 400 lb/yd^3. The volume required to store refuse to feed an incinerator processing 500 tons/day from 5 P.M. on Friday to midmorning on Monday would be about 6500 yd^3. A pit 30 ft deep by 50 ft wide by 120 ft long would be required. The length of the pit would provide sufficient unloading space to accommodate the approximately 100 loads that would be delivered. Plants handling larger tonnages will naturally require wider and deeper as well as perhaps longer pits. Width:depth: length ratios for these pits are not specified. The dimensions are determined by a variety of site and economic factors.

Slab storage. An alternative to pit storage that is preferred by many plant managers is slab storage. This consists of a concrete slab on which the collection truck dumps the refuse. One or more front-end loaders alternately load the refuse onto the conveyor feeding the processing system or pile the unneeded refuse to the side (see Figure 8.1). The slab is partially surrounded by a ''push wall.'' This is a reinforced concrete wall designed to withstand the force of a large front-end loader pushing refuse against the wall to load the bucket. The refuse in excess of that the processing line can handle accumulates in a pile. When the processing line has caught up with the incoming refuse, the loaders retrieve the stored material and load it onto the feed conveyor.

Slab storage is much cheaper than pit storage, especially when the required storage volume is moderate. It also has the advantage of operator control. The front-end loader operators are in a position to observe the quality of the refuse discharged from the trucks. They can be selective in what they introduce into the processing line. Elimination of massive steel, large truck tires, solvent cans, and so on, can provide a degree of protection for the processing equipment. The large storage requirements for the continuous operation of mass-burn systems may make the pit storage more economical. The height of the pile for slab storage is limited to about 15 ft by the reach of the front-end loader.

Atlas Bin storage. A third storage system that has been considered for storage of refuse is the Atlas Bin. This device is used extensively for storing wood chips and similar materials. It was logical to assume that it would work on refuse. It became clear in a short time that raw refuse could not be stored in such a unit. Some success has been obtained with processed refuse that has more uniform characteristics. There are still problems that must be addressed.

The Atlas Bin consists of a metal silo in the shape of an inverted cone. The processed refuse is conveyed to the top (the apex of the cone) and dropped into the bin. The refuse forms a pile in the shape of a cone. The angle of the bin walls is designed to approximate the angle of repose of the refuse pile. Under these conditions, the wall pressure exerted by the refuse is minimal. The costs are reduced from normal bin storage because of the lighter construction required for the walls.

The retrieval mechanism is illustrated in Figure 8.4. The refuse pile grows from the center since the incoming stream drops down from the center of the inverted cone. There are several sweep bucket chains that are attached to a rotating pull ring on the periphery of the circular base. As these chains are pulled, they tend to migrate toward the center. They drag along the bottom of the pile, pulling the refuse toward the exit conveyor, located in a channel under the floor. The refuse pulled by the chain drops onto this conveyor and is conveyed to the desired process.

One of the major problems with this system is the maintenance costs. Processed refuse contains quantities of grit, glass, and metal. This material is very abrasive and causes extreme wear on the bottom of the bin. Protection of the concrete is necessary to prevent damage to the bin bottom. The abrasion also requires frequent replacement of the chain and buckets. Another problem occurs when the refuse is allowed to accumulate in the bin. The retrieval mechanism works on a last-in first-out basis. If the bin is not completely emptied at frequent intervals, some of the refuse may remain in the bin for extended periods of time. If the moisture content is adequate, biodegradation will occur, causing odor problems. The refuse can also compact sufficiently over a period of time so that the drag buckets do not dislodge it as they move around the bottom.

The extreme of this problem was illustrated by a reported event that occurred with the Atlas Bin at the Baltimore, Maryland, gasification plant. It appeared that some hot metal in the shredder ignited refuse being transported to the bin, where it proceeded to burn. The resulting fire was extinguished using a considerable

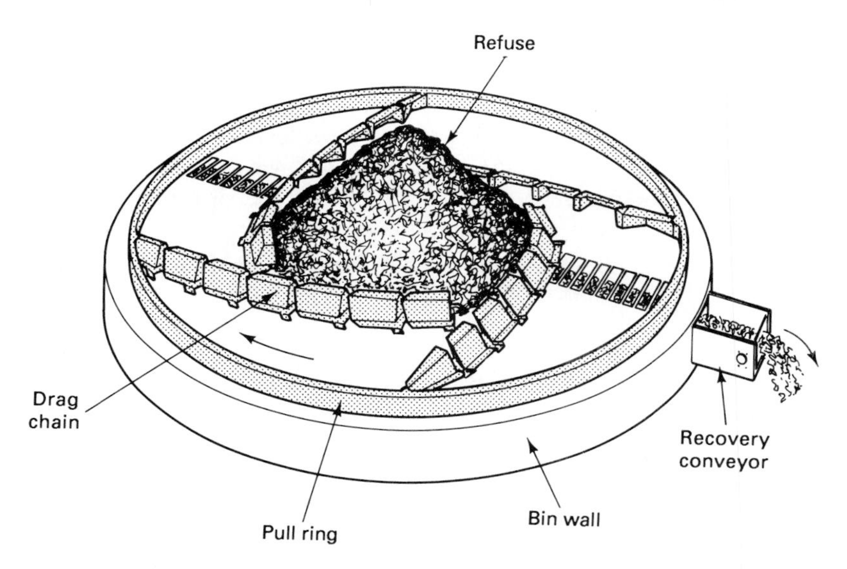

Figure 8.4 Refuse retrieval from an Atlas Bin. (From: *Waste Age*, July 1985, p. 54.)

quantity of water. The plant was shut down during this time and after the fire was out, they decided not to resume operation until they figured out what to do with the wet material. The gasification plant was a thermal process and could not process the wet refuse. After several days of planning, a solution was devised that required removal of the refuse from the bin. As they tried to retrieve the material, they were surprised that the wet paper had dried on the surface of the pile, encasing it with a thick coating of papier-mâché. As you know, dry papier-mâché can be hard and reasonably resistant to attack. How to break this tough skin? Dynamite! A few sticks were strategically placed around the pile and exploded. The explosion did damage to the bin, but the papier-mâché was still intact.

The relative economic advantages of these storage systems are difficult to quantify. Many of the costs are site specific or process specific. The best solution is to determine the operational advantages each system may offer. The cost of the storage is not a dominant cost factor in the total system cost. Therefore, operational advantages usually decide which storage system is used.

REFUSE CONVEYING

Conveyor Types

The movement of refuse through any processing facility will involve a number of conveying systems. There are four systems that have application in short-distance transport of refuse. Each has unique characteristics that determine the appropriate application. With the exception of the pneumatic conveyor, the other conveyors will probably be found in every refuse processing plant. The purpose of conveyors is to elevate the refuse so that it can be gravity fed into a unit process or a storage facility.

Pan or apron conveyors. The pan or apron conveyor is the first conveyor employed in processing raw refuse. It consists of a continuous loop of heavy steel plates pinned together and supported by rollers. It is enclosed with stationary vertical side plates and the conveyor serves as the bottom of a receiving bin for loading raw refuse onto the processing line. The steel plates are necessary to withstand the impact of large dense objects that may be dropped from the height of a front-end loader or similar equipment. The conveyor is slow moving, and therefore it must be wide to handle any quantity of refuse. The slow speed provides an opportunity for the loader operator to achieve a relatively uniform depth of refuse on the conveyor. The conveyor is driven by a variable-speed drive that enables the operator to adjust the feed rate to meet the capacity of the processing equipment. Figure 8.5 shows an inclined pan conveyor used as an in-feed conveyor for a vertical shaft hammer mill.

Belt conveyor. The belt conveyor is a continuous belt stretched between a head and a tail pulley (see Figure 8.6). These pulleys are supported by a structural frame designed to span the distance between supports. The rubber-impregnated fabric belt is carried by a series of top rollers supported by the structural frame. The rollers are as closely spaced as needed to support the load carried by the belt. They are positioned to have the belt assume a concave shape. This increases the capacity of the belt by providing sides. It is also possible to place vertical metal plates on the sides of the belt to increase the depth of material on the belt.

The belt returns on the underneath side of the frame. A small number of horizontal rollers is provided to support the belt on the return trip. Because the belt is made of a fabric material, it is subject to stretching. A free-floating idler pulley is positioned on the return run to maintain tension in the belt. When these belts are subject to wind and rain (exterior runs), they can be covered. This cover is hinged to allow access to the belt. The angle of inclination of these conveyors is limited by the ability to keep the refuse from sliding back down the belt. The maximum angle is about 30° from the horizontal.

Bucket conveyor. The bucket conveyor is designed to provide a nearly vertical lift. Since belt conveyors are limited in the angle of inclination, a different conveyor is necessary to obtain a vertical lift. There are a variety of designs for these conveyors, but the variation is mostly in the manner of support for the buckets. The buckets can be attached to a chain and lifted up a metal channel. A continuous rubberized belt that has the buckets fabricated as part of the belt is more common for refuse conveying. The capacity of the conveyor is determined by the volume of the buckets and the speed at which the buckets are elevated. These conveyors are not very satisfactory for raw refuse. The particle size is too large, requiring extremely large buckets. The application is almost exclusively for processed refuse.

Figure 8.5 Inclined pan conveyor feeding refuse into a vertical shaft hammer mill.

Pneumatic conveyors. As the name implies, these conveyors use a stream of air to transport the refuse. The application of such conveyors is limited because of the high energy requirements as well as the environmental problems associated with the air used for transport. Air is blown into a sheet metal pipe at a velocity sufficient to transport the refuse particles. Air-velocity requirements are in the range 3000 to 5000 ft/min. Depending on the diameter of the duct, airflow rates of 10,000 to 20,000 ft^3/min are experienced. The power required to transport this quantity of air at these speeds is substantial, an order of magnitude greater than the power required for the conveyors above.

Refuse is fed through an air lock into the airstream. The refuse is recovered by passing the suspension through a cyclone. The cyclone is designed to remove only the larger particles; therefore, the air must be cleaned further with a baghouse (fabric filter) before being released into the atmosphere. The normal application of pneumatic conveyors is in conjunction with air classification systems where air is used for other purposes. Since the material is already suspended in an airstream, it becomes practical to use the pneumatic conveyor. The substantial maintenance cost of these conveyors is due to the sandbasting effect of shattered glass and grit con-

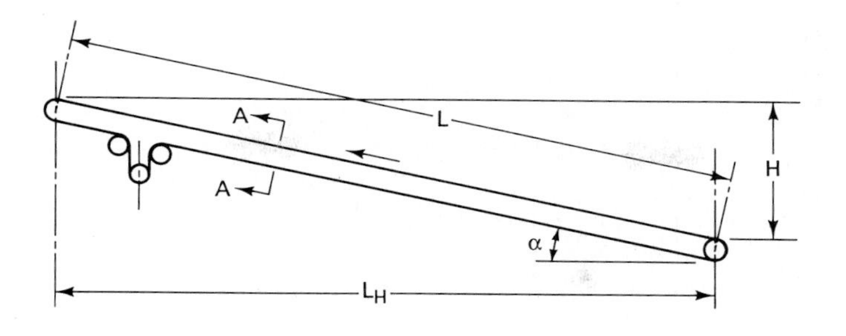

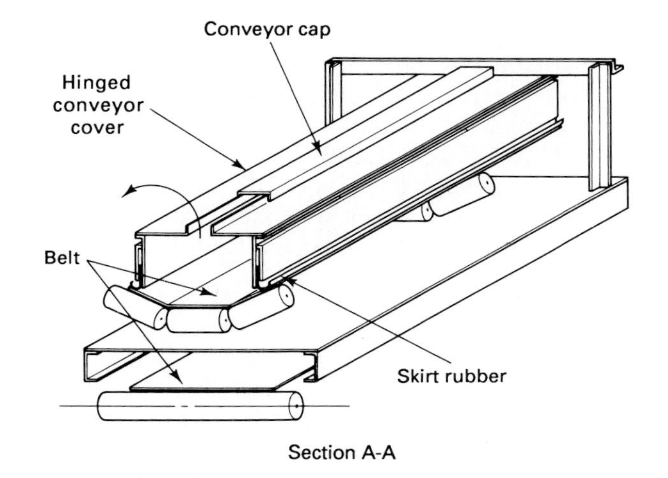

Figure 8.6 Schematic representation of a belt conveyor.

tained in the refuse. This problem is particularly acute at bends. Reinforced replaceable wear plates are included at these locations to assist in maintenance.

Sizing of Conveyors

Belts, pans, and bucket conveyors. The capacity of a conveyor is a function of the cross-sectional area and the speed of the conveyor. The width is fixed by the positions of the side plates or edge of the belt. The depth is limited to a degree by the height of the side plates or the shape of the belt and the angle of repose of the solid waste. These two dimensions determine the cross-sectional area. The following equation can be used to calculate the volume flow rate:

$$Q = AV \tag{1}$$

where Q = flow rate in ft^3/min, A = cross-sectional area in ft^2, and V = belt speed in ft/min. Mass flow rate can be determined by knowing the density of the refuse on the conveyor. As with other examples previously discussed, the density can be expected to be less than the density of the refuse in the collection vehicle, perhaps 300 to 400 lb/yd^3 (10 to 15 lb/ft^3) for raw refuse. Processed refuse density, especially shredded refuse, will be closer to 5 lb/ft^3 (150 lb/yd^3).

The speed of a pan conveyor is in the range 10 to 20 ft/min. The belt conveyor operates at a higher speed, generally about 200 ft/min (2 to 3 mph). Higher speeds are possible, but the wind resistance can become a problem with refuse. Plastic film is especially prone to wind action and can literally form a plug on the belt if it is covered, or blow off an uncovered belt. The bucket conveyors move at a higher speed than the belt conveyor because the refuse is contained in the bucket and is not as susceptible to wind action.

Pneumatic conveyors. Sizing of pneumatic conveyors is an art and requires experience with the material being conveyed. There are theoretical equations for calculating the velocity required to transport a particle with a specific size, shape, and density. However, the variation in size, density, and shape that occurs in processed refuse render these equations useless. In addition, these equations are valid only for a single particle. When multiple particles are transported, particle interaction greatly complicates any technique for calculation of the required air velocity or flow rate.

PARTICLE SIZE REDUCTION

All refuse processing systems must, as a first step, consider the condition of the refuse as it is received at the processing facility. The initial processing step is the liberation of the refuse contained in bags and boxes. These containers must be opened in order for the refuse particles to be available for the processing steps involved in separation of wanted and unwanted components. The mechanical processing entailed in this initial step can be gentle or harsh. It may be as simple as passing the refuse through a rotating drum that contains ''shark's teeth'' that can break open the bags and boxes as they tumble in the drum, or it may involve passing it through a shredder that reduces the particles to some prescribed size. How well the subsequent processing steps work will depend on this initial step and the effects on the individual constituents in the solid waste.

Particle Size Distribution of Raw Refuse

The constituents in raw refuse have characteristic particle sizes that can be used to assist in their separation. Figure 8.7 shows typical particle size ranges for the various components in unprocessed solid waste. This figure is idealized. There will likely be a very small percentage of every component at the smaller sizes. Paper and paper products are the largest components in the refuse stream and also have

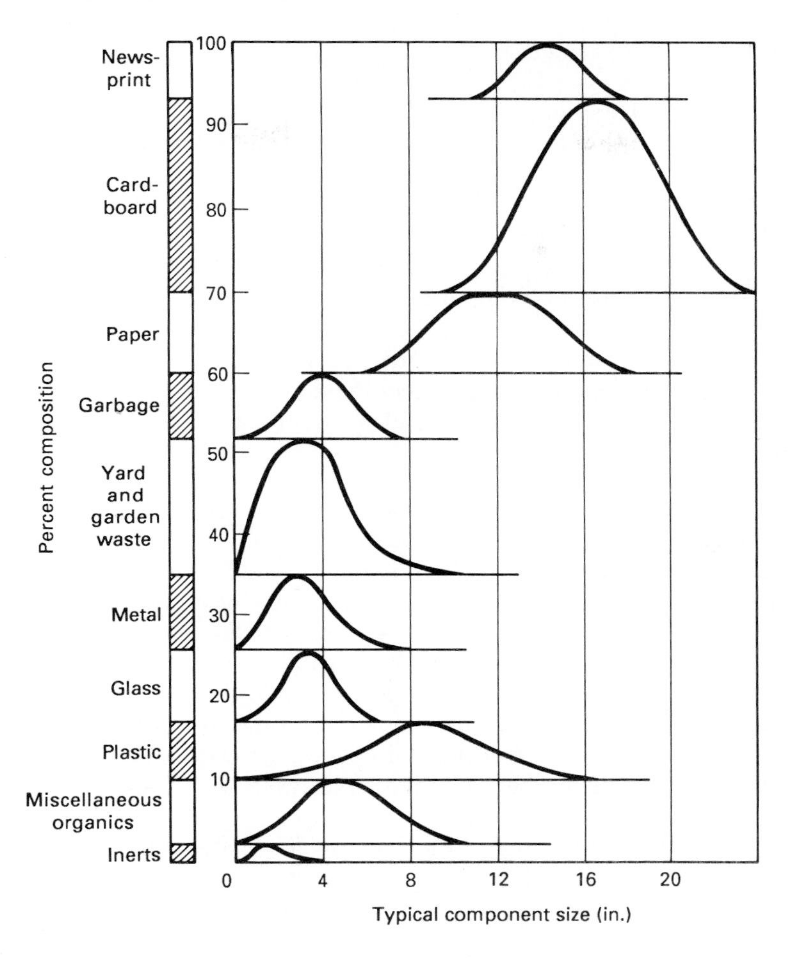

Figure 8.7 Typical size distribution of raw refuse.

the largest particle size. From this figure, a particle size cut at about 8 to 9 in. would produce a stream containing only paper and corrugated, with some plastic film.

The 4- to 8-in. size range contains the garbage and yard waste, rubber and wood, plus the miscellaneous organic materials. Glass and ferrous and nonferrous metal tend to be less than 4 in. The inert residue (ashes, dirt, etc.) will generally be less than 2 in. Unfortunately, none of these distributions is absolute. For example, grass clippings will be less than 2 in. However, these size distributions do provide an approach for obtaining streams that are highly enriched with certain refuse components. Simple separation of the refuse according to size will provide an opportunity to apply a variety of technologies to the recovery of specific materials from refuse. As the composition of refuse changes, so will these size distributions.

Solid Waste Shredding

Shredding has been used as the initial step for processing solid waste since the early attempts to produce a useful product from this material. The original concept was massive size reduction. Under this scenario, it is a battle of survival. Which will wear out first, the refuse or the hammers in the mill? This is especially acute when the refuse contains difficult-to-shred items such as rubber tires, bowling balls, engine blocks, large pieces of carpet, and so on. This intensive shredding was partly in response to the need to produce a relatively uniform material for the production of RDF. As more experience with shredding was obtained, the emphasis on massive size reduction has changed. Alternative devices have been evaluated in this role and found more effective than the power-intensive hammer mills. The following description of shredders will identify some of the advantages and disadvantages of the various shredding systems.

Flail mill. The flail mill is a well-established size reduction unit that has been employed in many industries. It has a low power requirement, commensurate with the limited particle size reduction achieved. The refuse receives only one exposure to the hammers or heavy chains (in which case it is called a chain mill) swinging from the rotating shaft. As it passes through the mill, the hammers strike the refuse and knock it against the anvil plate (see Figure 8.8). If the particle is small enough, it will pass through the mill without being reduced in size. Small items such as glass and metal containers have a low probability of being struck by the hammers, so they will probably pass untouched. Most plastic items and paper will also pass untouched. Only the large containers, bags, boxes, cans, and so on, will be impacted by the mill.

There are three primary advantages associated with the flail mill: low power requirements, low maintenance requirements, and minimum size reduction. The first two advantages are obvious in that the operational costs are much lower per ton of refuse processed than for the more destructive mills. The third advantage is

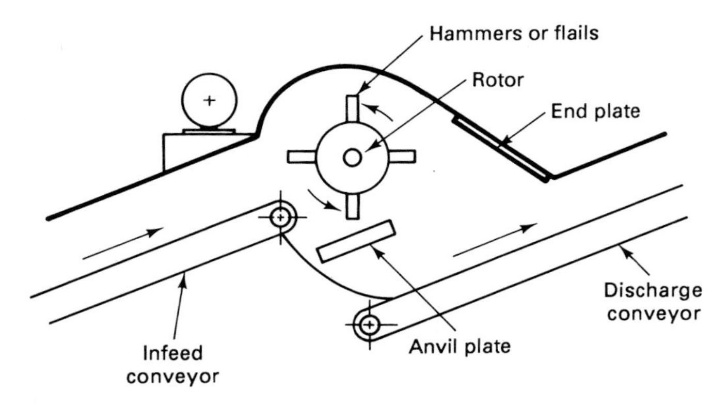

Figure 8.8 Schematic of a flail mill.

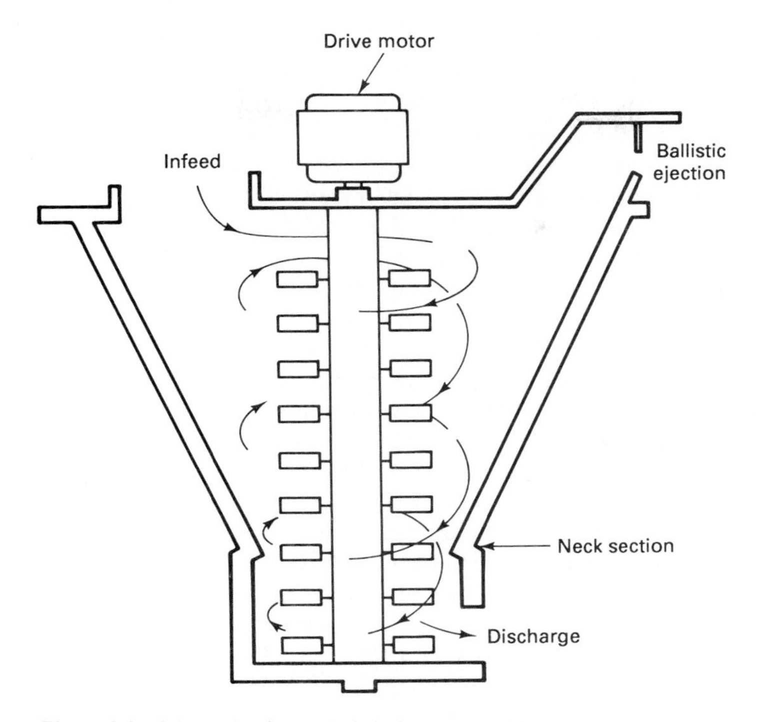

Figure 8.9 Schematic of a vertical shaft hammer mill.

associated with the quality of the output from the mill. If it is desired to recover a useful product from the refuse, separation of the various components will be necessary. Since particle size reduction is limited, the various constituents in the refuse remain intact and retain their characteristic particle size. For example, the glass does not shatter and become embedded in the organic fraction. The recovered organic fraction is of higher quality without the embedded glass. Can recovery, both ferrous and aluminum, is more efficient if whole cans are available rather than small pieces. Consequently, the flail mill offers a better size-reduction unit process for a material recovery processing system.

Vertical shaft hammer mill. Hammer mills are classified as to the orientation of the rotor or shaft. The vertical shaft has a drive shaft mounted vertically. Figure 8.9 is a representation of the vertical shaft mill. This type of mill, originally known as the Tollemache Mill, was developed in Great Britain for the shredding of solid waste. Massive steel hammers are pinned to this shaft. When rotating at high speed, the centrifugal force of these swinging hammers causes them to strike the refuse that is being fed into the mill. The hammers are enclosed in a massive steel housing that serves as a counter force to the hammers. The diameter of the housing decreases from top to bottom, decreasing the space between the hammer tips and the housing. For the refuse to pass through the mill, the particle size must be small enough to permit passage through this space.

The material is fed from the top. The fan action of the rotating hammers sets up an airflow from top to bottom which, along with gravity, pulls the material into the mill. As the housing narrows at the neck section, material that has not been reduced in size will not pass. If this material is very difficult to mill, the continued impact from the hammers will impart a centrifugal motion that will cause the object to move to the ballistic ejection port. Items such as bowling balls or massive steel sections will normally be ejected rather than allowed to remain in the mill and destroy it.

Since the object of these mills is to obtain significant size reduction, there needs to be an understanding of what controls the particle size. The retention time in the mill and the number of hammer impacts are important factors. The spacing between the hammers and the housing in the lower part of the mill assists in regulating the time of passage through the mill. The number of hammers in the lower portion of the mill determines the number of impacts, which in turn, determines the particle size. Therefore, by changing the number and location of hammers in the mill, the particle size can be changed. With time, the tips of the hammers wear and become rounded, changing the distance between the hammer and the housing. Since this wear can occur after a few hours of operation, the product particle size can also change significantly in a relatively short time.

The advantage of the vertical shaft hammer mill is the ability to achieve a high degree of size reduction, even with refuse that may contain difficult to mill material. The primary disadvantage is the high energy cost and high maintenance cost. Also, in the process of shredding raw refuse, the contamination of the organic material with the shattered glass and other inorganic components significantly reduce the value of the organic product. The most effective use of this type of mill is on a stream of solid waste that has undergone processing to remove the inorganic constituents.

Horizontal shaft hammer mill. There are two primary differences between the vertical shaft and the horizontal shaft hammer mill illustrated in Figure 8.10. First, the rotating shaft is horizontal. Steel hammers are pinned to this rotor and centrifugal force causes them to impact the refuse in the mill. The second difference is the presence of a grate that has specific-size openings. Only those particles smaller than the opening will be passed. This provides a more positive control over the maximum particle size. It can also be a potential problem. For example, if a small piece of steel too large to pass through the grate openings enters the mill, it will remain there until it is beaten to a smaller size. If this requires a long time, the continued impact of the hammers will heat the steel to a sufficient temperature to ignite the refuse, resulting in a fire. Rejection ports exist, but they work only if there is a significant mass to the rejected material.

These mills were developed for and are used extensively for the crushing of ores and stone. The application to solid waste shredding has not been very success-

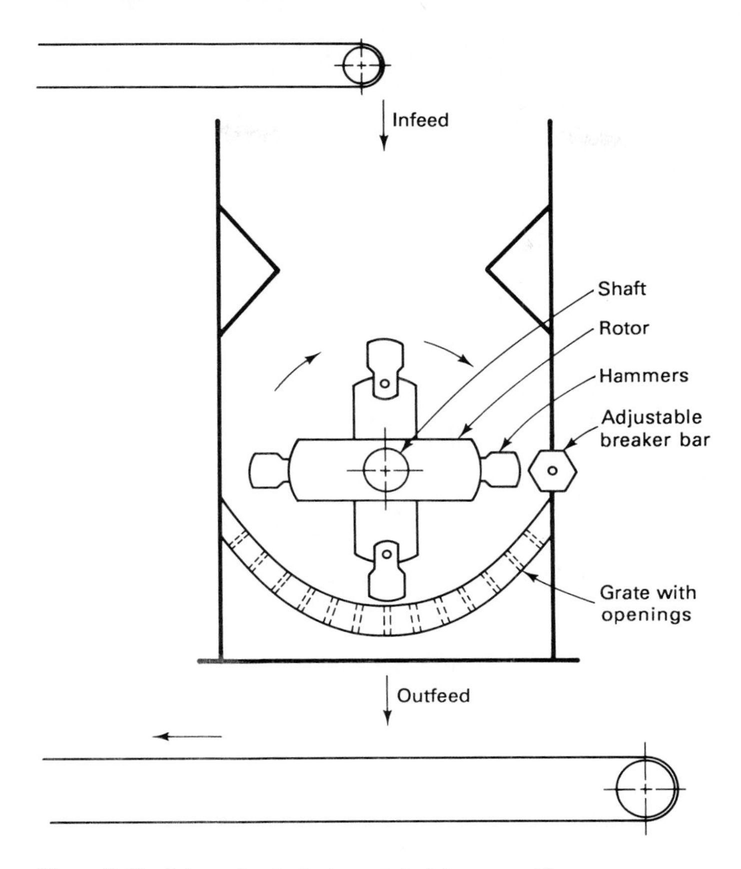

Figure 8.10 Schematic of a horizontal shaft hammer mill.

ful. Several manufacturers still sell these mills for refuse processing. They have made some modifications, but the solid waste just does not respond like the brittle ores and stone. Plastic and rubber products are considerably more resilient and are much more difficult to mill. Because of the grate, this mill cannot pass these materials like the vertical shaft mill. If a sheet of plastic or textile aligns properly in the vertical shaft mill, it can pass down the sides with little contact with the hammers. This is not true with a horizontal mill; everything must be reduced to a size that will pass through the screen openings.

Shear shredder. A recent addition to the equipment that can be used for size reduction in solid waste is the shear shredder. A schematic of this unit is presented in Figure 8.11. The size reduction accomplished by the hammer mill is from the impact of the hammers. A shear shredder achieves the size reduction by cutting actions. Two slow-speed counter-rotating shafts support cutters that literally cut the material as it becomes entrapped between the cutting teeth. The cutters reduce the

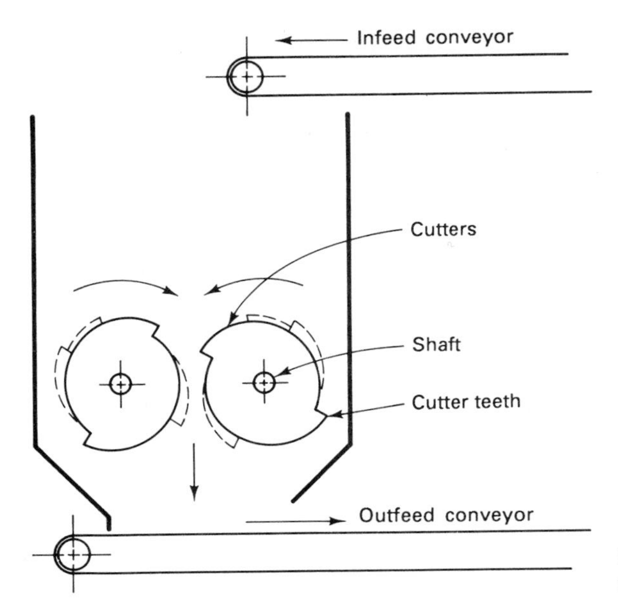

Figure 8.11 Schematic of a shear shredder.

size sufficiently for the material to pass between the shafts and cutters. These shafts are hydraulically driven at speeds of about 100 rpm. If the material jams the cutters, the rotation is automatically reversed to free the jam. After a couple of revolutions in reverse, the shafts resume normal rotation. This sequence will be continued until the material is finally passed through the shredder or the system is shut down and the offending material is removed from the mill. Shutdown is not normal because even objects such as railroad ties and large truck tires can be processed by these mills.

Particle size is controlled by the spacing and orientation of the cutters on the shafts and the spacing between the shafts. Nominal particle sizes can range from perhaps as low as 1 in. to 8 to 10 in. A major advantage of this mill is the lack of destructive force. Glass bottles and metal cans will pass the mill with little damage if large spacings are used for the cutters. Consequently, there is good control over the large particles, while at the same time, the components with a characteristic smaller size can pass the mill with little destruction. This is a definite advantage for the subsequent separation and material recovery.

The operating costs are more reasonable. The power cost is less than that for the hammer mills because of the slow speeds used. The maintenance costs are much lower because of the cutting action of the mill. The cutters are made from extra-hard steel and are not significantly impacted by organic material and lightweight metal. Massive steel will be a problem and care is exercised to remove material of this type before it reaches the mill. Also, concrete materials are not acceptable. However, such materials are not found in any quantity in normal residential and commercial refuse. Construction and demolition debris would not be accepted into this unit.

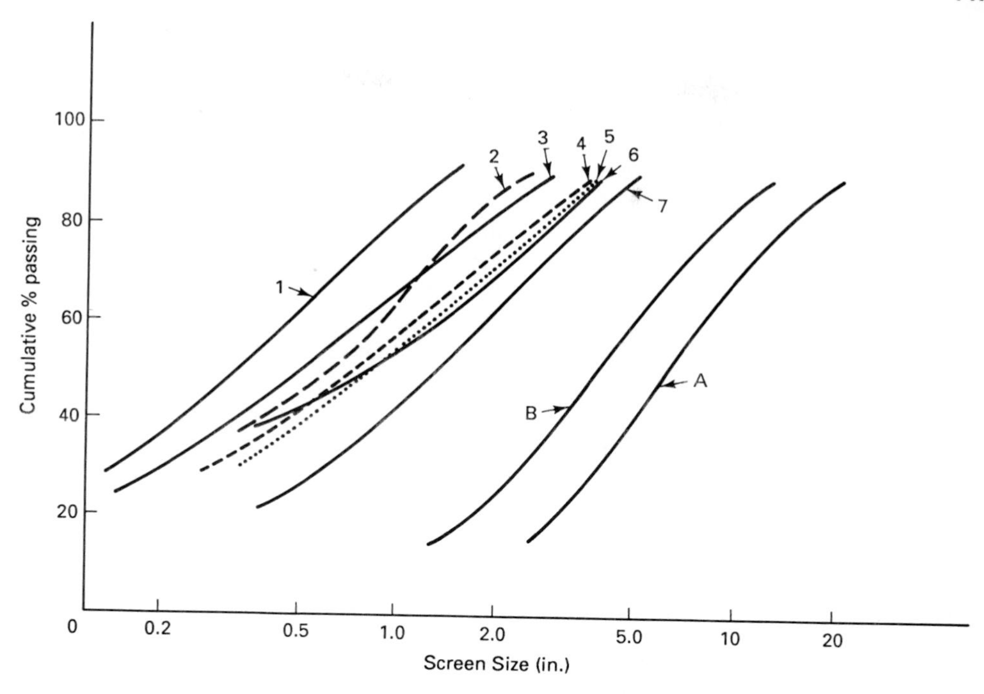

Figure 8.12 Particle size distribution of shredded refuse: A, raw commercial; B, raw residential; 1, Ames, Iowa, secondary; 7, Ames, Iowa, primary; 2, Great Falls, Maryland; 3, Cockneysville, Maryland; 4, Odessa, Texas; 5, Tilton Falls, New Jersey; 6, Appleton, Wisconsin. (From *Solid Waste Management*, May 1980, p. 36.) Reprinted with permission from *Solid Waste Management*, May 1980. Copyright Communication Channels, Inc. 1980 Atlanta, GA, USA.

Particle Size Distribution: Shredded Refuse

Since the objective of shredding is size reduction, it is interesting to know how particle size differs for different mills. The rating of a mill is on the basis of a "nominal" particle size. This is the screen size that will pass 90% by weight of the milled refuse. A mill that produces a nominal particle size of 2 in. will have a product in which 90% of the weight will consist of particle sizes less than 2 in. Some authors use a "characteristic" particle size, which is defined as the screen size passing 63.2% of the weight of particles. Why this particular number was selected is not apparent except that these authors report a better fit for the data.[1] Particle size distributions of shredded refuse from various mills are known by the manufacturer. However, as mentioned earlier, the particle size and size distribution can change significantly with hammer wear.

Figure 8.12 illustrates the size reduction achieved by several operating mills. The size distributions shown in this figure are for horizontal shaft hammer mills except for curves (2) and (5), which are for vertical shaft mills. The size distributions

[1]G. Trezek, G. Savage, and L. Diaz. *Solid Waste Management*, May 1980, p. 36.

for raw residential and commercial solid wastes are shown on the right side of the curve. The commercial refuse falls primarily in the size range 2 to 20 in., while residential refuse appears to be in the range 1 to 10 in. After shredding, the particle sizes are reduced significantly. The nominal particle size for all mills except for the secondary mill curve (1), appear to be between 2 to 4 in. Ames, Iowa primary and secondary mill, size distributions were generated by a two-stage shredding system. The discharge from the primary mill (curve 7) was, after some screening, the feed for the secondary mill. The discharge from the secondary mill (curve 1) had nominal particle size of approximately 1.5 in. It has generally been observed that multistage shredding is necessary to achieve a nominal particle size much below 2 in. One can design a single-stage mill that can produce the small particle, but the capacity of the mill is so restricted that it is not practical. After passing the first stage, the smaller particles can be removed with a simple screen. This significantly reduces the load to the second-stage shredder, improving the performance of this unit.

Energy Required versus Particle Size

Shredding is a very energy intensive step in the processing of refuse. Energy is required to destroy particle integrity; the more destruction, the more energy. There is a close correlation between particle size and energy requirements. Typical data for operating mills are shown in Figure 8.13. A nominal particle size of 4 in. will require about 7 kWh per ton of dry weight processed. As the nominal particle size is reduced to 2 in., the power demand increases to 12 kWh/ton; at 1 in., 27 kWh/ton; and at 0.5 in., 50 kWh/ton. With power cost in the range $0.05 to $0.10 per kilowatthour, the power costs for shredding refuse to small particle sizes can be substantial.

Maintenance Cost: Hammer Wear

A second operation and maintenance cost is the wear on the hammers and the housing or grates of the mills. There is substantial metal loss from both portions of the mill. Figure 8.14 demonstrates the magnitude of this problem. Hammer wear is expressed in pounds per ton of as-received refuse processed. The metal used for the hammers plays a significant role in the metal loss. A soft metal wears away rapidly, while a very hard metal will chip upon impact with some of the more resistant materials. An alloy hardness of about 50 (Rockwell C) appears to be the better range for the hammers. The loss due to chipping is not easily quantified since it tends to be isolated events. Upon inspection, one can visually observe the large chunks missing from the hammers. A mill processing 500 tons/day could be expected to lose about 22 lb/day of metal from the hammers. It is likely that similar amounts will be lost from the stationery portions of the mill.

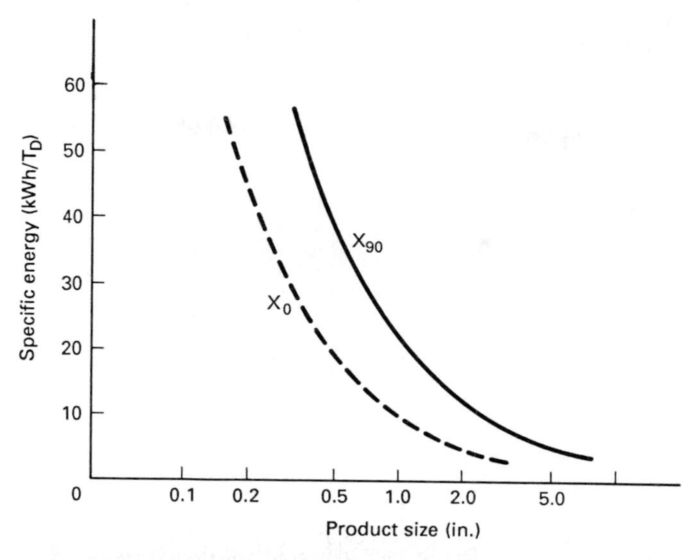

Figure 8.13 Energy requirements for size reduction. X_{90}, nominal particle size; X_0, characteristic particle size. Reprinted with permission from *Solid Waste Management*, May, 1980. Copyright Communication Channels, Inc. 1980 Atlanta, GA. USA.

The maintenance cost is the cost of replacing the metal on the hammers and the mill housing. This requires removing the hammers from the mill and replacing them with a new set. New tips are then welded onto the worn hammers. It takes a welder a significant period of time to use several hundred pounds of welding rods. When new metal is needed for the mill housing, it is necessary to take the mill out of service to allow welding of new metal to the interior of the housing. This maintenance can easily cost \$2 to \$4 per ton of refuse processed, depending on local conditions.

Explosion Hazard/Control

A constant concern with shredders, especially hammer mills, is the possibility of explosions. There are many possible causes of explosions, ranging from organic vapors to discarded dynamite to dust. There are a number of ways to handle explosions, but the industry is slowly coming to recognize that these techniques are not absolute. Consequently, the primary means of addressing explosion hazards is isolation of the shredder from the rest of the processing facility. The shredder is housed in a separate structure that has appropriate structural walls protecting the areas where personnel are expected to be located. The refuse is conveyed to the mill, which is operated by remote control. If a blast occurs, the personnel hazard is greatly reduced. This is the major concern because the damage likely to occur to

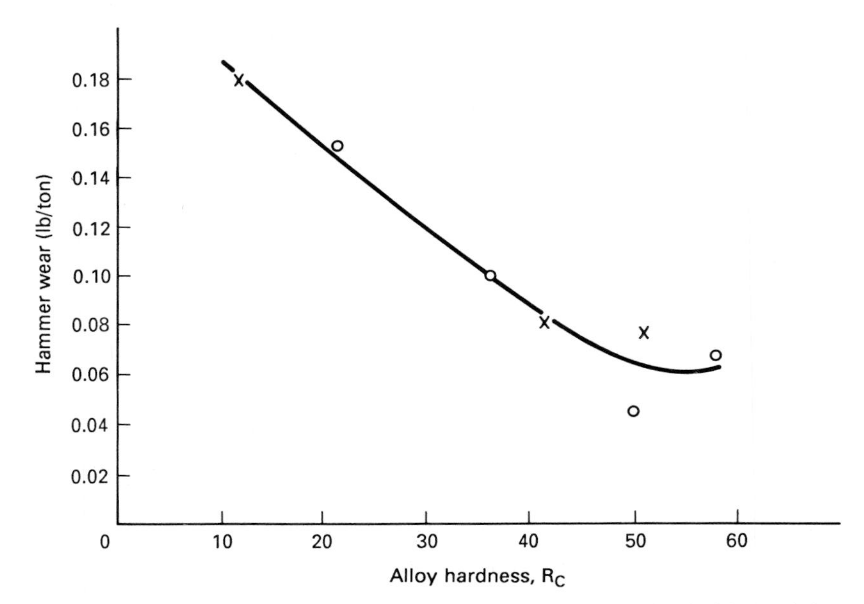

Figure 8.14 Hammer wear as a function of hammer properties. Reprinted with permission from *Solid Waste Management*, May, 1980. Copyright Communication Channels, Inc. 1980 Atlanta, GA. USA.

the mill is minimal. When 1000 hp is being dissipated inside a housing constructed of 1- to 2-in. steel plate, a tremendous amount of explosive energy will be required to damage the mill.

However, efforts should be made to protect against and reduce the effect of explosions even if there is adequate personnel protection. Time will be lost even for minor repairs to such components as the conveyors. The most common cause of explosions is likely to be dust. As the refuse is milled in any hammer mill, many fine particles are produced and suspended in the atmosphere, in both the shredder housing and the shredder building. A spark can ignite this suspension, causing a blast of major proportions. Dust control is essential. One common stop-gap measure is a water spray to keep the refuse wet and reduce the dust production. This may not be a practical solution, especially if the refuse is being prepared as RDF. Dust can be controlled by good ventilation. The location of exhaust fans that continuously pull dust-free air into the shredder will keep the dust level low. Of course, this air must be passed through filters to recover the dust.

Volatile solvents are frequently blamed for explosions, but with the new controls on the disposal of volatile organic compounds, the probability of finding a significant quantity of these materials in the refuse is greatly reduced. This is one reason for having an operator on the floor monitoring the waste stream being fed to the mill. If drums or cases of solvent cans are observed in the refuse, they can be removed manually before the refuse is sent to the shredder. A single empty solvent can is not enough to create a significant explosion. A can full of a volatile solvent can be a problem if the vapors are allowed to remain in the shredder. However, the ventilation that provides the dust control also removes the vapors.

Explosion suppression devices are available to reduce the force of any such event. When the explosion starts, there is an increase in internal pressure. Sensors strategically located sense this pressure change and release a gas to extinguish or suppress the explosion. An inert gas such as carbon dioxide or nitrogen will retard the propagation of a flame front by displacing the oxygen. These gases also prevent fires that may occur after the explosion. Halogenated hydrocarbons are suppressants that chemically extinguish the incipient explosion. The key to successful application of suppressants is the sensitivity of the pressure sensors and the response time. In a detonation (say, dynamite) the propagation of the flame front is instantaneous and a pressure rise occurs in microseconds, so the response time of the suppressant must be comparable. In a deflagration, the flame front propagates at a much slower velocity, and it may be 0.1 to 1.0 second before a pressure rise is noted. Fortunately, most explosions in shredders are of the deflagration type.

The final assist in reducing the damage due to these explosions is to provide an outlet for the pressure buildup. Blast doors are frequently located in the roof above the shredder and a blast duct is used to direct the force of the blast up through these blast doors, where it will do minimal damage.

PROCESSES FOR SEPARATION OF REFUSE COMPONENTS

After the refuse has received some size reduction, if only enough to open the bags and boxes, it can be subjected to a number of different unit processes that can recover specific components from the solid waste. Some of these processes are very simple and straightforward, while others are complex. Each process generally has a specific component that it will recover. A train of processes can be designed to recover the useful components found in solid waste. It is possible to vary the process train to fit the specific needs of the refuse being processed. Market value and refuse composition will be major factors in determining the desired material recovery.

Magnetic Separation of Ferrous Metals

Recovery of ferrous metal is easily accomplished with magnetic separation. The refuse is passed under a strong electromagnet that lifts the magnetic metal from the refuse stream. The efficiency of separation and the quality of the product is determined by the degree of preparation of the raw refuse and the specific recovery device used. Much of the ferrous metal is captive (i.e., enclosed in a plastic bag, a box, or other container). Liberation of the metal from these containers is necessary for recovery. Therefore, some degree of size reduction of the raw refuse is necessary to open the containers.

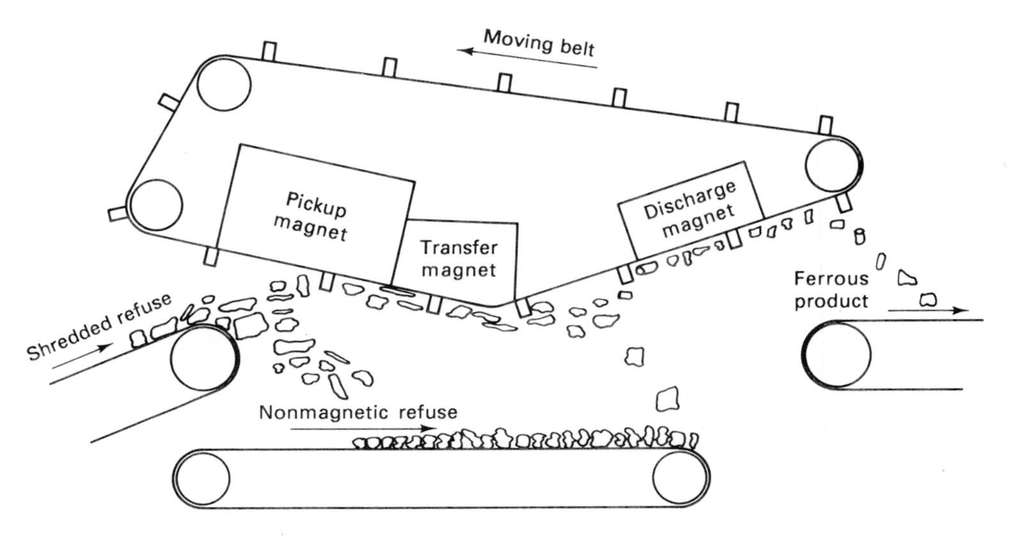

Figure 8.15 Magnetic belt for ferrous metal recovery.

The most simple recovery device is a magnetic head pulley. A permanent magnet is built into the head pulley of a conveyor belt. The velocity of the belt tends to impart an inertia to the material as it is discharged, causing it to continue for a short distance in the direction of the belt motion. The magnet in the head pulley binds the ferrous metal to the belt surface and the metal travels back under the belt as the belt begins the return trip. After a short distance, the metal leaves the magnetic field of the head pulley and drops into a recovery bin. Because of the very low cost of magnetic head pulleys, they can be located on a number of conveyor belts. A high ferrous recovery efficiency, 95%+, is possible because of the multiple exposures. The quality of recovered metal is low because of the contamination carried over with the metal. Any paper, plastic, or other material that is trapped between the metal and the conveyor belt will report to the recovered ferrous bin.

Because of the poor quality of recovered metal, a different magnetic separation system has been developed. This is a belt magnet, as shown in Figure 8.15. This belt is suspended above a normal conveyor belt that is transporting processed refuse. It consists of a strong electromagnet that has the power to recover relatively heavy pieces of ferrous metal. The magnets are covered by a belt that transports the metal to a recovery bin. As the metal is transported from the main magnet, the polarity of the magnetic field is reversed, causing the metal to rotate. As the polarity changes, the metal drops a small distance from the belt and rotates 180°. This movement provides an opportunity for the entrapped paper, plastic, and so on, to fall from the belt.

The ferrous product from this unit is considerably cleaner than that obtained from the magnetic head pulleys. The efficiency of recovery is only about 80 to 85%, and the cost of the belt magnet is sufficiently high to preclude multiple belts to enhance the recovery.

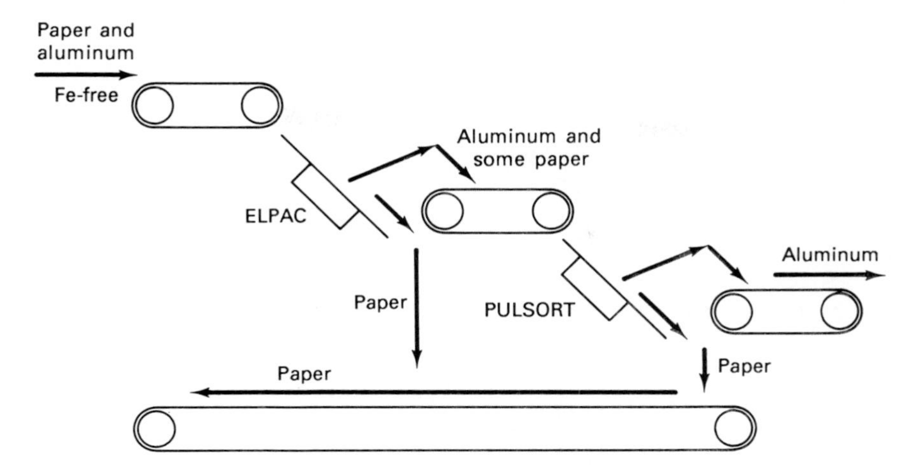

Figure 8.16 NRT aluminum recovery system.

The value of ferrous scrap is relatively low. The transportation costs to the steel mill may be more than the scrap value. It is, however, prudent to remove most of the ferrous metal early in the processing train. The properties of ferrous metal are such that it will be a cause of operating problems in all subsequent unit processes.

Aluminum Separation

The value of recovered aluminum is high and during the past 20 years, significant progress has been made in developing a recovery technology. A commercial system is now marketed by National Recovery Technologies, Inc. Figure 8.16 is a flow diagram of this technology. Partial separation of shredded raw refuse is necessary to produce a stream enriched with aluminum. The removal of ferrous metals, glass, and smaller organic and inorganic particles as well as removal of large (>6 in.) paper and plastic particles will make the process more efficient. The aluminum-enriched stream is passed down an inclined plane that contains a series of metal detectors spaced across the width of the plane. Each detector is connected to a high-pressure air jet located directly below it. When the detector senses metal, a short blast of air from the air jet blows the metal and any accompanying material onto another conveyor. This stream, although highly enriched with aluminum, still contains some paper and plastic. This initial step of the process is identified as ELPAC (electronic-pneumatic aluminum concentrator).

The second step of the process, PULSORT, separates the aluminum from the contamination carried over from the ELPAC concentrator. The quantity of material remaining in this stream is very small, less than 5% of the raw refuse input. Consequently, it is possible to consider discrete items as they cascade down an inclined plane. This plane contains a metal detector to sense when aluminum is present. The detector energizes a large magnet for a split second at the same time the aluminum item passes over it. As the magnetic field builds in the magnet, eddy currents are induced in the aluminum. The magnetic field that is associated with the eddy currents produces an oppositely directed force that acts on the large magnet to eject the aluminum object from the cascading stream. The resulting ejected stream is 100% aluminum.

Field tests on this unit show that it recovers almost 100% of the aluminum that is in the feed stream to the ELPAC. Small pieces of aluminum foil and metal may pass undetected. It is 100% efficient in recovery of aluminum cans from this stream. Any large aluminum objects or cans that are not released from bags and boxes will not be recovered. As with all recovery processes, the material to be recovered must be accessible to the recovery mechanism.

Air Classification

Principles of operation. Air classification is a process that has wide industrial application for separation of unwanted components from products. Attempts have been made, with some success, to apply this process to shredded solid waste. If a particle is introduced into the upper end of a chamber containing air under quiescent conditions, it will fall through the chamber, eventually exiting at the bottom. If there is an upward velocity to the air that resists the downward motion of the particle, one of three things can happen: the particle will fall, the particle will remain suspended in the airstream, or the particle will fly. The final response of the particle to the upward air velocity can be calculated from theoretical equations, including Stokes law. Unfortunately, these equations are for idealized particles that do not exist in shredded solid wastes.

These equations are useful to ascertain the variables that are important in understanding the operation of such a system. Particle density and air density are two primary variables. This density difference is the primary reason it was thought that this process would be successful in separating organic solids from inorganic solids. Particle diameter, along with the density, determines the particle mass. The downward force exerted by gravity is determined by the mass of the particle. Then, of course, the upward air velocity creates a drag force that counters the gravitational force. However, this drag force is very sensitive to particle shape, the one particle characteristic that will have the most variability. It is also sensitive to the Reynolds number. A careful analysis will make it obvious that it is impossible to analyze theoretically the design requirements for an air classification system for shredded solid waste. The equipment manufacturers resorted to pilot scale units to determine the appropriate design, and then proceeded to construct commercial units.

Figure 8.17 illustrates the concept of an air classifier that separates inorganic solids from organic solids (paper, plastic, garbage, textiles, etc.). The shredded refuse is introduced through a rotary valve (essentially an air lock) into the classifier. A fan creates a high upward velocity by pulling air out the top. The shape of the duct is designed to induce a degree of turbulence in the airflow. An upper baffle creates a zone near the top that has a high level of turbulence. This turbulence is

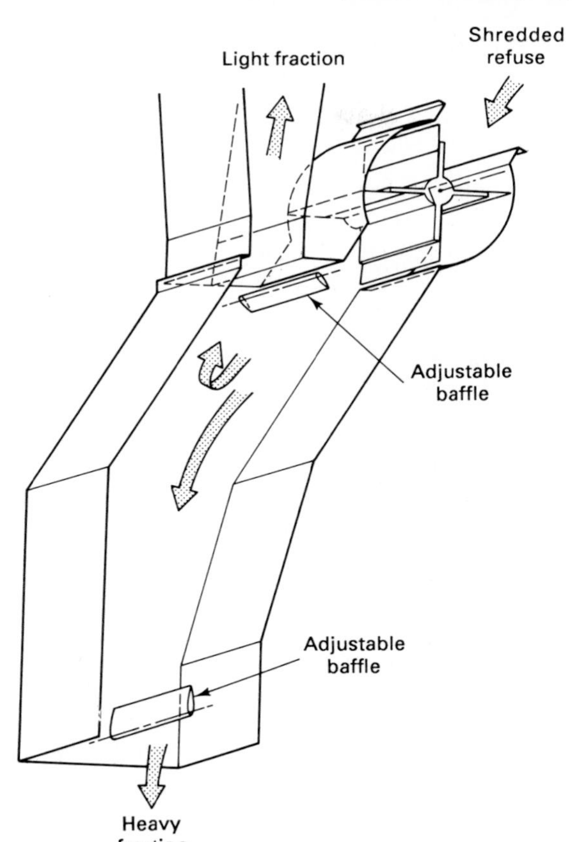

Figure 8.17 Schematic of the air classification concept.

intended to cause the particles to disperse in the airstream and approach the behavior of a discrete particle. The baffle at the bottom of the unit is intended to create a similar turbulence to disperse any clusters of solids that may have reached this level. Under ideal conditions, after the refuse is introduced into the separator, the particles will slide down along the back of the unit. In their downward trip, the particles are washed with a turbulent airstream that is intended to suspend the lighter particles and carry them out in the "light" fraction. The dense particles continue the downward trip and are discharged at the bottom as a "heavy" fraction.

Operational success. Commercial units are available from a number of manufacturers. Figure 8.18 is a cross section of a commercial air classifier. It operates similar to the system discussed above except for the air input and mechanical shaking. An irregular-shaped duct induces turbulence as the air passes through it. Blowers also introduce air into the underside of the classifier in an attempt to suspend the organic particles before they reach the outlet for the heavy fraction. Vibration of the entire unit is designed to free the organic from the inorganic

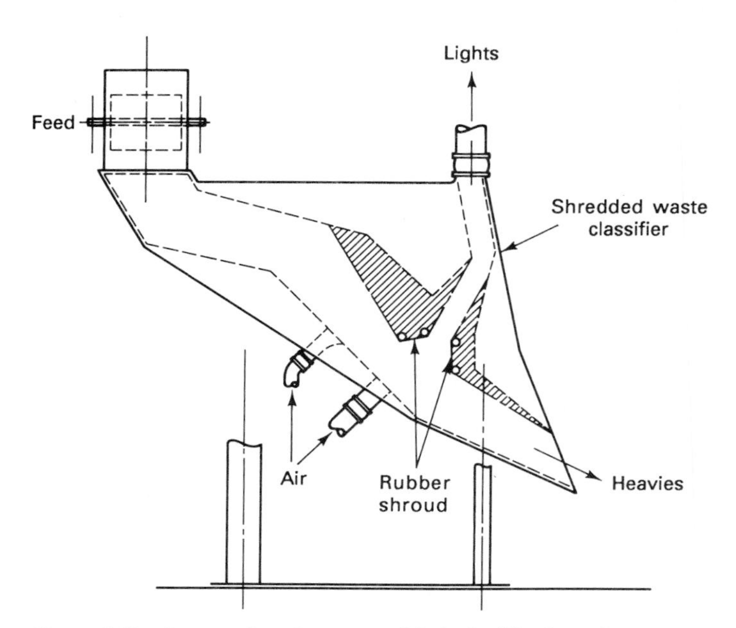

Figure 8.18 Cross section of a commercial air classification unit.

particles. When vibrated, dense particles tend to force particles of lower density to the top, where they have more exposure to the airflow. A high-volume blower draws air out with the light fraction, which passes through a cyclone to recover the particles. The air will then be filtered before discharge to the atmosphere. Air is introduced through the feed opening as well as at the outlet for the heavy fraction. These openings are baffled to control the distribution of air between them. A flexible rubber shroud in the classifier permits adjustment of the geometry to suit the specific field conditions.

The theory of air classification is sound, especially if applied to the idealized system. Unfortunately, shredded refuse is far from ideal. There are several key problems that make it impossible to use this system on such a heterogeneous stream. Perhaps the biggest variable is the moisture content. The moisture content of the entire waste stream may only change a few percentage points, but the added moisture may be in one component, say paper. Wet paper responds very differently from dry paper in an air classifier. Wet refuse particles tend to stick together, forming larger particles. Wet paper will attract dust and dirt, which will be carried over to the light fraction. When refuse is shredded to relatively small particles in a hammer mill, the shattered glass will become embedded in the paper and other organic constituents in the refuse. Of course, this glass is carried to the light fraction. The final factor that makes separation of the shredded refuse difficult is the aerodynamic properties of the particles, especially can lids. These metal items will "fly" to the light fraction.

Table 8.1 General recovery level for air classifier

Component	Percent to heavy fraction
Paper and corrugated	5–10
Plastic	15–20
Wood	30–35
Textiles	5–10
Nonferrous	90–95
Ferrous	95–100
Glass	95–100
Yard waste	30–40
Putrescibles	85–90
Rubble	95–100
Rubber and leather	95–100

Source: D. L. Murray, "Air Classification Performance and Operating Principles," *8th ASME National Waste Processing Conference,* 1978, p. 505.

It is possible to adjust the air classifier to perform reasonably well on a given shredded refuse. The typical adjustment will produce a light fraction that has a low ash content, but the recovery of the organic material will be low. The separation of the inorganic and organic solids is not very good, and as shown in Table 8.1, a significant portion of the organic material will remain in the heavy fraction. The converse effect can also be achieved. As long as the refuse characteristics do not change, it will continue to perform this manner. However, a change in moisture content or particle size will require readjustment. This in not practical on a daily basis because of the time required for adjustment. Several of these units have been installed, but the only ones operating to date are doing so because they are part of a pneumatic conveying system. The limited separation achieved is not worth the cost of operation.

Air knife applications. An especially successful application of the air classification concept is called an air knife. This device is restricted to a waste stream that has been subjected to a significant amount of separation. In particular, it is applied to a stream that has a narrow size range. Under these conditions, the density truly becomes the major difference between particles. Figure 8.19 illustrates the principle of an air knife. The material discharged from a conveyor has a nearly horizontal stream of air directed on it. The air current will carry the plastic and paper with it, while the dense material will fall more rapidly. The proper placement of a baffle to separate the two streams will result in a stream containing essentially lightweight material and a stream containing dense inert material. Because the particle sizes are relatively uniform, the separation based on density differences is very efficient.

The quantity of refuse passing through a specific air knife is small because of the classification and separation that has already occurred. Consequently, the

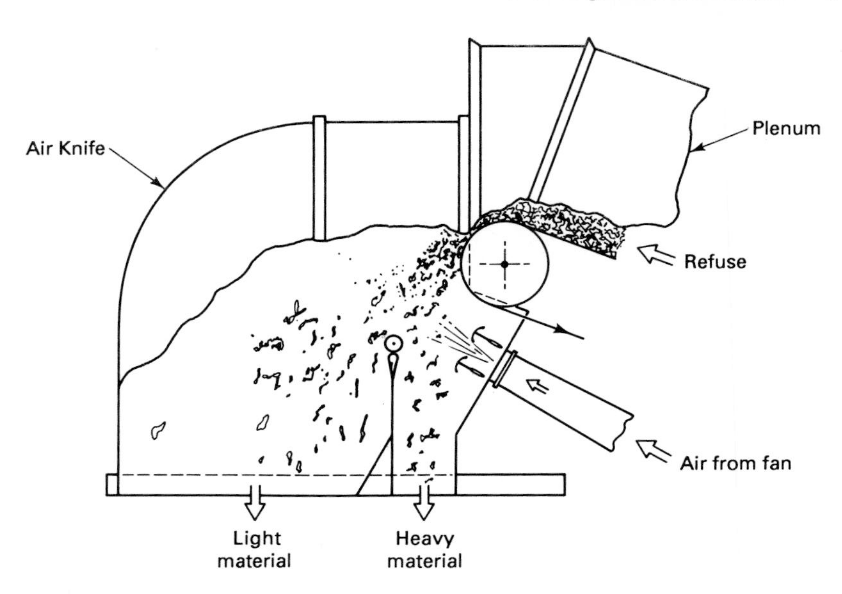

Figure 8.19 Schematic of air knife configuration.

quantity of air required is also low and since the particles are not suspended in the airstream, it is delivered at a relatively low velocity. There is no particular dust problem, as the air is contained in a hood, to allow for most suspended particles to settle. The fugitive dust leaving the hood can be handled by the building dust control system.

These units have low initial cost and low operating cost. It is practical to install an air knife on any separated stream to refine the quality of the reclaimed material, either to upgrade the inorganic content or the organic content of the stream.

Screening

Screens have a long history of use for separating materials according to particle size. As discussed previously, the various components in the refuse stream have characteristic size ranges. With proper screen sizes it should be possible to produce a variety of streams that are enriched with specific components of the raw refuse. Because of the overlap in the size range for some of the refuse constituents, 100% separation is not possible. Additional processing will usually be necessary to obtain a stream that contains only one constituent. Screens have also been found beneficial as the first processing step. Because many of the particles in the refuse are already of small size, they can be removed before shredding. This will significantly reduce the load on the shredder, which in turn reduces the cost of operating this unit.

Successful operation of screens involves two considerations. First, the size of the screen openings must be proper for the size of particles being separated. It is

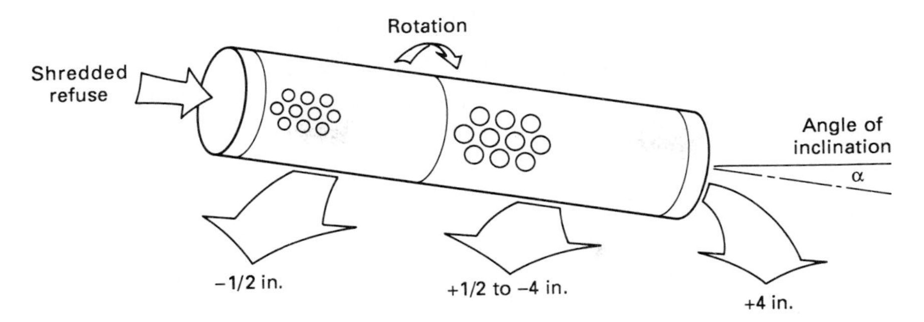

Figure 8.20 Schematic of rotary drum screen.

rather straightforward to determine the size of the openings required by conducting a particle size analysis of the feed material. Second, the particles must be exposed to the openings. A number of screening systems have been developed to accomplish this exposure. Two that have found the most acceptance for processing both raw and shredded refuse are the rotating drum screen (trommel screen) and the disk screen.

Rotary drum (trommel) screen. The trommel screen is a large-diameter drum positioned nearly horizontally. The diameter may range from 2 ft to perhaps 10 ft. As shown in Figure 8.20, the refuse is introduced into the elevated end. As the drum rotates, the particles are carried up the side of the drum until they reach a certain height, where they then fall to the bottom to repeat the cycle. It is the number of these cycles that determine the opportunity for the particles to contact the openings and pass through. The more cycles, the higher the separation efficiency. However, the throughput of the trommel will be reduced unless a large percentage of the particles are passing through the openings. There is a trade-off between capacity and separation efficiency.

The drum is inclined at a small angle to the horizontal. For each rotation, the particles move forward, depending on the angle of inclination and the distance the particles are carried up the side of the drum before they fall back. This distance can be calculated from the geometry of the drum and the rotational speed. Lifter bars are installed on the interior of the drum to prevent the particles from sliding back along the drum surface. The particles will be carried at least halfway up the side of the drum. Any additional distance will depend on the rotational speed of the drum. The centrifugal force will hold the particles to the drum side until gravitational forces exceed the centrifugal force.

If the drum rotates at too high a speed, the centrifugal force is such that the particles will be held against the drum surface and not fall back to the bottom. The "critical" speed is determined by the diameter of the drum and can be calculated as

$$r\omega^2 = 32 \tag{2}$$

This is the rotational speed where the centrifugal force equals the gravitational force. In this equation, r is the drum radius, in feet, and ω is the rotational speed,

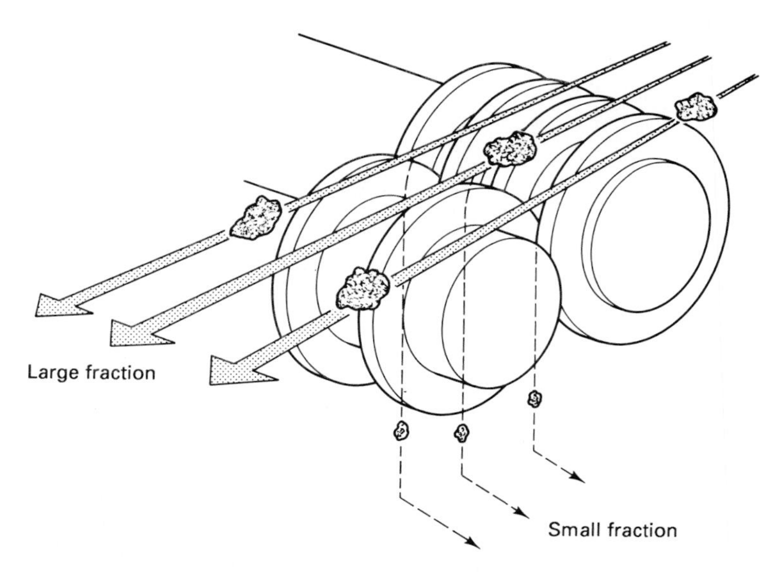

Figure 8.21 Schematic representation of a disk screen.

in rad/sec. The larger the drum, the lower the critical speed. Typical rotational speeds are in the range of 20 to 30 rpm.

Figure 8.20 is a representation of a compound trommel, a trommel that has two different-size openings. The first portion of the drum has 0.5-in. holes to pass dirt, grit, shattered glass, and other small particles. The quality of this ''undersize'' for recovery is poor, and the material would be taken to the landfill. The remainder of the drum has 4-in. openings. The material passing these openings is considered to be an acceptable product for subsequent processing to produce a refuse-derived fuel. The ''oversize'' is recycled back through a shredder to be reduced in size so that it will pass through the openings. The drum does not have to have different-size openings, but can be a simple trommel with only one hole size, as the application dictates.

Trommel screens have been successful in a number of installations. There are several manufacturers of the systems. In fact, they are so simple in construction that any fabrication shop could build one. Design of these units from a process consideration is more of an art than a science. Therefore, the experience of the manufacturer in processing solid waste is a very important consideration in sizing of the equipment to achieve a given objective.

Disk screen. The disk screen (Figure 8.21) consists of a bed of rotating shafts containing a number of disks. The shafts and disks all rotate in the same direction, so that the particles ''float'' down the length of the bed. The turbulent nature of the bed provides numerous exposures of the particles to the screen openings. The size selection is determined by the distance between the outer diameter of

Figure 8.22 Refuse separation by a disk screen.

the shafts and the spacing of the disks on the shaft. Particles having two dimensions equal to or less than these spacings have an opportunity to pass through the screen. Most particles tend to orient such that the two larger dimensions are in a horizontal plane. Therefore, the size separation is usually based on these two larger dimensions. Of course, it is possible for a particle to orient in such a way that the largest dimension is vertical and consequently, report to the undersize fraction. No process can guarantee 100% discrimination of size.

These screens can be constructed to provide a wide range of particle size cuts, ranging as small as 0.25 in. to 6 to 8 in. or larger. Figure 8.22 shows the refuse passing across a 5- by 3-inch disk screen. Only the large particles remain on the screen. The capacity of the unit is determined by the length and the surface loading rate. The number of exposures the particles have to the openings is the important consideration. Particle movement on the bed is excellent and causes separation of clumped particles. This results in good size separation.

Figure 8.23 illustrates the performance of a 3.5- by 1.0-in. screen. These curves show the percentage of the various particle sizes in the feed stream that report to the oversize fraction. It does not show the percentage of the individual components in the feed stream that are found in the oversize. The spacing between the shafts was 3.5 in. and between the disks was 1.0 in. The refuse had been passed through a shear shredder for minimal size reduction. There is a very clear separation of the glass and ceramics according to particle size. Since the size of the intact glass containers is larger than the openings in the screens, these

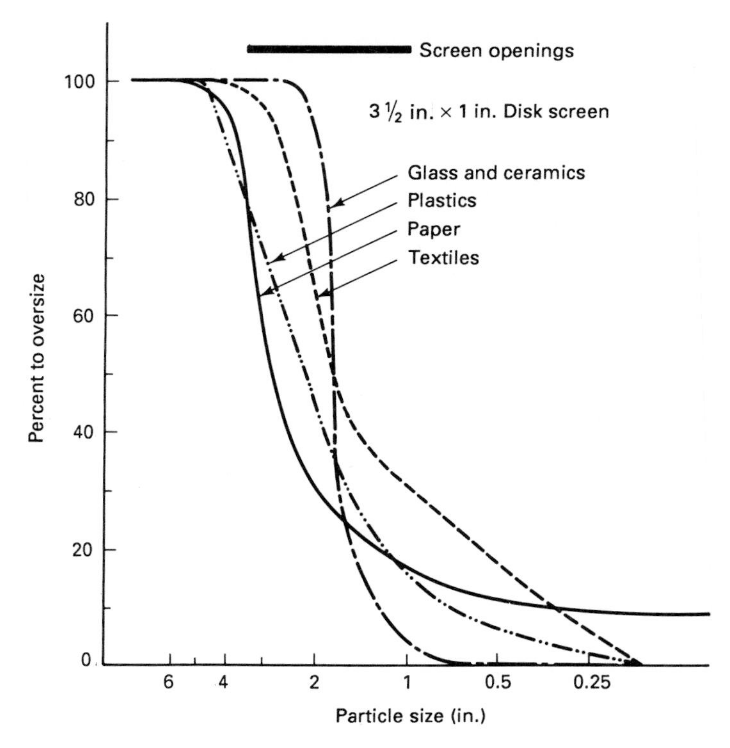

Figure 8.23 Performance of a disk screen.

containers are concentrated in the oversize, greatly facilitating recovery. Because the glass was not broken to any extent, most of it remained in the oversize fraction. However, all of the broken glass was found in the undersize; none of it reported to the oversize.

The separation of the plastic, paper, and textiles according to size was not as good. These materials have very irregular shapes and it is possible for the particle to be oriented such that the largest dimension is in the vertical direction. Under these conditions, the larger particles will pass the screens. Also, the smaller particles tend to attach to the larger particles and report to the oversize stream.

Disk screens will provide a reasonably distinct separation of particles according to size. Those refuse components that have a definite size, such as cans and bottles, can be isolated by simple screening. The stream will contain essentially all of the cans and bottles, with some other constituents, such as paper and plastic. However, once these concentrated streams are obtained, it is possible to use other techniques efficiently to further refine the glass and cans (i.e., the aluminum recovery system, air knife, etc). Consequently, a variety of screen sizes can be assembled to isolate the various refuse components in concentrated streams. This is the basis for some of RDF systems discussed later.

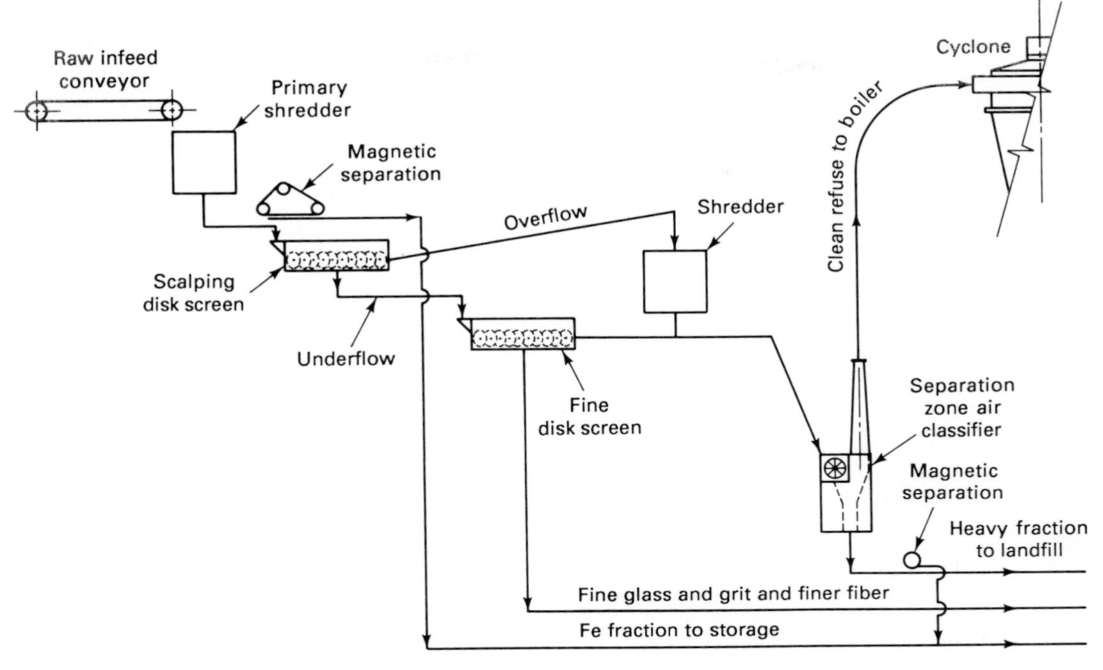

Figure 8.24 Schematic of the Ames, Iowa, RDF system. (From *Solid Waste Management*, May 1979, p. 40.) Reprinted with permission from *Solid Waste Management*, May, 1979. Copyright Communication Channels, Inc. 1979 Atlanta, GA. USA.

REFUSE-DERIVED FUEL SYSTEMS

Ames, Iowa

The Resource Recovery Act of 1970 funded a demonstration project designed to produce refuse-derived fuel (RDF) from solid waste. This project was undertaken by the city of St. Louis and the Union Electric Co. As a result of the apparent success, a full-scale plant was designed and constructed for the city of Ames, Iowa. This facility became operational in 1976. It consisted of a two-stage shredder and an air classifier. The discharge from the primary shredder was fed to the secondary shredder. The small particles produced by the second-stage shredder were passed through the air classifier to produce a light fraction for the RDF. The heavy fraction was sent to a material reclamation center for recovery of metals.

The system operated, but the maintenance costs were high and the quality of the RDF was poor. Radar Corp. investigated the use of disk screens to improve the overall performance of the system. These screens were installed and the system shown in Figure 8.24 was operated. There were three objectives for installing the screens. First, a scalping screen was installed to remove from the primary shredder discharge all particles less than 1.5 in. Only the larger particles were sent to the second-stage shredder. The second objective was to remove the glass from the system early, before it became pulverized and embedded in the refuse fibers as the material passed through the second-stage shredder. The glass crushed in the first-stage shredder was removed by the fine disk screen and sent to the landfill.

Finally, the disk screen was intended to improve the performance of the air classifier. The fine, less than ⅜ in., mostly inorganic refuse was removed by the fine disk screen. Glass greater than ⅜ in. was sent to the Radar air classifier, which was equipped with a secondary separation zone having a high removal efficiency for glass in excess of ⅜ in. Before installation of the disk screens, the glass and grit portion of the RDF ranged from 4.1 to 5.5%. Limited data show that this component of the RDF has been reduced and there has been an increase of about 15% in the Btu value of the RDF. The effect of the scalping screen on the operation of the second-stage shredder was significant. The less than 1.5 in. underflow from the primary shredder accounts for 59%, by weight, of this shredder output. This reduces the mass flow through the secondary shredder to only 41% of the previous value. Power consumption in the second-stage shredder has dropped from 125 amperes (with surges to 200) to 85 amperes (with rare surges to 100). The maintenance costs has also been reduced.[2]

National Resources Technology

National Recovery Technologies (NRT) has developed a processing system that incorporates a number of technologies to produce RDF and to recover various materials from refuse. A schematic of the system is shown in Figure 8.25. It consists of a large rotating drum where most of the separation occurs. The raw refuse, after perhaps passing a hand-picking station for recovery of corrugated paper products, is conveyed directly to the rotating drum. The bags and boxes are opened by the action of "shark's teeth" located in the first third of the drum (step 1). These are sharp metal triangular plates that protrude into the drum. When the bag or box hits the "tooth," it is ruptured and the contents are released. The second third of the drum has permanent magnets attached to the surface. These magnets attract the ferrous metals and carry them to the top of the drum, where they are scraped from the drum surface and fall onto a conveyor belt. In the final third of the drum (step 3), lifting bars, or better, buckets, are fastened to the drum wall. As the material is elevated with the drum rotation, the paper, plastic, and so on, will fall back into the drum. However, the dirt, grit, shattered glass, and so on, are retained by these lifting bars. Near the top of the drum, these materials fall onto a conveyor.

There are three streams exiting the drum: one stream containing the plastics, paper, other organic material, and nonferrous metals; one containing ferrous metals

[2] *Solid Waste Management*, May 1979, p. 40.

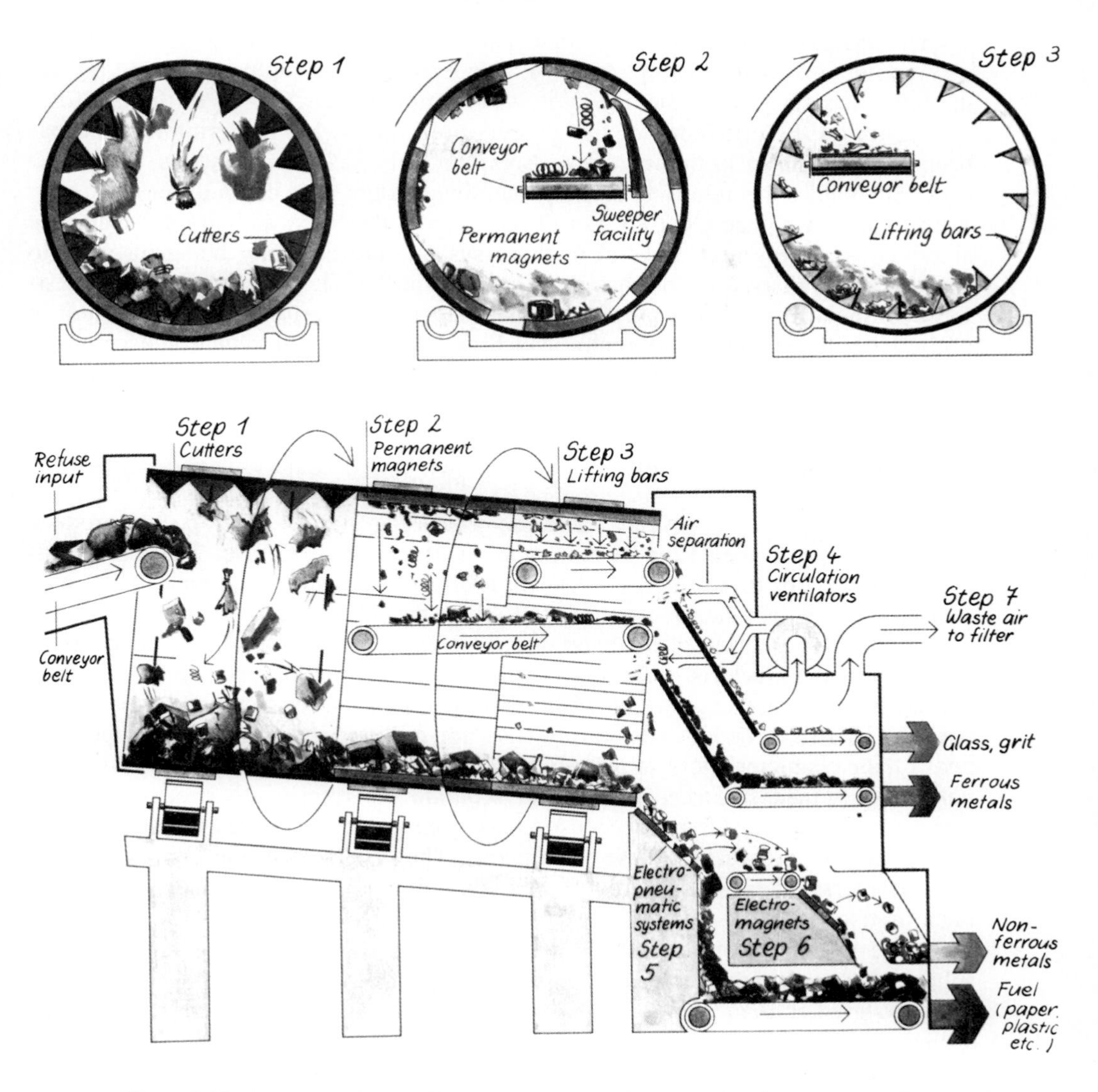

Figure 8.25 Schematic of the NRT separation system. (Courtesy of National Recovery Technologies, Inc., Nashville, Tennessee.)

plus organic contamination; and one containing the shattered glass, grit, dirt, and fine organic particles. The first stream passes on to the aluminum recovery system described earlier for recovery of the aluminum and production of RDF (steps 5 and 6). The stream intended for RDF production may be refined additionally to remove any large dense particles that pass through the drum. It may also be necessary to shred this material, depending upon the use for the RDF. The second and third streams are passed by an air knife that removes much of the fine organic material, such as lint and grass clippings (step 4). The ferrous metal can be recovered, as can the unbroken glass. The fine dirt, grit, and shattered glass will go to the landfill.

Data on the efficiency of this process are limited, but some information has been made available in the literature. Approximately 85% of the metals, glass, and noncombustibles are removed in the processing.[3] Aluminum can recovery is nearly 100%. Only those cans that are not liberated from the bags, and so on, escape the aluminum recovery unit. Several of these systems are currently being installed to produce RDF. Also, a system has been installed at the XL Disposal Corp. in Crestwood, Illinois, for the sole purpose of material recovery prior to hauling the refuse to the landfill. Except for aluminum and ferrous metals, the product recovery is by hand picking, including corrugated paperboard and various plastic materials.

RefCoM Separation System

RefCoM is an acronym given to a refuse-processing system developed to produce methane gas by fermentation of the organic material in the refuse. The refuse separation portion of this system produces a RDF that contains the organic fraction of refuse. This product can be used as RDF or as the feed for a biological processing system (i.e., production of methane or compost). Since the various components of refuse have a characteristic size, only minimal size reduction prior to screening is needed to open containers to allow the particles within to be exposed to the screen openings. This means lower operating costs for milling, and the refuse components retain their characteristic size and shape. This greatly facilitates separation, and a number of screen sizes can be used to separate the refuse into streams that are enriched with particular refuse constituents. This enrichment greatly enhances the opportunity for recovery of these constituents. The system shown in Figure 8.26 has been developed to produce a product that is a high-quality organic material suitable for a number of uses.

The essential components of this system are as follows. The raw refuse is subjected to a bag opener prior to passage over a coarse disk screen that rejects particles greater than about 5 in. After ferrous removal, this oversize stream is milled and then passed over another coarse disk screen for rejection of oversized particles, especially plastic containers and film. This oversized material can be processed through a paper separator for plastic and paper recovery if the shredding has not been excessive. It can also be hand-picked to recover plastic and paper. The material can also be returned to the shredder until it is finally reduced in size, or sent to a separate small shredder for size reduction. The undersize from both coarse disk screens passes to a can separation screen to produce an oversize stream rich in aluminum cans and glass containers. This stream goes to an aluminum recovery process for extraction of the aluminum, and the remaining material in this stream is

[3] *Waste Age*, Feb. 1988, p. 54.

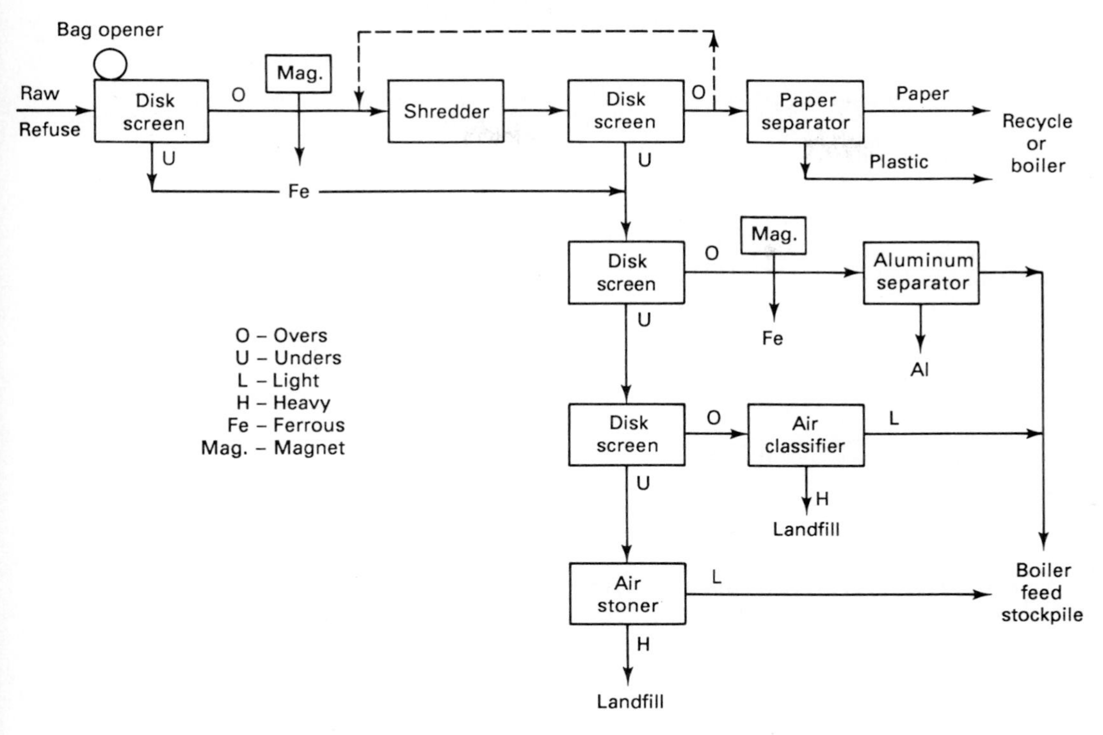

Figure 8.26 Schematic of the RefCoM processing system.

added to the RDF stockpile after passing through an air classifier (air knife) that removes the glass and other dense material. The undersize from the can separation screen is passed over a fine disk screen to remove the dirt, grit, shattered glass, and so on. The oversize goes to the stockpile after passing through an air knife that removes any remaining dense inorganic material. The undersize from the fine disk screen is processed through an air stoner for recovery of the fine organic material, such as grass clippings. The air stoner is a modification of an air knife designed to separate small inorganic and organic particles.

In addition to the aluminum and ferrous streams, two additional streams are generated from the separation processes: a stream rich in organic material that can be used as a RDF and an inorganic (inert) stream that goes to the landfill. A computer model of the separation processes has been developed that includes a variety of screen sizes, air knifes, air stoners and ferrous and aluminum recovery systems that can be assembled in any desired combination or sequence. The model calculates the input and output streams for each process in the flow scheme selected.[4] It can compute the fraction of the individual refuse components, such as paper or

[4]James J. Geselbracht, Black and Veatch, Pleasant Hill, CA, personal communication, 1987.

Table 8.2 Mass balance on the RefCoM separation system

Stream	Water	BDC	NBDC	Inerts	Total
Raw refuse	101.1	194.6	54.9	49.4	400.0
Fe recovered	0	0	0	15.0	15.0
Feed to paper separator	3.1	50.5	26.2	1.9	82.1
Feed to can reject screen	97.9	144.1	28.4	32.5	302.9
Can screen oversize	15.9	78.1	21.3	7.5	122.8
Can screen undersize	82.0	66.0	7.1	25.0	180.1
Fine screen oversize	13.6	28.2	5.0	16.3	63.2
Fine screen undersize	68.4	37.7	2.0	8.8	116.9
Air classifier light fraction	12.6	28.2	1.7	12.6	55.4
Air classifier heavy fraction	1.0	0	3.3	3.7	8.1
Air stoner light fraction	57.1	31.6	2.0	0	90.7
Air stoner heavy fraction	11.3	6.1	0	8.8	26.2
Feed to Al recovery	15.9	78.1	21.3	3.8	119.1
Fe removed from feed to Al recovery	0	0	0	3.7	3.7
Al recovered	0	0	0	1.0	1.0
Rejects from Al recovery unit	15.9	78.1	21.3	2.8	118.1

plastics, or can present the categories based on the activity of the material in the various processes (see Table 8.2). In this case, biodegradable solids (BDC) refer to paper, food wastes, and yard wastes. The combustible category (NBDC) is plastic, wood, rubber, textiles, and so on, and the inert materials constitute all of the noncombustible components. The biodegradable solids are also considered to be combustible, but have a potential for uses other than combustion.

Table 8.2 presents a lot of numbers that are probably of little interest to the reader. The intent is to illustrate the extent of the data base that some manufacturers and firms working in this field have accumulated. This same level of detail is probably available on most refuse unit processes. It is not always publicly available because of the desire of equipment suppliers to control use of the equipment. As mentioned earlier, some of these units are very simple mechanically and could be fabricated in any metal shop. The protection retained by the developer of the technology is the result of the many field tests conducted. This information is required if the unit is to be properly utilized.

STUDY QUESTIONS

1. What are the essential elements of a receiving area of a refuse processing system?
2. Differentiate between pit, slab, and bin storage of refuse. Discuss the advantages and disadvantages of each.
3. What are the four types of conveyors considered for refuse transport? Where is each type applied in the processing train?

4. Calculate the capacity in tons per hour, of a 36-in. flatbelt conveyor with 12-in. metal skirts. Assume that the angle of repose for the processed refuse is 35° and that the conveyor can be loaded so that the side depth is equal to the 12-in. skirt height. The density of the refuse is 5 lb/ft^3. The belt speed is 100 fpm.
5. How can the particle size of the unprocessed refuse be used to enhance the separation of the refuse into individual components?
6. What is the negative effect of shredding raw refuse to a small size on the separation of individual constituents?
7. Distinguish between hammer mills, flail mills, and shear shredders.
8. What is the relationship of shredded particle size to operating and maintenance costs for shredders?
9. What is the difference between a magnetic belt and a magnetic head pulley for removal of ferrous metal from refuse?
10. How does an aluminum ''magnet'' work?
11. What is the principle of operation for air classification? What are the problems associated with the application of this process to urban refuse?
12. Why is the ''air knife'' concept an improvement in this process?
13. What property of refuse is exploited by the use of screens for component separation?
14. How does a rotary drum (trommel) screen operate? What is the critical speed of a trommel screen?
15. How does a disk screen operate?
16. Calculate the critical speed of a trommel screen 6 ft in diameter.
17. Draw a schematic diagram of the Ames, Iowa, RDF system.
18. Draw a schematic diagram of the RefCoM RDF system.
19. How does the quality of refuse change by passing it through the RefCoM RDF system?

CHAPTER 9

Combustion Principles and Mass-Burn Design

PRINCIPLES OF COMBUSTION

Definition of Combustion

Combustion is a chemical reaction involving the rapid combination of oxygen with the combustible components in a fuel. This is a reaction where the elements in the fuel are oxidized. When there is adequate oxygen for the oxidation to be complete, the elements will be in their most stable oxidized state. The major elements in the fuels are carbon, hydrogen, and oxygen. Some fuels have significant quantities of sulfur and, to a lesser extent, nitrogen. Carbon is oxidized to carbon dioxide, hydrogen to water, and of course, oxygen does not react with atmospheric oxygen. Sulfur will oxidize to sulfur dioxide, an undesirable product. Nitrogen is generally thought to be converted to a nitrogen molecule (gas). Because of the large amount of nitrogen present in the combustion air, it is difficult to determine where the oxidized forms originate. Since only a small percentage of the atmospheric nitrogen is oxidized to nitrogen oxides in the combustion process, it is assumed that the same small percentage of fuel nitrogen is oxidized beyond molecular nitrogen.

Conditions Necessary for Combustion

Since this is a chemical reaction, it follows the laws of chemical equilibrium, chemical kinetics, and thermodynamics. Following is a brief presentation of the important conditions that affect the combustion reaction. A simple way to remember these conditions is by the initial letters, T, T, T, O.

Time. Because this is a chemical reaction, adequate time must be available for the reaction to go to completion. The time required for complete combustion is related to the following conditions.

Temperature. The temperature at which the combustion occurs has several effects. First, a higher temperature results in a higher reaction rate, allowing the reaction to reach completion more quickly. Second, many compounds have an ignition temperature that must be reached before the energy level is sufficient to initiate the reaction. Finally, the temperature will determine the equilibrium concentration of some of the combustion products, especially the oxides of nitrogen.

Turbulence. This condition is important because the reactants must have intimate contact if the reaction is to proceed to completion rapidly. Mixing in the furnace is essential to mix the combustion air (oxygen) with the combustible gases and vapors.

Oxygen. Having this reactant in excess will drive the reaction to completion more rapidly. If a stoichiometric quantity of oxygen is available, the reaction will eventually reach completion, but excess oxygen will increase the rate at which completion is achieved.

Combustion Zones in a Solid Fuel

When a solid fuel is introduced into a furnace, combustion will proceed through a series of steps. Understanding these steps is important in understanding the design of a furnace.

Distillation zone. The fuel is introduced into the primary combustion chamber and subjected to high temperatures. Oxygen is generally not sufficient to provide for rapid and complete oxidation of all of the fuel. There is sufficient oxygen to cause enough burning to maintain relatively high temperatures. As the fuel's temperature increases, volatile materials present will be driven off as gases. The volatile matter determination of the proximate analysis indicates the significance of this volatilization. A continued increase in temperature will cause many of the organic compounds in the fuel to thermally "crack" and form gaseous products.

Solid waste has a high volatile matter content, and there is a rapid release of combustible volatile compounds that burn in the gas phase. Refuse will respond differently than a fuel such as coal, which has a much lower volatile matter content. In coal-fired power plants, the coal is crushed to very fine particles and blown into the furnace like a gas. Because of the very large surface area of the small particles, the combustion approaches a gaseous combustion. Since refuse has a high volatile matter content, small particles are not required. However, all of the refuse particles must be exposed to the high temperature if volatilization is to occur. This is achieved by designing the grates to turn the material over as it progresses through the furnace.

Incandescent zone. After the volatile compounds and the compounds that are thermally cracked have been driven from the fuel, only fixed carbon and ash remain on the grates. When the temperature reaches the ignition temperature of carbon, 1300°F, the fixed carbon, essentially charcoal, is ignited and burns on the surface. To achieve good ''burnout'' (i.e., destruction of all the combustible material), it is necessary to obtain at least 1300°F in the refuse bed and to have oxygen available to burn the fixed carbon. The oxygen is supplied as the ''underfire air,'' air that is blown in under the grates and passes up through the refuse retained on the grate. Because surface burning is slow, the retention time of the residual material on the grates must be long, to ensure good burnout.

Some of the heat released by burning of the fixed carbon is used to evaporate the moisture in the incoming fuel and to provide the heat energy for the many thermal cracking reactions that are endothermic (i.e., heat is required for the reaction). The heat balance becomes an important consideration for these first two phases. If sufficient heat is not available, the combustion will be poor at best.

Flame zone. During the first stage of combustion, most of the combustible solids are converted to gaseous components, either by volatilization, thermal cracking, or partial oxidation. These hot gases form a flame when mixed with oxygen. The combustion, oxidation, is rapid. Very high heat release rates are experienced and high temperatures can be encountered in this combustion zone. The flame zone operates with a great deal of turbulence so that the gases are intimately mixed with the oxygen added with the secondary air supply. The combustion occurring at a high temperature with excess oxygen will be complete in 1 or 2 seconds. This is a critical phase in refuse combustion. If properly designed, the concentration of non-oxidized organic compounds will approach zero. All of the toxic and hazardous compounds produced during the early combustion phases will be destroyed.

Combustion Temperature

In recent years there has been a tendency to increase the maximum combustion temperature. In early refuse incinerators, it was generally conceded that 1200°F was sufficient to minimize the odors produced during incineration. This was an important consideration for two reasons. First, the lower temperature was more benign in the furnace; less maintenance was required. Second, during certain periods of the year, the refuse would be wet and auxiliary fuel would be required to maintain a higher temperature. When the refuse was dry, the temperature would be much higher than 1200°F. Under these conditions, a lot of excess air was vented through the furnace to control the temperature at this lower level. Air pollution control was not required, and the air was free, so it was freely used.

As the profession became wiser (i.e., developed the ability to measure the multitude of compounds that were present in these gases), it was clear that it would be necessary to improve the combustion efficiency. Also, the application of emission standards to incinerators made it impractical to use the high airflow rates to cool the furnace when burning dry refuse. As much as 300% excess air would be used. The air pollution control device would have to be more than twice the size required for normal airflows. These changes caused a reevaluation of the role of incineration in refuse disposal. The high temperatures resulting from the restricted airflow necessitated an alternative means for cooling the hot combustion gases. The added costs for construction and operation of these units needed another income stream to keep the costs reasonable, so the idea of steam recovery and finally the production of electricity with steam turbines became a reality. Once in the power generation business, it was desirable to reach as high a temperature as practical so that a high boiler efficiency was obtained. More of the refuse energy was turned into steam.

As more information is developed, the role of temperature in determining the degree of combustion becomes apparent. Many constituents will not even ignite at some of the temperatures previously used. Examples of some of the ignition temperatures of organic materials are presented in Table 9.1. Ignition temperatures are unique for each substance. These are intended only as examples of the wide temperature ranges encountered. These are the temperatures at which the ignition occurs. The rate of combustion is still a function of the temperature. Therefore, higher temperatures would be advantageous for increasing the rate of combustion. These are some of the considerations for establishing furnace temperature standards for refuse incineration.

Berthelot's Second Law of Combustion

This law describes the relationship between the initial reactants and the final products of a combustion reaction. "In a furnace, with work equal to zero, the heat evolved from the union of combustible elements with molecular oxygen depends

Table 9.1 Ignition Temperatures of Selected Compounds

Substance	Ignition temperature (°F)
Carbon disulfide	248
Wood	350
Paper	350
Acetone	1042
Benzene	1097
Carbon monoxide	1128
Methane	1170
Cyanogen	1562

Source: R. H. Perry and C. H. Chilton, Eds., *Chemical Engineers Handbook,* 5th ed., McGraw-Hill, New York, 1973, p. 9–19.

upon the ultimate products of combustion and not upon any intermediate combinations which may result.''

Example

Consider the reactions of carbon with oxygen:

Two-step conversion:

$$C + \tfrac{1}{2}O_2 \rightarrow CO \qquad \Delta H_R = -10{,}100 \text{ Btu} \tag{1}$$

$$CO + \tfrac{1}{2}O_2 \rightarrow CO_2 \qquad \Delta H_R = -\ 4{,}000 \text{ Btu} \tag{2}$$

One-step conversion:

$$C + O_2 \rightarrow CO_2 \qquad \Delta H_R = -14{,}100 \text{ Btu} \tag{3}$$

This law simplifies the combustion analysis. It is not necessary to consider all of the possible reactions and intermediate products. It is only necessary to know the composition of the reactants and the final products.

Unfortunately, refuse is not a well-defined chemical compound. However, this law applies if the elemental composition (ultimate analysis) of the fuel is known. The products will be the oxidized form of the elements, CO_2 and H_2O primarily. Sulfur will be oxidized to SO_2, and nitrogen is assumed to be elemental nitrogen, N_2. The other elements present in the combustible fraction will be insignificant in the combustion reactions. The cans, glass, ceramics, and so on, do not enter into the combustion analysis. The heat of reaction, ΔH_R, of refuse must be determined from the bomb calorimeter test discussed previously. The value obtained is the HHV (high heating value) at the standard temperature of 77°F.

HHV (ΔH_R) Determined by the Oxidation State of Carbon

Because of the nature of refuse, it is very difficult to determine the HHV from theoretical considerations. In addition to the ash content, the oxidation state of carbon is quite variable, having a significant impact on the HHV. This effect is illustrated in Table 9.2. The oxidation state of carbon in most natural fibers, such as cellulose (paper), is the same as that in elemental carbon. The average oxidation number for carbon in an organic compound can be determined from the following relationship:

$$\text{oxidation number} = 2(\text{number of oxygen}) - (\text{number of hydrogen}) \tag{4}$$

Plastic polymers are generally more saturated with hydrogen and the carbon has an oxidation number closer to that of methane. Fats and oils also contain more saturated carbon compounds, which cause these materials to have a higher HHV.

Volumetric and Gravimetric Analysis

Avogadro's law. ''At the same temperature and pressure, equal volumes of all perfect gases contain the same number of molecules (i.e., 6.02×10^{23}).''

Table 9.2 Effect of carbon oxidation state on HHV

Compound	Carbon oxidation number	ΔH_R	
		Btu/lb-mol	Btu/lb C
CH_4	−4	382,786	31,899
CH_3OH	−2	310,333	25,861
C	0	169,183	14,099
CO	+2	121,666	10,139
CO_2	+4	0	0

Source: R. H. Perry and C. H. Chilton, Eds., *Chemical Engineers Handbook,* 5th ed., McGraw-Hill, New York, 1973, p. 3–145.

Mole. A mole is defined as that quantity of a substance that has a weight equal to the molecular weight. A gram-molecular weight is the weight in grams of 6.02×10^{23} molecules of a substance. For example, the molecular weight of methane is 16, so a g-mol of methane weighs 16 g. A lb-mol simply references everything to the pound (i.e., everything is 454 times larger than the g-mol). This is a convenience only for use with the Imperial System of units.

Volume occupied by 1 mol. The ideal gas law can be used to determine the volume occupied by a gas:

$$PV = NR_uT \tag{5}$$

where N = number of moles
R_u = ideal gas constant
P, V, T = pressure, volume, and temperature, respectively

This equation stipulates that at a constant pressure and temperature, the volume will depend only on the number of moles. The volume occupied by 1 mol of any gas is the same as long as they are at the same temperature and pressure. For any gas,

1 lb-mol occupies 359 scf (32°F and 1 atm).
1 g-mol occupies 22.4 L at STP (0°C and 1 atm).

Gravimetric/volumetric analysis. The analyses of solid fuels are reported in percentage by weight (gravimetric analysis), and the gaseous products of combustion are reported in percentage by volume (volumetric analysis). The relationships above are important in the conversion of weight percents to volume percents. When a solid fuel composition is presented, it is necessary to convert this composition into a mole percentage since all combustion reactions occur on a molar basis. It is also necessary to know the moles of gaseous products, since this determines the volume of gaseous combustion products. Table 9.3 illustrates the relationships between these expressions of composition. The unifying parameter is the molecular weight.

Table 9.3 Conversion of volumetric analysis to gravimetric analysis

Constituent	Volumetric analysis [% (mol/100 mol)]	Mol. wt	Weight of individual gas (lb/100 mol)	Gravimetric analysis [% (lb/100lb)]
H_2	26	2	52	100(52/2052) = 2.5
CO	68	28	1904	100(1904/2052) = 92.8
CH_4	6	16	96	100(96/2052) = 4.7
	100		2052	100.0

The reverse calculation can be accomplished by expressing the gravimetric analysis percent as lb/100 lb. The number of pounds of a specific compound or element is then divided by its molecular weight to determine the number of moles of this constituent in 100 lb. This is done for each constituent to find the number of moles in the 100 lb of fuel. The number of moles of individual constituents is divided by the total number of moles per 100 lb and multiplied by 100% to obtain the volumetric analysis.

Complete Combustion Is Necessary

All calculations associated with the mass balance for combustion reactions must be based on the complete combustion of the elements in the fuel (i.e., carbon is oxidized to carbon dioxide, hydrogen to water, and sulfur to sulfur dioxide). The correctness of the calculations can be checked by knowing that the mass of the reactants must equal the mass of the products. The same is *not true* for the volume of the reactants and the products. The following equations illustrate these relationships:

$$C + O_2 \rightarrow CO_2 \tag{6}$$

$$12 \text{ lb C} + 32 \text{ lb } O_2 \rightarrow 44 \text{ lb } CO_2 \tag{7}$$

$$1 \text{ mol C} + 1 \text{ mol } O_2 \rightarrow 1 \text{ mol } CO_2 \tag{8}$$

For this reaction, the mass of reactants, 44 lb (12 + 32) equals the mass of the products, 44 lb [equation (7)]. However, the 2 mol of reactants yield only 1 mol of product, as illustrated in equation (8). A mass balance should be completed on any type of combustion computations.

Theoretical Combustion Air

The theoretical or stoichiometric oxygen requirements can be determined from the chemical reactions for the oxidation of the combustible elements in the fuel. The gaseous products of combustion will generally be CO_2, H_2O, and SO_2. If air is used as a source of the oxygen, these combustion products will contain the nitrogen associated with the oxygen. However, since the stoichiometric amount of oxygen is used, there will be no oxygen remaining in the combustion gases. In all combustion systems, it is desirable to have an excess of oxygen to drive the reaction to completion more rapidly. The excess air employed is usually 25 to 50% of the theoretical air. In older incinerators, values as high as 300% were used primarily to cool the furnace when the refuse was dry. Because of the cost of air pollution control, the excess air is minimized. Only enough excess oxygen is supplied to ensure complete combustion.

It is necessary to know the composition of air since nitrogen will be present in all mass and energy balance calculations. For this application, air is assumed to contain only oxygen and nitrogen. The volume percentages of oxygen and nitrogen are 20.9% and 79.1%, respectively. The weight percentages are 23.15% and 76.85%, respectively. The following numbers are important to have for conducting mass balance computations.

To supply 1 lb of O_2 requires (1/0.2315) = 4.32 lb of air:

$$1 \text{ lb } O_2 + 3.32 \text{ lb } N_2 = 4.32 \text{ lb air} \tag{9}$$

To supply 1 mol of O_2 requires (1/0.209) = 4.78 mol of air:

$$1 \text{ mol } O_2 + 3.78 \text{ mol } N_2 = 4.78 \text{ mol air} \tag{10}$$

$$1 \text{ scf } O_2 + 3.78 \text{ scf } N_2 = 4.78 \text{ scf air} \tag{11}$$

MASS BALANCE FOR COMBUSTION OF A FUEL GAS

Given the volumetric composition of a fuel gas, the following procedure can be used to compute a mass balance for the combustion of this fuel (Table 9.4). The mass and volume flow rates are necessary to size the furnace components and the air pollution control equipment and to compute an energy balance. The tabulation is simply an attempt to assist in keeping the numbers straight, as there are many ways

Table 9.4 Mass balance computations: oxygen requirements

Constituent	Vol %	Combustion reaction	Moles O_2 required
H_2	9	$H_2 + \frac{1}{2}O_2 \rightarrow H_2O$	$9(\frac{1}{2}) = 4.5$
CO	24	$CO + \frac{1}{2}O_2 \rightarrow CO_2$	$24(\frac{1}{2}) = 12.0$
CH_4	2	$CH_4 + 2O_2 \rightarrow CO_2 + 2H_2O$	$2(2) = 4.0$
CO_2	6		
O_2	3		(−3.0)
N_2	56		

Theoretical oxygen requirements = 17.5 mol/100 mol
50% excess air = 8.75 mol/100 mol

in which the numbers can be manipulated. The first step involves determination of the theoretical oxygen requirements. This requires writing the combustion reactions to determine the moles of oxygen required for each combustible element. These equations also provide the gaseous combustion products. The residual excess oxygen and the nitrogen in the total combustion air will also be part of the combustion products. It is important to recognize the oxygen present in either a solid, liquid, or gaseous fuel. The availability of this oxygen in a fuel gas may be obvious, but in the other fuels it represents the amount of oxidation of the carbon that has already occurred, and must be deducted from the oxygen requirements.

Combustion Air Requirements

The required combustion air can be calculated as follows:

$$\text{0\% excess air:} \quad \frac{17.5 \text{ mol } O_2}{100 \text{ mol fuel}} \times \frac{4.78 \text{ mol air}}{\text{mol } O_2} = \frac{83.7 \text{ mol air}}{100 \text{ mol fuel}} = 0.837 \text{ scf/scf}$$

$$\text{50\% excess air:} \quad \text{combustion air} = 1.5(0.837) = 1.26 \text{ scf/scf}$$

Combustion Products

The combustion equations also provide the combustion products in Table 9.5. Gases associated with the combustion air will be in the stack gases. The table shows how these numbers are handled. Nitrogen is an important component of any combustion system that uses air as the oxygen source.

Table 9.5 Mass balance computations: combustion products

			Combustion product (mol/100 mol)			
Constituent	Vol %	Combustion reaction	CO_2	H_2O	N_2	O_2
		Theoretical air: 0% excess air				
H_2	9	$H_2 + \frac{1}{2}O_2 \rightarrow H_2O$	0	9	0	0
CO	24	$CO + \frac{1}{2}O_2 \rightarrow CO_2$	24	0	0	0
CH_4	2	$CH_4 + 2O_2 \rightarrow CO_2 + 2H_2O$	2	4	0	0
CO_2	6		6	0	0	0
O_2	3		0	0	0	0
N_2	56		0	0	56	0
			32	13	56	0
		Combustion air	0	0	66.2	0
		50% excess air	0	0	33.1	8.75
			32	13	155.3	8.75

Table 9.6 Volumetric analysis of the dry products of combustion

	0% excess air			50% excess air		
Constituent	mol/100 mol fuel		Vol %	mol/100 mol fuel		Vol %
CO_2	32	100(32/154.2) =	20.7	32	100(32/196.1) =	16.3
O_2	0		0	8.75	100(8.75/196.1) =	4.5
N_2	122.2	100(122.2/154.2) =	79.3	155.3	100(155.3/196.1) =	79.2
	154.2		100	196.1		100

Composition of the Dry Products of Combustion

The composition of the gases exiting the stack is used for operational control as well as for emission standards. The oxygen or carbon dioxide content of these gases is frequently used as a baseline for determining the concentration of allowable contaminants. Sampling and analysis of these gases require a dry sample. It is impossible to extract a sample from a stack and retain a constant moisture. As the gas cools, the water condenses. Also, many of the analytical instruments require a dry sample for analysis. Therefore, calculation of the dry products of combustion (dpc) is necessary. The combustion products listed in Table 9.5, less the water, are used in this calculation, as shown in Table 9.6.

Calculation of the Weight of Total Products of Combustion

Mass flow rates are important to size the motors required for the fans and blowers used to move the combustion gases. These weights are also useful in making estimates of the energy balances on a combustion unit. This will be discussed later. The weight of the gases can be computed from the number of moles and the molecular weight. This is shown in Table 9.7.

Table 9.7 Weight of the total products of combustion (tpc)

		0% excess air			50% excess air	
Constituent	Mol. wt	mol/ 100 mol		lb/ 100 mol	mol/ 100 mol	lb/ 100 mol
CO_2	44	32	(× 44)	1408	32	1408
H_2O	18	13	(× 18)	234	13	234
O_2	32	0		0	8.75	280
N_2	28	122.2	(× 28)	3420	155.3	4348
				5062		6270

To compute the pounds of combustion gases per pound of fuel, the weight of 100 mol of the fuel gas must be known. This number can be calculated in the same manner as in Table 9.7. It is 2650 lb/100 mol of fuel.

$$\text{At 0\% excess air:} \quad \frac{\text{lb tpc}}{\text{lb fuel}} = \frac{5062}{2650} = 1.91$$

$$\text{At 50\% excess air:} \quad \frac{\text{lb tpc}}{\text{lb fuel}} = \frac{6270}{2650} = 2.37$$

The weight of water in the combustion gases is used to compute the heat loss associated with water vapor and to determine the potential for condensation in various parts of the furnace.

$$\frac{\text{lb } H_2O}{\text{lb fuel}} = \frac{234}{2650} = 0.089 \text{ lb/lb fuel}$$

The specific heat of the water vapor is almost twice the specific heat of the dpc. When using the weight of combustion products to estimate heat loss, it is necessary to separate the weight of water from the weight of dry gases.

$$\frac{\text{lb dpc}}{\text{lb fuel}} = \frac{\text{lb tpc}}{\text{lb fuel}} - \frac{\text{lb } H_2O}{\text{lb fuel}} \tag{12}$$

$$\text{At 0\% excess air:} \quad \frac{\text{lb dpc}}{\text{lb fuel}} = 1.91 - 0.089 = 1.82$$

$$\text{At 50\% excess air:} \quad \frac{\text{lb dpc}}{\text{lb fuel}} = 2.37 - 0.089 = 2.28$$

Weight of Combustion Air per Pound of Fuel

To complete the mass balance, it is necessary to determine the weight of combustion air used per pound of fuel. This is shown in Table 9.8.

Table 9.8 Weight of combustion air

	0% excess air			50% excess air		
	mol/ 100 mol		lb/ 100 mol	mol/ 100 mol		lb/ 100 mol
O_2	17.5	(× 32)	560	26.25	(× 32)	840
N_2	66.2	(× 28)	1853.6	99.3	(× 28)	2780.4
			2413.6			3620.4

$$\text{At 0\% excess air:} \quad \frac{\text{lb air}}{\text{lb fuel}} = \frac{2413.6}{2650} = 0.91$$

$$\text{At 50\% excess air:} \quad \frac{\text{lb air}}{\text{lb fuel}} = \frac{3620.4}{2650} = 1.37$$

The mass balance can be used to check the accuracy of the computations as follows:

At 0% excess air: 1 lb fuel + 0.91 lb air = 1.91 lb tpc

At 50% excess air: 1 lb fuel + 1.37 lb air = 2.37 lb tpc

MASS BALANCE FOR THE COMBUSTION OF A SOLID FUEL

The mass balance computations for a solid fuel are similar to those for a fuel gas. There are a few differences in the initial fuel compositions that must be recognized. First, the ultimate analysis of a solid fuel is a gravimetric analysis (i.e., a weight percent basis). All combustion reactions occur on a mole basis. The fuel gas composition, a volumetric analysis, can be translated directly to molar units, since the volume percent is simply a mole ratio expressed as a percentage. With a solid fuel, the weight of the individual fuel components must be converted to moles in order to use the combustion equations. If the weight percent is considered to be based on 100 lb of fuel, the percentage numbers are the weight of the individual constituent per 100 lb of fuel. The number of moles per 100 lb can be calculated by dividing the weight of the element by its molecular weight. The computation continues as for the fuel gas; only the results are based on 100 lb rather than 100 mol.

The ultimate analysis is based on a dry sample. If the percentage combustible elements sum to 100%, the analytical results are also presented on an ash-free basis. The combustion air and the combustion products are based on an "as received" fuel (i.e., the fuel will contain the moisture and the ash). Therefore, it is necessary to adjust the weight percent (ultimate analysis) to reflect the actual contents of the elements in the fuel as it is burned. If one is given the solid fuel analyses shown in Table 9.9, they can be used to construct Table 9.10.

Table 9.9 Solid fuel analysis

Proximate analysis	(wt %)	Ultimate analysis	(wt %)
Moisture	9.7	Carbon	78.3
Volatile matter	19.6	Hydrogen	5.4
Fixed carbon	60.0	Oxygen	8.7
Ash	10.7	Sulfur	6.0
	100.0	Nitrogen	1.6
			100.0

Table 9.10 Mass balance computations: oxygen requirements

Constituent	Wt %	Mol. wt	mol/100 lb	Combustion reactions	O_2 required (mol/100 lb)
S	4.8	32	0.15	$S + O_2 \rightarrow SO_2$	0.15
H_2	4.3	2	2.16	$H_2 + \frac{1}{2}O_2 \rightarrow H_2O$	1.08
C	62.3	12	5.20	$C + O_2 \rightarrow CO_2$	5.20
N_2	1.3	28	0.05	No reaction	
O_2	6.9	32	0.22		(−0.22)
Ash	10.7				
H_2O	9.7	18	0.54		
	100.0			Theoretical oxygen required =	6.21
				50% excess =	3.11

The carbon content in the fuel can be computed as follows:

$$\%C = 78.3(1.00 - 0.097 - 0.107) = 62.3\%$$

This same relationship is used to compute the percentage of each element in the fuel. With these corrections, the calculation of the oxygen requirements for a solid fuel can be made as illustrated in Table 9.10. The gaseous products of combustion are calculated as before (see Table 9.11). The nitrogen in the fuel is assumed to report as N_2. Because the quantity of nitrogen oxides formed in the furnace is very small compared to the amount of nitrogen present in the combustion gases, it is reasonable to ignore these compounds.

Table 9.11 Gaseous combustion products from solid fuel

	Combustion product (mol/100 lb)				
Constituent	Co_2	H_2O	N_2	SO_2	O_2
S				0.15	
H_2		2.16			
C	5.2				
N_2			0.05		
O_2					
Ash					
H_2O		0.54			
	5.2	2.70	0.05	0.15	0
Combustion air			23.50		
50% excess air			11.75		3.11
	5.2	2.70	35.30	0.15	3.11

ENERGY BALANCE ON COMBUSTION PROCESSES

Theoretical Considerations

The first law of thermodynamics states that the heat transferred (Q) to any system is equal to the change in the internal energy (ΔU) plus work (W) done by the system. This law can be applied to any system, whether or not a chemical reaction occurs. A statement of this law is

$$Q = \Delta U + W \tag{13}$$

If a reaction occurs such that there is no mass transferred across the system boundary (e.g., a closed system), the system remains at constant pressure, and the only work done is due to volume change, equation (13) becomes

$$Q = \Delta U + p\Delta V = \Delta H \tag{14}$$

H is a property of the system that represents the heat content and is defined as enthalpy. In combustion reactions, the change in enthalpy may be due to changes in the temperature or due to chemical reactions taking place at a constant temperature. To distinguish between the two phenomena, let ΔH_R denote the enthalpy change due to the chemical reactions and ΔH_T denote the enthalpy change due to temperature change. This relationship is

$$Q = \Delta H = \Delta H_R + \Delta H_T \tag{15}$$

ΔH_R is commonly referred to as the heat of reaction and, for refuse, is determined by the bomb calorimeter test. This test is run at a constant temperature of 77°F, which is the reference temperature (T_0) that will be used in subsequent energy balance calculations.

Figure 9.1 illustrates the relationship defined by equation (15). For a steady-flow process between states 1 and 2 in which the changes in kinetic energy and potential energy are zero and the only work done is volume change, the heat flow is determined by the enthalpy change. The change in enthalpy can be determined by computing the difference in the enthalpy at point 1 and point 2:

$$\Delta H = H_2 - H_1 = (H_2 - H_x) + \Delta H_R + (H_y - H_1) \tag{16}$$

In this equation, H_x and H_y are the enthalpies of the products and reactants, respectively, at the reference temperature. For convenience, the sign on the last term of equation (16) can be changed to a negative, to give

$$\Delta H = H_2 - H_1 = (H_2 - H_x) + \Delta H_R - (H_1 - H_y) \tag{17}$$

However, the reactants and the products are composed of a number of specific substances. $(H_1 - H_y)$ represents all of the reactants, and $(H_2 - H_x)$ represents all of the products. It is necessary to consider the number of moles of each substance (N) and the molar enthalpy of that substance (h). Therefore, the following is a

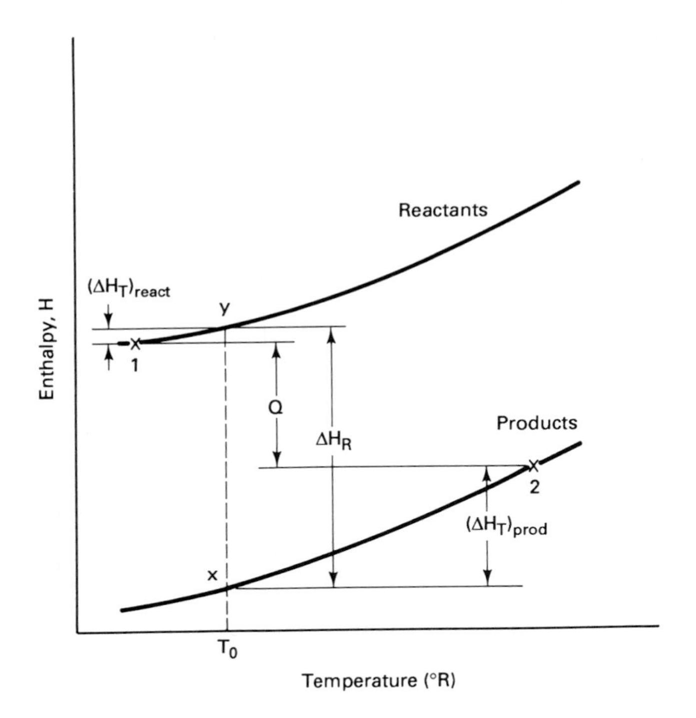

Figure 9.1 Relationship between enthalpy of reactants and products.

general equation that can be used to compute the enthalpy change considering each individual substance involved in the reaction:

$$H_2 - H_1 = \sum_{\text{prod}} N_i(h_2 - h_0) + \Delta H_R - \sum_{\text{react}} N_j(h_1 - h_0) \tag{18}$$

where $(h_1)_j$ = molar enthalpy of the reactants, j, at the temperature they enter the furnace

$(h_0)_j$ = molar enthalpy of the reactants at the reference temperature, 77°F

$(h_2)_i$ = molar enthalpy of the products at the final temperature

$(h_0)_i$ = molar enthalpy of the products at the reference temperature

When $T_2 = T_0$ and $T_1 = T_0$, $h_2 - h_0 = 0$ and $h_1 - \text{h}_0 = 0$. Consequently,

$$Q = \Delta H_R \tag{19}$$

A sign convention is used throughout to determine if heat is added or released from the reactions. When ΔH_R is negative, Q is negative, indicating that the reaction is *exothermic* (i.e., heat is released by the system). If ΔH_R is positive, the reaction is *endothermic* and it is possible to achieve the indicated end products only if heat is added to the system.

To compute the energy balance, information is needed for the enthalpy of the different reactants and products as a function of the temperature. The enthalpy can be determined from the heat capacity of the material as indicated by

$$H_T - H_0 = \int_{537}^{T} C_p \, dT \quad (20)$$

where H_T and H_0 are the enthalpy at the elevated temperature and the reference temperature of 537°R (77°F) in Btu/lb-mol and C_p is the heat capacity at constant pressure. Integration of this equation produces a general equation for the computation of the enthalpy change for various materials. The values for constants in this general equation can be found in thermodynamic properties tables.[1] These constants have been determined experimentally or computed from theoretical considerations.

$$CO_2: \quad H_T - H_0 = 10.34T + 1.37 \times 10^{-3}T^2 - 1.96 \times 10^5/T - 5583 \quad (21)$$

$$CO: \quad H_T - H_0 = 6.6T + 1.6 \times 10^{-3}T^2 - 3717 \quad (22)$$

$$O_2: \quad H_T - H_0 = 8.27T + 1.29 \times 10^{-4}T^2 - 1.88 \times 10^5/T - 4129 \quad (23)$$

$$N_2: \quad H_T - H_0 = 6.50T + 5.0 \times 10^{-4}T^2 - 3635 \quad (24)$$

$$H_2: \quad H_T - H_0 = 6.62T + 4.05 \times 10^{-4}T^2 - 3672 \quad (25)$$

$$H_2O: \quad H_T - H_0 = 7.14T + 8.1 \times 10^{-4}T^2 - 4.0 \times 10^{-8}T^3 - 4062 \quad (26)$$

These equations are cumbersome and it is easier to have the enthalpy displayed in tabular or graphical form for ease of use, especially when considering a variety of substances at several different temperatures. Table 9.12 contains such values as computed from equations (21) through (26).

Water requires special consideration. First, the enthalpy values calculated from equation (26) and presented in Table 9.12 are only for the water vapor. It is assumed that the water evaporates at the standard reference temperature and that only the vapor is subject to the temperature increase. The C_p for liquid water is 1 Btu/lb-°F, while water vapor has a specific heat of about 0.5 Btu/lb-°F. This is a valid assumption since the humidity in the furnace gases will be very low until the water evaporates and the combustion water appears. Since the values in Table 9.12 and equation (26) do not consider any change of state, the heat of vaporization of water must be included in the heat-loss calculations.

Energy Balance Computations

Heat loss with the combustion gases. The mass balance computations conducted in Tables 9.10 and 9.11 provide the necessary flow rates for the reactants and the products. The initial and final conditions must be specified and the HHV of the fuel (refuse) must be known. The enthalpy value of a fuel gas such as methane is available, but the enthalpy value for a solid fuel is not easily determined since the

[1]R. H. Perry and C. H. Chilton, Eds., *Chemical Engineers Handbook,* 5th ed., McGraw-Hill, New York, 1973, p. 3–119.

Table 9.12 Enthalpy of ideal gases

	Enthalpy (Btu/lb-mol)					
Temperature (°R)	CO	CO_2	H_2	H_2O	N_2	O_2
0	0	0	0	0	0	0
300	2,076	2,572	2,022	2,214	1,995	1,866
400	2,736	3,865	2,713	2,983	2,680	2,859
500	3,450	5,121	3,411	3,768	3,375	3,791
537	3,717	5,583	3,672	4,062	3,635	4,128
600	4,176	6,371	4,118	4,567	4,080	4,695
700	4,914	7,629	4,832	5,381	4,797	5,584
800	5,664	8,904	5,555	6,210	5,520	6,464
900	6,426	10,198	6,286	7,053	6,255	7,340
1000	7,200	11,514	7,025	7,910	7,000	8,211
1100	7,986	12,854	7,772	8,781	7,755	9,082
1200	8,784	14,217	8,527	9,665	8,520	9,953
1300	9,594	15,606	9,290	10,563	9,295	10,824
1400	10,416	17,021	10,062	11,474	10,080	11,697
1500	11,250	18,467	10,841	12,398	10,875	12,570
1600	12,096	19,929	11,629	13,334	11,680	13,445
1700	12,954	21,422	12,424	14,282	12,495	14,321
1800	13,824	22,942	13,228	15,243	13,320	15,196
1900	14,706	24,489	14,040	16,216	14,155	16,079
2000	15,600	26,062	14,860	17,200	15,000	16,962

specific compounds in the fuel are not known. It is convenient to assume that the fuel is introduced into the furnace at the reference temperature (77°F). It is possible to estimate the heat required to elevate the fuel temperature to the reference temperature by using the specific heat of the solid fuel. Typical values for organic compounds are 0.3 to 0.35 Btu/lb-°F. Therefore, if the solid fuel is not at the reference temperature, the heat requirement can be calculated from $C_p|\Delta T|W$, where W is the pounds of dry fuel. For more precision, a correction can be made for the water (moisture) present in the fuel using this same approach and a C_p value of 1.0 Btu/lb-°F.

Given the following information (see Tables 9.10 and 9.11), determine the amount of heat that can be recovered from the combustion of a solid fuel.

Combustion reactants	Gaseous combustion products
O_2: 9.32 lb-mol/100 lb	O_2: 3.11 lb-mol/100 lb
N_2: 35.25 lb-mol/100 lb	N_2: 35.30 lb-mol/100 lb
T = 140°F	CO_2: 5.20 lb-mol/100 lb
	SO_2: 0.15 lb-mol/100 lb
Fuel at 77°F	H_2O: 2.70 lb-mol/100 lb
ΔH_R = −9000 Btu/lb	T = 540°F

Step 1. $\Sigma N(h_1 - h_0)$ of combustion air at 140°F.

Constituent	lb-mol/100 lb	h_1	h_0	$N(h_1 - h_0)$
O_2	9.32	4695	4128	5,284
N_2	35.25	4080	3635	15,687
				20,971

Step 2. $\Sigma N(h_2 - h_0)$ of gaseous products of combustion.

Constituent	lb-mol/100 lb	h_2	h_0	$N(h_2 - h_0)$
O_2	3.11	8,211	4,128	12,698
N_2	35.30	7,000	3,635	118,785
CO_2	5.20	11,514	5,583	30,841
SO_2	0.15	10,088	13,504	−512
H_2O	2.70	7,910	4,062	10,389
	$N(\Delta H_v)_{H_2O}$ = 2.70(18,926 Btu/lb-mol) =			51,100
				223,301

Step 3. Compute the heat of reaction, ΔH_R.

$$\Delta H_R = -9000 \text{ Btu/lb(lb dry solids/100 lb fuel)} = -9000(100 - 9.69)$$
$$= -812{,}790 \text{ Btu/100 lb fuel}$$

Step 4. Compute Q as follows:

$$Q = \sum_{\text{prod}} N_i(h_2 - h_0) + \Delta H_R - \sum_{\text{react}} N_j(h_1 - h_0) \tag{27}$$

$$= 223{,}301 + (-812{,}790) - 20{,}971 = -610{,}460 \text{ Btu/100 lb of fuel}$$
$$= -6105 \text{ Btu/lb of fuel}$$

Additional heat losses. The analysis above represents only the heat balance on the combustion gases. Additional heat loss can be attributed to the ash fraction and to radiant heat from the furnace. The heat loss associated with the ash will depend on the temperature at which the ash exits the furnace and the heat capacity of the ash components. Most incinerators have an airstream called the underfire air that passes over the ash and extracts some of the heat. Ash temperatures will generally be about 400°F as it drops into the ash quench tank. A ΔT value of about 350°F would be reasonable. Multiplying this temperature change by a representative C_p and the pounds of ash generated will yield the heat loss with this residue.

The following are specific heats for some of the common residues found in refuse incineration.

Specific heats for inorganic materials	
Ash	0.2 Btu/lb-°F
Glass	0.17 Btu/lb-°F
Aluminum	0.21 Btu/lb-°F
Iron	0.1 Btu/lb-°F

In the example above, the fuel contained 10.68% ash, or 10.7 lb/100 lb of fuel. Assume that the final temperature of the ash is 400°F and the initial temperature is 77°F:

$$\text{heat loss with ash} = (10.7\ \text{lb}/100\ \text{lb})(0.2\ \text{Btu/lb-°F})(400 - 77)$$
$$= 691\ \text{Btu}/100\ \text{lb of fuel}$$

If this heat loss is compared to the approximately 200,000 Btu/100 lb of fuel lost with the combustion gases, it is apparent that it is insignificant. If the ash content doubled, the heat loss would still be minimal.

The radiant heat loss from the furnace will also be insignificant in a modern design. Because of the size of these units, it would be impossible to dissipate the heat if the loss was a significant percentage of the furnace load. A system burning 1000 tons per day of refuse releases about 400 million Btu/hr. Even a 1% loss would be 4 million Btu/hr. A sizable ventilation system would be necessary to remove the heat from the building. The insulation of these furnaces is substantial to minimize the radiant heat loss.

Estimating Heat Loss from Average Combustion Product C_p

An alternative technique for approximating the heat loss with the combustion gases can be used. This assumes an average C_p for the dry products of combustion. Over the temperature range in Table 9.12, the C_p for these gases will average between 0.23 and 0.25 Btu/lb-°F, with nitrogen having a value of 0.25. Since nitrogen is the largest component of the combustion gases, a value of 0.25 Btu/lb-°F would be appropriate. Table 9.13 presents the computations that are necessary to obtain the pounds of combustion products.

The exit temperature of the combustion gases is 540°F, so the temperature change is (540 − 77) = 463°F.

$$\text{Heat loss:}\quad \text{dpc} = (1327\ \text{lb}/100\ \text{lb})(463°\text{F})(0.25\ \text{Btu/lb-°F})$$
$$= 153{,}600\ \text{Btu}/100\ \text{lb}$$
$$(H_2O)_v = (48.6)(463)(0.5) = 10{,}251\ \text{Btu}/100\ \text{lb}$$

Table 9.13 Pounds of combustion gases

Gas	lb-mol/100 lb	Mol. wt	lb/100 lb
O_2	3.11	32	99.5
N_2	35.33	28	988.4
CO_2	5.20	44	229.0
SO_2	0.15	64	9.6
		Σ dpc =	1326.5
H_2O	2.70	18	48.6

$$(\Delta H_v)_{H_2O} = (48.6)\left(\frac{18{,}926}{18}\right) = 51{,}100 \text{ Btu/100 lb}$$

$$\text{Total heat loss} = 153{,}600 + 10{,}251 + 51{,}100 = 214{,}951 \text{ Btu/100 lb}$$

This estimation of the heat loss with the combustion gases compares with 223,301 Btu/100 lb as computed in step 2 on page 187. The loss is underestimated by about 4%, which is sufficiently accurate for an initial assessment of the energy balance on an incineration system. Slightly more precision is obtained by using the actual enthalpy values for the product gases.

DESIGN CONSIDERATIONS FOR MULTIPLE-CHAMBER INCINERATORS

Components of a Multichamber Incinerator

Figure 9.2 is a cutaway of a simple in-line incinerator. One can compare it with Figure 9.4, an illustration of a modern large-scale incinerator, and recognize the general flow path. Figure 9.2 is more detailed and identifies the various component parts. There are three major compartments in this unit: the ignition or primary combustion chamber, the mixing chamber, and the secondary combustion chamber. There are two parameters that dictate the geometry of the chambers, retention times and gas flow velocities. The retention times are associated with the combustion rate, while the gas velocities control the mixing that is achieved.

Geometry of the Combustion Chambers

Incinerator design is still an art, but there are some general guidelines that must be considered to have a successful design. One of these is the size and geometry of the furnace chambers. The size of the primary combustion chamber is controlled by grate loading and arch height. The grate loading is an empirical relationship that is unique for the type of grate employed. The stationery grate illustrated in Figure 9.2 has loading defined by

$$L_G = 10 \log R_C \tag{28}$$

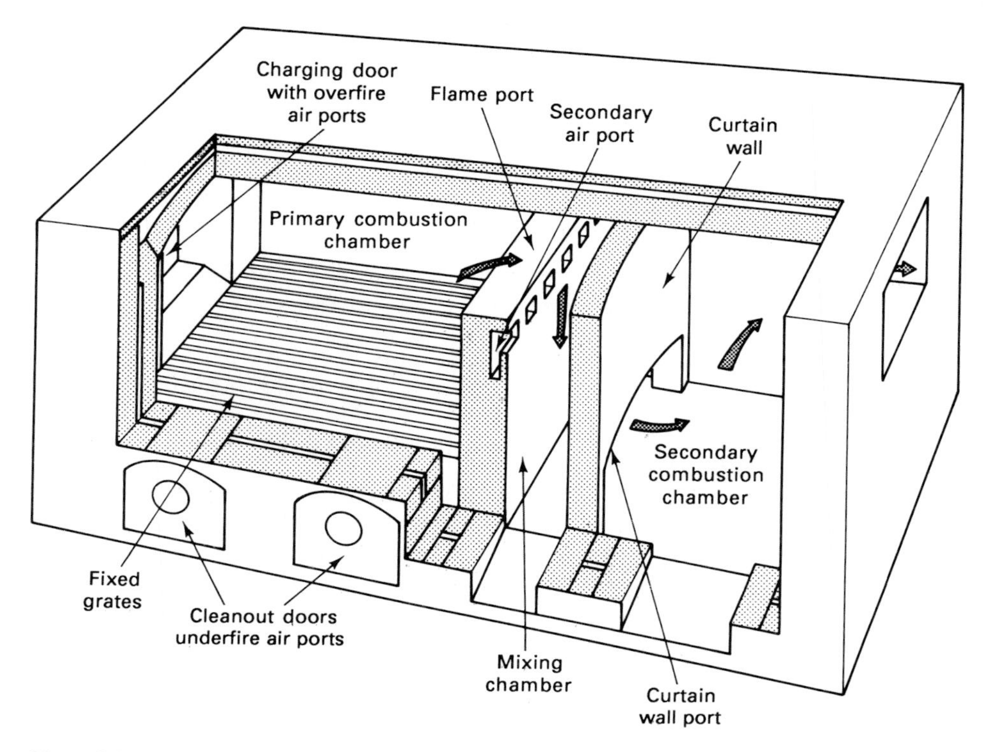

Figure 9.2 Cutaway of an in-line multichamber incinerator. (From *Air Pollution Engineering Manual*, U.S. Department of Health, Education, and Welfare, Public Health Service, PHS Publication No. 999-AP-40, Washington, DC, 1967.

where L_G is the grate loading rate in lb/hr-ft^2 and R_C is the refuse combustion rate in lb/hr. The grate area (A_G) is computed by dividing the refuse combustion rate by the allowable loading rate. This area fixes the horizontal dimensions of the primary combustion chamber. There are a number of different types of grates; many of them are mechanical. Each grate will have a different allowable loading rate as dictated by field evaluations of the grate performance.

For a small incinerator that is burning 1000 lb/hr of refuse, the grate loading would be 30 lb/hr-ft^2 of as-received refuse. The heating value of the as-received refuse would be 4000 to 5000 Btu/lb, depending on the moisture and ash content. The heat release rate of this grate would be between 120,000 and 150,000 Btu/hr-ft^2. This is a typical rate for a fixed grate. Mechanical grates can be loaded at higher rates, resulting in higher heat release rates. Figure 9.3 illustrates some additional grates available for burning solid fuels. The applications illustrated in this figure employ a spreader stoker that mechanically throws (from the feed hopper on the right side of the grate) the fuel across the width of the grate. The stationary grate in Figure 9.3(a) requires the ash to fall through the grate openings without

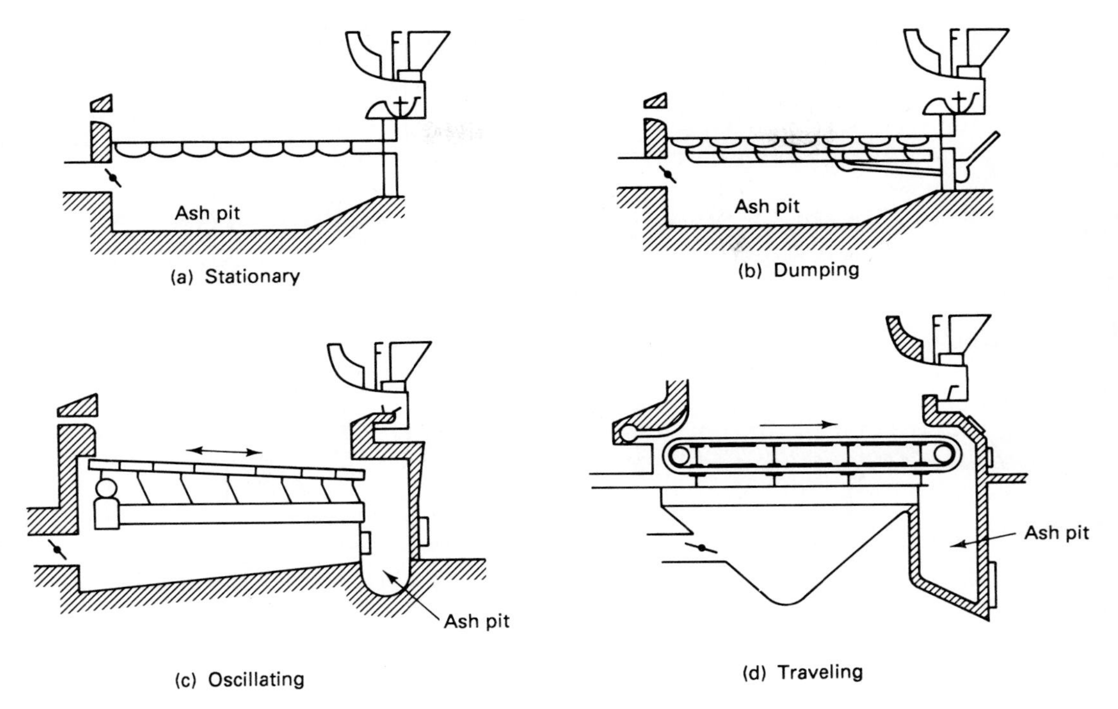

Figure 9.3 Types of grates available for combustion of solid fuels.

any movement of the grate. In Figure 9.3(b), the grates are mechanically rotated. This dumps the ash into the ash pit underneath the grates. Neither of these grates provide any mixing of the fuel on the grate. Consequently, heat release rates would be low, less than 200,000 Btu/hr-ft^2.

The grate in Figure 9.3(c) moves the ash to the side by oscillating (vibrating). It is sloped to assist in the movement of the ash to the ash hopper. Figure 9.3(d) shows a traveling grate that is an endless metal belt. It moves slowly to the right and carries the ash to the ash hopper. Both of these grates can also accept fuel from a feed conveyor or ram. The fuel would be introduced from the left side of the grate as presented in these figures. This would be termed a crossfeed, as contrasted to an overfeed, which occurs with a spreader stoker. The manner in which the fuel is fed to the furnace determines the maximum heat release rate. For example, a traveling grate with a crossfeed of fuel has a maximum heat release of about 300,000 Btu/hr-ft^2, while this same grate fed with spreader stoker has a maximum heat release rate of 1,000,000 Btu/hr-ft^2.

Except for the stationary grate, the grates above could be considered for use in the combustion of RDF. However, the characteristics of unprocessed refuse are so variable that these grates are not very satisfactory. Because of the bulky nature of the refuse, some grate action is necessary to mix the refuse to expose it to the high

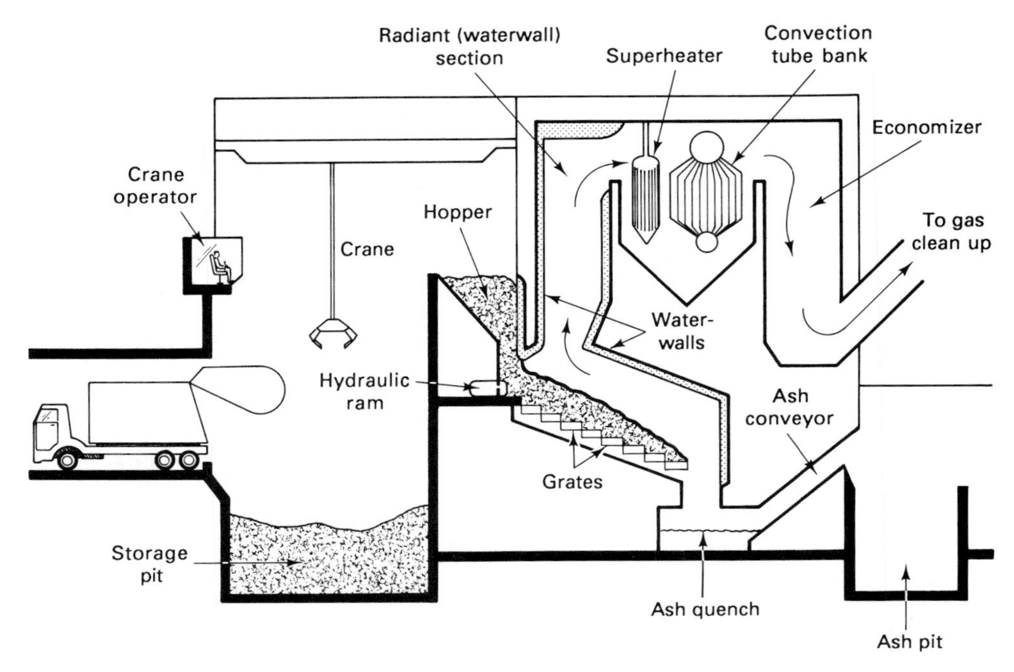

Figure 9.4 Mechanically fired incinerator with reciprocating grates.

temperatures and combustion air present in the primary combustion chamber. Figure 9.4 illustrates the reciprocating grate. The grates are arranged similar to stair steps, with alternate steps being stationary. The movable steps are slowly moved back and forth, gradually pushing the refuse to the ash hopper. Each time the refuse tumbles from one step to the next, it is mixed to expose more of the unburned material to the proper environment. There are other types of grates that achieve the same objective. It is important to provide this mixing capability when burning unprocessed refuse.

The other dimension required is the height of the primary chamber, or the arch height. Retention time of the combustion gases in this chamber is one factor that determines the chamber height. Again this is defined by an empirical relationship unique for each furnace. For the unit in Figure 9.2, the arch height (H_A) can be calculated from

$$H_A = \tfrac{4}{3}(A_G)^{4/11} \tag{29}$$

An additional factor that is considered is the length-to-width ratio of this chamber. This ratio is considered in relationship to the remaining portion of the furnace.

When the combustion gases exit the primary chamber, they pass through a flame port into a mixing chamber, where the secondary air is mixed with the combustion gases. In these two components of the incinerator, gas velocity is the controlling factor in generating turbulence, as the gases change directions of flow. The gas flow rate and the cross-sectional area determine this velocity. For the incinerator in Figure 9.2, the recommended velocity for the flame port is 55 ft/sec when the gas temperature is 1000°F. Since the volume of a gas is temperature dependent, the 1000°F is a reference temperature. If a higher temperature is encountered in the burn, the gas velocity will increase. The recommended downward velocity for the mixing chamber is 25 ft/sec at 1000°F, and the velocity through the combustion wall port is about 70% of the mixing-chamber velocity. The height and width of the mixing chamber are set by the dimensions of the primary combustion chamber.

The gas velocity in the secondary combustion chamber is much slower since the mixing should have occurred in the previous chambers. The horizontal velocity should be less than 10 ft/sec, preferably in the range 5 to 6 ft/sec. Since the height and width of the secondary chamber are fixed by the primary chamber, it may be necessary to adjust the length-to-width ratio of the primary chamber. The length of the secondary chamber is determined by the desired retention time.

Combustion Air Distributions

There are three locations where it is necessary to introduce combustion air. The underfire air is necessary to maintain a good burnout. The air passing through the ash on the grates will cool the grates and ash, thereby reclaiming some of the heat. It will also provide the oxygen necessary to completely oxidize the residual organic material in the ash. There is a limit to the underfire air. Too much will suspend a large fraction of the ash in the combustion gases, greatly increasing the air pollution problem. Underfire air will be about 10 to 20% of the total combustion air.

The majority of the air is introduced into the primary chamber to obtain high temperatures and good burning. About 40 to 50% of the air required is added in this chamber as overfire air. There are small incineration units that operate with only enough over- and underfire air to have an effective gasifier in the primary chamber. The lower air velocities reduce the amount of entrained ash particles found in the combustion gases. The primary heat release is in the secondary combustion chamber. The balance of the required air is added as secondary air. The excess oxygen in the secondary combustion chamber and the high temperatures force the combustion to completion as long as the retention time is not too short.

Air distribution is controlled by dampers on the air ducts or ports that conduct the air to the furnace. The capacities of these air-handling systems are overdesigned by as much as 50% to permit flexibility in the distribution of the air through the furnace. The head loss is kept low, about 0.1 in water gage. If forced-air or induced-draft airflow is used, the head loss should not be significant. The power required for the fans, or stack height or temperature differential required for the induced draft, can become excessive if this head loss is not low. A stack draft (D_T) of 0.5 in. of water gage may be required to overcome the head loss in the furnace. The relationship between the stack draft, height, and temperature differential is

$$D_T = 0.52PH\left(\frac{1}{T} - \frac{1}{T_i}\right) \tag{30}$$

where P = atmospheric pressure, psi
H = stack height, ft
T = ambient temperature, °R
T_i = stack temperature, °R

Modern incineration systems have high stacks, but the designs include fans to ensure positive control over the airflow. The temperature of the combustion gases is reduced to low levels as a result of the energy recovery systems associated with mass-burn or RDF facilities. Because of the reduced temperature differential, little natural draft will be generated. The stack height is not intended to provide draft but is used to discharge the combustion products higher into the atmosphere.

Current Design Guidelines and Objectives

There is no set of design standards that can be followed for the design of an incinerator. An incineration system design is the sum of the designs of a host of individual components, ranging from the ash-handling conveyor to the steam condenser on the steam turbine to the dry scrubber for air pollution control. Each furnace is unique and the design is based on the cumulative experience of the manufacturer in the design and operation of many units. There are some specific objectives in the design and operation of an incinerator that reduce the air pollution potential, especially from unburned organic materials. As illustrated in Table 9.14, these objectives are broken into three categories: design, operation/control, and verification. The physical design of the furnace should achieve these goals.

Table 9.14 presents the proposed goals for the design and operation of mass-burn incineration systems. Similar goals have been proposed for RDF facilities and starved-air incinerators. These guidelines may be revised at any time by the U.S. EPA. The major consideration in the design element is being able to achieve the temperature of 1800°F and to have good control of the distribution of the combustion air. During operation, excess oxygen is necessary to achieve the oxidation of the organic materials and to keep the concentration of CO at low levels. Combustion is assumed to be reasonably complete when the concentration of the CO is low. Also, the ability to add auxiliary fuel to obtain operating temperatures during startup and at any time the temperature drops or the CO concentration increases beyond acceptable levels is a requirement. Finally, verification of the design success is achieved by monitoring the oxygen and CO in the flue gas. The furnace temperature should achieve the required 1800°F and the mixing of the air with the combustion gases is verified by a CO profile of the furnace.

An additional constraint that is not shown in this table is the retention time in the furnace and in a postcombustion chamber. Retention times that are being considered and actually imposed in some permits are a minimum furnace retention time of 1.5 seconds. This may increase to 2.0 seconds. As discussed presently, a postcombustion chamber with a retention time of 2.0 seconds may be necessary to ensure the destruction of the trace organic compounds formed during combustion.

Table 9.14 Minimizing trace organic emissions from mass-burn incinerators

Element	Component	Target
Design	Temperature at fully mixed height	1800°F 1 m above overfire jets
	Underfire air control	At least four separately controlled plenums over depth of grate per grate module
	Overfire air capacity	40% of total air
	Overfire air injector design	That required for penetration and coverage of furnace cross sections
	Auxiliary fuel capacity	60% of design load
Operation/ control	Excess air	6–12% excess oxygen
	Turndown restrictions	80–110% of design or as set by CO profiling
	Startup procedures	On auxiliary fuel to design temperature
		On auxiliary fuel for high CO or low furnace temperatures
Verification	Oxygen in flue gas	6–12% by volume of dry gas
	CO in flue gas	100 ppm on 8-hr average
		400 ppm on 30-min average
		800 ppm on 10-min average
	Furnace temperature	Minimum of 1800°F at fully mixed height across furnace
	In-furnace CO profiling for air distribution	CO variation allowed 50% in plane after fully mixed height

Source: Combustion Control of MSW Incinerators to Minimize Emissions of Trace Organics, U.S. EPA, Washington, DC, 1987.

This chamber may be considered as the secondary combustion chamber if the combustion is nearly complete in the primary combustion chamber.

AIR POLLUTION CONTROL REQUIREMENTS FOR MASS-BURN INCINERATION

State Permit Requirements

Each incinerator—mass-burn, RDF, or starved-air—requires a discharge permit. The stack discharge emission guidelines imposed by the states are variable, but Table 9.15 presents two of the more complete listings. As of 1987, these were only proposed guidelines, but these numbers have been known to move from guidelines to specific permit requirements.

In addition to the normal air pollutants, these guidelines include acid gas, HCl, and hydrocarbons. The levels are generally specified in concentrations, but in some cases, a percent reduction is also included. No standard has been proposed for dioxin and furan emissions. There is still considerable discussion regarding a "safe" level. The risk issue has been clouded by the extreme concern for the apparent chronic toxic and mutagenic effects of these compounds. In 1987, an incinerator in Essex, New Jersey, was issued a permit with a level of 5 lb/hr for 2,3,7,8-TCDD.

Table 9.15 State emission guidelines for mass-burn systems

Pollutant	New Jersey, 7% O_2	Illinois, 12% CO_2
Particulates	0.015 gr/dscf[a]	0.010 gr/dscf
HCl	50 ppm/hr[b]	30 ppm/hr[b]
SO_2	50 ppm/hr[c]	50 ppm/hr[d]
NO_x	300 ppm/hr	100 ppm/hr
Hydrocarbons	70 ppm/hr	
CO	400 ppm/hr 100 ppm/4 days	100 ppm/hr
Temperature		
Design	1800°F	1800°F
Minimum	1500°F	1500°F
Minimum furnace residence time	1.0 sec	1.2 sec
Minimum lime injection		100 lb/hr
Certified operator	No	No
Minimum excess O_2	6%	

[a]Grains/dry scf.

[b]90% reduction.

[c]80% reduction.

[d]70% reduction.

Source: Waste Age, Jan. 1988, p. 83.

Air Pollution Control Systems

Particulates. Particulates can be very important contaminants, depending on their origin. They can result from solid materials that are present in the refuse or they can condense from organic vapors or metal fumes. Solid particles form under normal combustion conditions in a furnace by the noncombustible materials released from the refuse in the form of fly ash. These solids consist of inorganic oxides, including heavy metals and are emitted over a wide range of particle sizes. Condensation particles occur when partial combustion occurs in the furnace, producing organic vapors. When the combustion gas cools, these vapors condense to form particles that are organic compounds. Certain metals, and metal oxides, can become vapors at high furnace temperatures. Upon cooling, these particles condense to form particles containing these metals.

There are two primary devices for control of particulate emissions from incineration: electrostatic precipitators (ESP) and fabric filters. ESP uses an electrical field to remove the particles that acquire a charge when passing through this field. Conventional ESPs are about 99.8% efficient for removal of particles greater than 2 μm. However, removal of smaller particles is more critical because particles of less than 2 μm are respirable (i.e., can be inhaled deep into the lungs). The ESP efficiency is only about 93% for this size range. Newer designs that have a larger collection plate area per unit of gas flow have improved this efficiency, but at a significant cost. There is still some concern that the ESP may not be able consistently to meet the discharge limit of 0.01 gr/dscf.

For new incinerators fabric filters will probably be considered the choice for control of particulate emissions. These filters are closely woven fabrics that trap the particles as the gas stream is pulled through the filter. The fabric is formed into tubes or "socks" and a large number are suspended in a "baghouse." A large filter area can be obtained with this arrangement. The filtration mechanism is a "deep bed" filtration. Only the larger particles are trapped on the surface, where they form a "bag cake." The fine particles actually penetrate the bag cake and the filter cloth. This filtration mechanism exhibits a high efficiency for the particles below 2 μm. Mass collection efficiencies of 99.99% can be expected. The stack discharge consistently meets the 0.01 gr/dscf, and discharges as low as 0.001 gr/dscf have been obtained.

The bags are manufactured from a variety of materials suited to the specific chemical nature of the gas being cleaned. Two environmental conditions can increase the operating problems of the bags: high temperature and high moisture. The bag material deteriorates rapidly at high temperatures, so it is necessary to reduce the temperature of the combustion gases before entering the baghouse. A steam boiler can lower the temperature from 1800°F to about 300°F. If the gas stream contains excessive amounts of moisture and hygroscopic particles, the bags can become clogged with wet particles. If a wet process is used for emission control prior to the filters, the particles must be dry before they reach the filter fabric.

There is a substantial pressure drop across the filter, considerably more than with an ESP. Power costs for the fans are a significant operating cost that must be considered.

Acid gas control. Acid gases, HCl, HF, SO_2, and NO_x have limits imposed by some discharge permits. HCl and HF are of primary concern with refuse incineration. The combustion temperatures are generally not high enough (< 2000°F) to cause significant NO_x production, and the sulfur content of refuse is low, so SO_2 removal alone would probably not be required. However, the combination of the three acid gases creates both operational and environmental problems. Because these compounds are gases, they must be converted to a solid by precipitation or be ad/absorbed by a solid or liquid particle. The resulting particles are removed by an appropriate removal device. The three processes applicable to acid gas removal are wet scrubbers, spray-dry scrubbers, and dry scrubbers.

A wet scrubber consists of a spray chamber where a fine spray of water is mixed with the combustion gases. Gases that are soluble in the water spray will be absorbed and removed with the water droplets. A dilute base solution is used to improve the capture of the acid gas. With a base present, the acid is neutralized to a salt that is soluble in the water. The water droplets are removed by a cyclone or demister before the gases reach the baghouse. These droplets cannot be removed by the fabric filters because they rapidly clog the filter medium with a wet cake. There is a temperature drop with the wet scrubber, due to evaporation of some water. The scrubbers are 95% efficient in removal of HCl and HF and about 85% efficient in removal of SO_2. NO_x removal is not appreciable with these scrubbers.

The water continues to be circulated with the addition of more base. The concentration of resulting salts increases to the limits of solubility and precipitate. Unfortunately, the chloride and fluoride salts are reasonably soluble and will accumulate to high concentration. The solution can be very corrosive. In addition, the CO_2 in the combustion gas will react with the base and since the base ion is normally calcium, precipitate as calcium carbonate. The SO_2 forms $CaSO_3$ which has a limited solubility. The sludge produced from these scrubbers may also contain some of the metal particles present in the combustion gases. Disposal of the wet sludge is a problem, especially if contaminated by heavy metals.

A spray-dry scrubber sprays a concentrated base solution into the combustion gas to absorb the acid gas. The removal mechanism is the same as for the wet scrubber. The water content is only sufficient to effect the absorption and then evaporates before the gas reaches the baghouse. When the water evaporates, it leaves behind a solid residue that can be removed by the filter. As contrasted with the wet scrubber, the residue is dry and has much less volume. The heat loss with the evaporation of the water will reduce the gas temperature. The removal efficiencies are the same as for the wet scrubbers.

A dry scrubber injects a base such as lime (CaO) directly into the primary combustion chamber. The lime will react with the acid gas to produce $CaCl_2$, CaF_2, and $CaSO_3$ particles. The hydrogen in the HCl and HF will form water with the oxygen from the lime. These particles are removed by the particle removal unit on the stack gas. The dry residue is less expensive to dispose of, due to the much smaller volume of dry rather than wet residue. Also, the absence of water greatly reduces the corrosion problems associated with a wet system. Temperature loss from water evaporation does not occur, so more energy is available for recovery.

Control of heavy metals. Heavy metals are present in the combustion gases as suspended particles of fly ash and as metal fumes. If the temperature in the combustion chamber is high, the metals will be present as a vapor. Table 9.16 lists the boiling or sublimation temperatures for some of the heavy metals common to refuse. When the temperature in the refuse reaches 1800°F, all of the metals except lead will be present as a vapor. Salts of some of the metals have lower boiling points. For example, lead chloride will boil at about half the temperature of lead

Table 9.16 Boiling temperatures for some heavy metals

Metal	Boiling temperature (°F)
Lead	3164
Lead chloride	1742
Zinc	1664
Cadmium	1402
Arsenic	1135 (sublimation)
Mercury	673

metal. Mercury is a major problem because of the low temperature at which it boils. The metals in the ash remaining on the grates will not be exposed to the high temperature of the secondary combustion chamber. Consequently, only metals with low boiling temperatures will evaporate from the ash. Metals that are carried with the fly ash to the high-temperature zones can be expected to vaporize.

The success for control of the heavy metal emissions is (1) an efficient particle removal process and (2) cooling the gases to below the condensation point of the metals before reaching the particle removal system. The use of a steam boiler will generally not cool the gases sufficiently. An "economizer" is used to recover additional heat from the gases after they have passed through the boiler. This unit can reduce the gas temperature to less than 400°F. Passage of the cool gas through an efficient filter unit will yield a 99% reduction in all heavy metals except mercury. Mercury remains a problem. It is difficult to cool the gas sufficiently to condense all of the mercury vapors. Tests at Quebec City found 0% mercury removal when the gas temperature was 400°F, while operating at 280°F achieved a 90% reduction.[2] Temperatures below 300°F coupled with an efficient scrubber/baghouse particle control unit can be expected to achieve a 90% reduction of mercury.

Dioxin control. In general, control of dioxin and furan have concentrated on the combustion efficiency. As long as the efficiency is high, with low levels of CO and good temperatures and oxygen concentrations, the level of dioxin should be acceptable. Since there is no consensus as to an acceptable standard, the approach has been to use the best available technology (BAT). Success in Italy with postcombustion chambers may indicate a future direction for dioxin control.[3] These chambers have been added to conventional incinerators to provide better control of the temperature and the oxygen level. The chambers provide a retention time of 2.0 seconds. The temperature is maintained at 1700 to 1800°F, and the oxygen is 6% at the chamber exit.

The effectiveness of this technique was demonstrated on an existing facility that was retrofitted with a postcombustion chamber. Before the retrofit the total PCDD was 3×10^{-2} μg/dscf. After the retrofit, the level of total PCDD was reduced to 3×10^{-4} μg/dscf, or a reduction of 99%. It may be that similar results will be obtained by increasing the time in the area of the furnace where the temperatures are highest and excess oxygen is present. If complete oxidation of the PCDD precursors can be achieved, there is no reason to expect them to reform.

Operational Strategies for Emission Reductions

Recent evaluations of incineration systems have clearly indicated that proper operation can be a major factor in minimizing the emissions. Also, the preparation of refuse as RDF can remove many of the contaminants before they are in the gas stream.

[2] *Waste Age*, Jan. 1988, p. 92.

[3] *Waste Age*, Jan. 1988, p. 87.

Control of refuse composition. Understanding the source of the undesirable components in refuse can be important in evaluating the opportunity for removal prior to combustion. Some of the materials can easily be removed in preprocessing of the refuse.

Cadmium. Cadmium batteries account for about 60% of this metal in the refuse. An additional 25% can be found as an additive to certain plastics.

Chromium. About 40% of the chromium is found in rubber and leather. An additional 40% is associated with metals.

Mercury. Batteries (alkaline) account for about 60% of this metal. Additional sources include plastic (10%) and paper (13%).

Lead. About 20% of the 600,000 tons per year of the lead used in batteries end up in the refuse stream. The other major source of lead is the solder in electronic components.

Acid gases. The primary sources of HCl, HF, and SO_2 are plastic, especially the chloride in PVC, textiles, and rubber.

The metals that are present in batteries, appliances, electronic items, and so on, have a high probability of being removed in the various separation processes used to prepare the RDF. However, the contaminants contained in the paper, plastic, textiles, and rubber may end up in the fuel. The benefits of separation can be seen from the results of the NRT separation process (Figure 8.21). In addition to recovery of the aluminum, ferrous metals, glass, and other inert materials, the process resulted in a reduction of stack emissions of 52% for lead, 64% for chromium, and 73% for cadmium, compared to burning unprocessed refuse in the same facility.[4] It was also observed that complex hydrocarbons were reduced by 75% and the CO reduced by 63%. This improved combustion was a result of the uniformity of the fuel and the ability to better control the combustion.

Operational control of the combustion process. If a system is designed to provide operational flexibility, considerable reduction in the emissions is possible by careful operation of the incinerator. Not only is it necessary to have sufficient combustion air-handling capability, but it is also necessary to have the ability to distribute this air at varying rates to different sections of the furnace. A uniform feed rate is important in maintaining the proper oxygen level and temperature in the combustion zones. With this uniform feed rate, the airflows can be set more precisely. Moderate temperatures can be maintained in the primary combustion chamber. This will reduce the metal volatilization and the NO_x formation. Preparation of the refuse to provide a homogeneous fuel with reduced inert material can greatly improve operation. The NRT process is one example. Separation of the metals re-

[4]*Waste Age*, Feb. 1988, p. 55.

duces the metal content in the combustion gases, but the homogenization resulting during the processing makes it possible to provide a more uniform Btu feed rate to the furnace.

Complete oxidation of the organic compounds released from the refuse can be achieved if excess oxygen and high temperatures are maintained in the secondary combustion chamber. The temperature should be in excess of 1800°F to oxidize completely the precursors for PCDDs. A 99% destruction of pentachlorophenol (a precursor) was observed with the following secondary combustion chamber conditions:[5]

Percent O_2	Temperature °F
20	1350
2.5	1500
0	1800

At the highest temperature, complete oxidation of this compound was achieved at the stoichiometric combustion air. If an oxygen level of 6% is maintained at this temperature, dioxin formation will not be a problem. The oxygen should not be permitted much higher because of the cooling effect of the excess air. A maximum temperature of 2000°F will minimize the formation of the NO_x.

ASH DISPOSAL

The bottom ash and the fly ash from an incineration system will contain all of the metals present in the refuse, assuming that the emission control devices are removing the metals from the combustion gases. The composition of these ashes is poorly documented in the literature. In most cases, the ash has simply been hauled to a landfill for disposal. The U.S. EPA has proposed to categorize these ashes as hazardous if they fail the EP toxicity test. There is considerable opposition to such a classification. This test is an elaborate procedure that passes a weak acid solution through the ash under specified conditions to determine what elements and compounds are leached from the ash. If any exceed the regulatory limit, the ash is declared a hazardous waste, requiring disposal at a hazardous waste site or at least a landfill designed for incinerator ash, a ''monofill.''

The data in Table 9.17 show the mobility of heavy metals that may be found in incinerator ash. The results of the EP toxicity test will depend on the characteristics of the ash as determined by a variety of factors, including additives such as lime, temperature in the primary combustion chamber, degree of slagging in furnace, and degree of refuse preprocessing. Therefore, it is necessary to evaluate each incinerator ash independently.

[5]*Waste Age*, Dec. 1987, p. 162.

Table 9.17 Summary of EP Toxicity Tests on Incinerator Ash

Metal	Regulatory limit (mg/L)	Fly ash		Bottom ash		Combined ash	
		No.	Ave. (mg/L)	No.	Ave. (mg/L)	No.	Ave. (mg/L)
As	5.0	10	<1.0	7	<1.0	11	<1.0
Ba	100	11	2.4	6	0.5	8	0.6
Cd[a]	1.0	23	21.7	7	0.3	33	0.7
Cr	5.0	10	<0.1	6	<0.1	12	<0.1
Pb[b]	5.0	25	31.0	7	2.1	36	5.4
Hg	0.2	10	<0.01	6	<0.02	8	<0.02
Se	1.0	10	0.03	6	<0.02	12	<0.01
Ag	5.0	10	0.05	6	0.03	8	

[a]23 fly ash and 3 combined ash samples exceeded limit.

[b]21 fly ash and 16 combined ash samples exceeded limit.

Source: Waste Age, Sept. 1987, p. 89.

The data in this table are interesting in that they suggest that it may be advantageous to provide for separate disposal of the bottom ash from the fly ash. Fly ash tended to exceed the regulatory limits for cadmium and lead, whereas the bottom ash was acceptable for all metals. When the bottom and fly ash are combined, the lead standard is still violated on the average. In some samples, the combined ash also violated the cadmium standard. The quantity of bottom ash is generally considerably more than the fly ash, so separation of the two ashes may result in a lower ash disposal cost, especially if one ash is designated as hazardous.

PRODUCTS AND CAPITAL COSTS OF INCINERATION (MASS-BURN) SYSTEM

Historically, the only objective of incineration was the destruction of the combustible fraction of refuse. As the environmental standards required more air pollution control, the operational philosophy had to be changed to accommodate these standards. One major problem was dissipation of the excess heat generated by the combustion process. Because of the air pollution control equipment, it was not feasible to use the massive quantities of excess air for cooling the furnace. Heat recovery by the installation of boilers was instituted. Marketing of the steam was difficult because of the inability to transport steam any significant distance. Several plants constructed close to a major steam user were able to sell the steam for use in manufacturing processes. As the number of plants increased, it became more difficult to find the steam markets. The prospects of generating electricity from this steam was investigated and markets have been developed. The passage of the Public Utility Regulatory Policy Act (PURPA) required the electric utilities to buy any electricity produced at a price reflecting their avoided cost (i.e., the cost of the fuel). This is

Table 9.18 Status of waste-to-energy facilities: 1986 versus 1988

Status	Number of facilities 1986	1988
Conceptual	75	139
Advanced planning	72	63
Construction	25	28
Shakedown	7	4
Shutdown		
Temporary	8	5
Permanent	19	27

Source: Waste Age, Nov. 1988, p. 61.

about half of the price paid by the consumer. In 1988, these prices ranged from \$0.0369 in the southern region to \$0.0719 per kilowatt hour in the northeastern region of the country.[6]

Despite problems with air emissions, incineration or waste-to-energy plants continue to be considered as a solution to the refuse disposal problem for many urban areas. Table 9.18 compares activity in the development of new facilities in 1986 and 1988. These data indicate a strong commitment to this technology. The capital required for construction of sufficient capacity to process all of the refuse is substantial. The projected expenditures for all the facilities in the conceptual and advanced planning stage would be approximately \$20.7 billion.

Table 9.19 presents capital costs for waste-to-energy facilities. The data vary considerably because of the variety and sizes of systems considered. Also, some of the systems have been constructed for several years and were not subject to the tighter emission standards. Today, a waste-to-energy plant will be in the range of \$200,000 per ton/day capacity. For example, a 1000-ton/day facility will have a capital cost in the range of \$200 million. In addition to the amortized capital cost,

Table 9.19 Capital cost of waste-to-energy facilities

Mass burn (158 plants)	Capital cost[a] (10^6 \$)	Capacity (tons/day)	\$1000 / Capacity (tons/day)
Total	14,485	131,684	
Mean	92	833	110
Range	2–445	36–4800	29–219

[a]Capital costs are 1990 dollars.

Source: Waste Age, Nov. 1986, p. 27.

[6]*Waste Age,* Nov. 1988, p. 61.

an operation and maintenance cost of perhaps $30 per ton must be considered. Despite these high capital and operating costs, the tip fees projected for new facilities are in the range of $40 per ton. The revenue from the energy sales is sufficient to reduce significantly the revenue needed from the tip fees. It is worth noting that many of the projected tip fees increase substantially by the time the facility is on line.

STUDY QUESTIONS

1. Define "combustion."
2. What four general conditions are necessary to achieve good combustion?
3. What occurs in each of the three combustion zones associated with combustion of a solid fuel?
4. What factors have been responsible for the increase in furnace temperatures of refuse combustion systems?
5. If complete combustion is ultimately achieved in a furnace, what is the effect of the formation of intermediate combustion products during combustion on the total heat release?
6. What is the effect of the oxidation state of carbon in a fuel on (a) the high heating value, and (b) the low heating value.
7. Compute the average oxidation number of the carbon in these materials: C_6H_{14}, $C_6H_{12}O_6$, and $C_6H_{12}O_2$.
8. What is the difference between a gram-mole (g-mol) and a pound-mole (lb-mol)?
9. If 1 lb-mol occupies 359 scf at 32°F and 1 atm, what is the volume at (a) 1800°F and 0.5 atm, and (b) 1800°F and 1 atm?
10. Convert the following volumetric analysis of dry products of combustion to a gravimetric analysis: CO_2, 8%; O_2, 6%; N_2, 86%.
11. Convert the following gravimetric analysis of a solid fuel to a volumetric analysis: C, 39%; H, 7%; O, 52%; N, 2%.
12. What is the purpose of providing excess air for combustion systems?
13. A prepared RDF has been found to have the following composition:

Proximate analysis		Ultimate analysis	
Moisture	22%	Carbon	51.5%
Inorganic	19.5%	Hydrogen	7.5%
Organic	58.5%	Oxygen	41.4%

Calculate the theoretical air requirements as (a) lb/lb of RDF and (b) scf/lb RDF.

14. For the fuel in Question 13, calculate the composition of the dry gaseous combustion products (% by volume and % by weight) with 50% excess air.
15. For the fuel in Question 13, calculate the volume and mass of *total gaseous* products of combustion if the combustion gases are cooled to 340°F by a steam recovery boiler.
16. What is the need for computing (a) the mass of the combustion products, and (b) the composition of the dry products of combustion?
17. What is enthalpy? How does enthalpy change in a combustion reaction?
18. What is ΔH_R, and how is it determined for a solid fuel?
19. How is the enthalpy of a substance determined at a temperature other than the reference temperature?
20. For the fuel in Question 13, calculate total heat recovery with the refuse and combustion air at 77°F and the gaseous combustion products at 340°F. The dry refuse has a calorific value of 6500 Btu/lb.
21. For Question 20, calculate the total heat recovery with the combustion air at 40°F rather than 77°F.
22. When calculating heat balances, what special considerations are required for both the fuel moisture and the combustion water?
23. Given the following gaseous products of combustion, calculate the heat loss per pound of fuel associated with these products using the average values for the C_p and a final temperature of 350°F: CO_2, 1.1 lb; O_2, 0.45 lb; N_2, 4.44 lb; H_2O, 0.59 lb. The fuel and air temperature is 77°F.
24. What factors dictate the size and geometry of an incineration system?
25. What type of grate is most suitable for the combustion of unprocessed refuse? Why?
26. What are the design and operating objectives of an incinerator to minimize air pollution problems?
27. What are the air pollutants of primary concern? What are the procedures for controlling these pollutants?
28. How does the refuse processing associated with the preparation of RDF reduce the air pollution potential of solid waste?
29. What concerns are involved in the disposal of incinerator ash?

CHAPTER 10

Bioconversion of Urban Refuse: Fuel Gas Production

SOLID WASTE AS A SUBSTRATE FOR BIOPROCESSES

The organic matter present in refuse presents a significant disposal problem while providing a large, renewable source of organic material that has the potential to be converted into useful chemicals and fuels. The approximately 150 million tons of refuse generated each year has an energy content of 1.5 Q (quadrillion Btu). This is the equivalent of 75 million tons of coal if all of the refuse is incinerated. However, much of the organic matter, mainly paper, food waste, and yard and garden debris, can be converted biologically to a number of products, including ethanol, methane (CH_4), and carbon dioxide (CO_2).

Equation (1) describes the first biological transformation that occurs with the complex biodegradable fibers present in solid waste. Cellulose is the major fiber in various refuse constituents. In the initial reaction the cellulose is ultimately hydrolyzed to glucose. Cellulose hydrolysis is a very slow process that limits the rate of biological activity in the refuse.

Hydrolysis Reaction

$$(C_6H_{10}O_5)_x + x\,H_2O \rightarrow x\,C_6H_{12}O_6 \qquad (1)$$

Equation (2) shows the reaction for the fermentation of the glucose to ethanol (ethyl alcohol). This is but one of the intermediate reactions that can occur. With selection of the proper organism, the product of the fermentation can be specified. Microorganisms can produce a variety of chemicals, including several short-chain organic acids and alcohols. These products have commercial value in the industrial chemical markets.

Fermentation reaction

$$C_6H_{12}O_6 + 2H_2O \rightarrow 2CH_3CH_2OH + 2HCO_3^- + 2H^+ \quad (2)$$

A number of researchers has investigated the prospects for production of commercial quantities of these chemicals. Technologies have been developed for the production of both chemicals and fuels from the biodegradable organic fraction in refuse. However, the costs are much greater than when a well-defined substrate is used. The cost of processing the refuse to obtain a consistent-quality substrate is very high. Because of the solubility of the fermentation products, recovery from the fermentation liquor adds considerable cost to the processing. To date, these processes have not been economically practical on solid waste. Certain industrial residues have been used for these fermentations. In 1989, about 800 million gallons of ethanol were produced by the grain milling industry for use as an octane enhancer in gasoline. Much of the carbohydrate used as raw material was from spoiled or contaminated grain and residues from the milling process. The economic success of these fermentations was, in part, based on special tax incentives provided by the federal government during the energy crisis of the 1970s.

The most promising process for recovery of a useful product from refuse is methane fermentation. This process occurs naturally in a sanitary landfill and can be incorporated into an engineered system that produces a fuel gas. A brief outline of the reactions occurring during methane fermentation is presented below. In methane fermentation, complex organic fibers such as cellulose are converted anaerobically by a consortium of bacteria into CH_4 and CO_2. Methane fermentation can best be described by a three-stage scheme. In the first stage, the polymers are hydrolyzed and fermented by hydrolytic and fermentative bacteria to primarily fatty acids and alcohols [equations (1) and (2)]. These fermentation products are then oxidized to acetic acid and CO_2 by obligate H_2-producing acetogenic bacteria [equation (3)]. Finally, methanogenic bacteria produce CH_4 from the reduction of CO_2 with the H_2 produced from the reaction shown in equation (3) and the cleavage of acetic acid [equations (4) and (5)].

Acetogenic reaction

$$2CH_3CH_2OH + 2H_2O \rightarrow 2CH_3COO^- + 2H^+ + 4H_2 \quad (3)$$

Methanogenic reactions

$$4H_2 + HCO_3^- + H^+ \rightarrow CH_4 + 3H_2O \quad (4)$$

$$CH_3COO^- + H_2O \rightarrow CH_4 + HCO_3^- \quad (5)$$

The CO_2 that is produced by the metabolism of the substrate will either remain in the aqueous phase or escape into the product gas together with the methane. The amount of CO_2 absorbed is influenced by a number of factors, such as temperature, partial pressure of CO_2 in the gas phase, pH, bicarbonate/carbonate alkalinity, and ionic strength. The equilibrium solubility of CH_4 and CO_2 may be described by Henry's law and aqueous CO_2 dissociation equations.

FUEL GAS PRODUCTION FROM REFUSE

A number of research studies was conducted during the late 1960s and the 1970s on techniques to utilize this natural methane fermentation process to recover a fuel gas from solid waste. These studies were sufficiently promising, and with the Middle East oil embargo of the mid-1970s, encouraged federal agencies to fund a proof-of-concept experiment. In May 1975, a contract was awarded to Waste Management, Inc., for installation of this facility at their Pompano Beach, Florida, Solid Waste Reduction Center.

Initial Pilot Plant Evaluation

The site selected for construction of the Refuse Conversion to Methane (RefCoM) project was an operating sanitary landfill. RefCoM, with a nominal capacity of 10 tons/hour, received a slipstream from a vertical shaft hammer mill rated at 62.5 tons/hour that was being used to prepare refuse for landfill cover. The primary shredded refuse passed through a trommel screen that removed fine inorganic material such as shattered glass, sand, and ash. Then a second-stage horizontal hammer mill reduced the size of the particles to pass through 3-in. grate openings, producing a relatively uniform particle size. From the shredder the refuse was conveyed to an air classifier, which produced a "light" fraction consisting of low-density organic material for feed to the fermentation reactors. A "heavy" fraction containing high-density inorganic material was landfilled.

This "light" fraction was conveyed to the premix tank, where the fermenter feed slurry was prepared. Appropriate quantities of makeup water, recycle liquor, chemicals, and in some cases, sewage sludge were added to prepare the desired feed slurry. Steam was also injected at this point, to heat the feed slurry to obtain the desired fermentation temperature. Two 50-ft-diameter mechanically mixed reactors, each approximately 45,000 ft^3, were constructed. The fermenters were operated under a specified experimental program. The overflow from the tanks flowed by gravity to the vacuum filter, where the slurry was dewatered. The filtrate from the vacuum filter was used as makeup water to slurry the incoming dry refuse. This recycle eliminated the need for the disposal of a large quantity of contaminated water, which would require treatment at a significant cost prior to discharge.

Revised Facility Design

The severe operational problems that were encountered with the original refuse separation processes necessitated a complete reevaluation of this technology. The air classification process was totally unsatisfactory for preparing the feed for the fermentation tanks. Experience in Europe and in the United States with rotary drum screens (trommel screens) for the separation of refuse components was very positive. The concept of screening was appealing since separation is on the basis of particle size and not density. Therefore, the screens are less susceptible to moisture content and aerodynamic properties of the particles. In addition, the power required to operate a screening system is about an order of magnitude less than that required for the air separation system. The initial RefCoM refuse separation subsystem was replaced with a combination of disk screens of different sizes that produced an ideal feed for the methane fermentation system. The screening produced a feed that was free of cans and bottles and had particles between $\frac{5}{8}$ and 2 in. Only a portion of the shredded refuse stream from the Solid Waste Reduction Center was used. The balance was diverted to the landfill.

Methane Fermentation System

The methane and carbon dioxide produced by this process are a result of the activity of the microorganisms present in the anaerobic fermenters. These reactors are designed and operated such that the optimum conditions necessary for growth of these organisms are maintained. These conditions can be separated into two general areas as follows:

1. Physical environment
   (a) Aqueous homogeneous suspension
   (b) Thermophilic fermentation temperature
   (c) Gastight reactor
2. Chemical environment
   (a) Neutral pH
   (b) Adequate nutrients

Physical environment. The product from the refuse separation subsystem is relatively dry, with a typical moisture content of 25 to 35%. Uninhibited microbial activity generally occurs at moisture contents greater than 75%. An equally important consideration is the physical properties of a concentrated slurry. Because of the paper content of the feed solids, the viscosity of the slurry increases dramatically as the moisture content decreases. As the moisture content drops below 90%, the slurry begins to lose its fluid properties, significantly increasing the difficulty of transporting and mixing the slurry. Consequently, water is added to the feed from the stockpile to obtain a slurry solids concentration compatible with the reactor system. This water is obtained by recycle of the filtrate generated from dewatering of the fermented slurry.

Mixing plays an important role in the successful operation of this reactor. The degree of mixing must be sufficient to maintain a relatively homogeneous slurry.

Because of the long retention time of the solids in these reactors, a high rate of mixing is not necessary. Mass transfer is not a consideration. However, it is important that zones of stagnation be prevented from developing. Heavy inorganic solids can accumulate in these zones and reduce the effective reactor volume. The design objective of the mixer is to move heavy solids to the withdrawal point on the bottom of the tank, where they can be removed from the reactor. Similar problems can also be encountered with floating material, especially plastics (Styrofoam). Design of a mixer with sufficient power input to keep these materials from separating is not practical or necessary. Alternating slurry withdrawal between the top and the bottom of the reactor can prevent the accumulation of heavy, inert solids and floating plastics.

The rates of methane production are substantially higher at temperatures in the thermophilic range (130 to 140°F, 55 to 60°C) compared to operation in the mesophilic range (95°F, 35°C). The traditional drawbacks of operating at thermophilic temperatures do not exist with this system. The feed slurry has a high solids concentration, 12 to 15%, greatly reducing the amount of water to be heated. A major portion of this water is warm, recycled filtrate, so little heat is required to achieve the desired slurry temperature. However, heat transfer must be considered carefully since the heat-conducting properties of the slurry are very poor. Direct steam injection into the feed stream or the reactor is the recommended technique for heating the reactor contents.

The reactor must be designed to be gastight. The gas produced by the fermentation must be captured if it is to be marketed. Any gas loss will mean a decline in this revenue stream and will create a potentially explosive mix of air and methane. It is equally important to prevent air from entering the reactor, for two reasons. If sufficient air enters the reactor to develop a methane content between about 5 and 15% by volume, an explosive mixture exists. Also, the methanogenic bacteria are sensitive to the oxidation-reduction potential (ORP) of the reactor slurry. If enough oxygen enters the reactor to elevate the ORP, the methane fermentation will cease.

Chemical environment. The methanogenic bacteria are sensitive to pH and are inhibited when the pH drops below 6.6. Refuse is traditionally deficient in the natural alkalinity needed to control pH. In addition, the oxidation state of the carbon in carbohydrates is such that the gas produced by this fermentation is 50% carbon dioxide. This high partial pressure of CO_2 in the gas phase will depress the reactor pH and require a higher level of alkalinity to maintain a neutral pH. Since the alkalinity formed by the decomposition of the substrate is not adequate, additional alkalinity must be added to the reactor. Lime is generally the most economical source.

The lime requirements are significantly reduced by recycle of the filtrate. This recycle also reduces the amount of nitrogen and phosphorus required to satisfy the nutritional requirements of the microorganisms. As a result of the filtrate recycle, only intermittent additions of nutrients are required. Micronutrients such as sulfur and trace metals are available in sufficient quantities from the refuse.

Table 10.1 Fermenter loading conditions

Test runs	Average feed rate (tons/day)		Loading (lb vol. solids/ ft^3-day)
	Classified lights	Volatile solids	
1	5.9	4.3	0.25
2	14.7	7.6	0.54
3	9.4	4.4	0.31
4	5.4	2.9	0.20
5	10.0	5.1	0.33
6	7.1	3.3	0.19
7	14.3	6.5	0.38
8	11.8	6.7	0.61
9	9.5	5.1	0.49
10	17.6	7.9	0.49

RefCoM pilot plant fermenter performance. After modifications to the refuse processing subsystem were made, data were collected on the fermenter performance. Table 10.1 shows the loading conditions on the reactors during the test runs. The classified lights fed are the weights of the processed refuse, including the moisture as measured by the weigh belt on the feed to the premix tank. The volatile solids represent the dry ash-free solids in the classified lights. Table 10.1 also shows the volumetric loading on the reactor. Loadings as high as 0.61 lb of volatile solids per cubic foot per day were achieved.

The data in Table 10.2 are representative of the more traditional operating parameters for the reactors. Retention times investigated ranged from a low of 6.4 days to a high of 26.6 days. During the entire period of operation, the temperature was maintained in the thermophilic range of 135 to 140°F (57 to 60°C). The previous laboratory studies and the subsequent economic analysis strongly support thermophilic operation. The ability to feed a concentrated slurry was demonstrated. A

Table 10.2 Fermenter operating conditions

Test run	Retention time (days)	Feed slurry concentration (%)	Fermenter solids (%)	
			Total	Volatile
1	10.0	4.5	2.99	2.19
2	6.4	6.5	4.37	2.97
3	8.6	5.0	3.09	2.26
4	12.6	4.5	2.68	1.93
5	9.8	6.5	4.05	2.95
6	26.6	9.3	2.67	2.00
7	13.5	9.4	3.60	2.76
8	10.1	10.3	6.33	4.65
9	9.0	7.5	4.79	3.64
10	10.5	9.0	4.02	2.91

Table 10.3 RefCoM gas production data

Test run	Retention time (days)	Volatile solids destruction (%)	Gas production (scf/lb vol. solids fed)	
			Total	CH_4
1	10.0	43.6	5.9	3.1
2	6.4	45.6	6.2	2.7
3	8.6	46.3	6.2	3.2
4	12.6	52.1	7.0	3.8
5	9.8	42.6	5.8	3.1
6	26.6	75.1	10.1	5.5
7	13.5	66.3	8.9	4.8
8	10.1	52.4	7.0	3.7
9	9.0	47.1	6.3	3.4
10	10.5	63.2	8.5	4.5

slurry with solids concentrations as high as 10.3% was fed into the fermenters, resulting in a reactor slurry solids concentration of 6.33%.

Table 10.3 shows the gas production data for the test periods indicated. Frequent data on the influent and effluent solids were obtained so that it was possible to compute solids balances on the reactors as well as the volatile solids destruction in the reactor. Since the dominant biodegradable organic material in the refuse was carbohydrate, gas production was computed based on a gas production of 13.5 scf/lb volatile solids destroyed. The methane production was determined from the measured gas composition. The gas composition was relatively consistent during periods of "steady-state" operation. The methane content ranged between 50 and 54%. The pH was maintained about 7.0 by the addition of lime. The RefCoM data are better than the data on gas production obtained from prior thermophilic laboratory studies.[1]

A belt-filter press with a stainless steel belt was obtained to evaluate the dewaterability of the fermenter slurry. The process was used in a production mode and was never operated under optimum conditions. Under these conditions, the dewatering system was able to reduce the moisture content of the fermenter slurry to below 70%. It has been demonstrated that the dewatering characteristics of the RefCoM digested solid wastes are much superior to those of digested sewage sludge. Chemicals were not needed to condition the slurry prior to filtration. The concentration of total solids in the filtrate from the belt press stabilized at approximately 2.5%. The belt was a very porous weave that passed a substantial quantity of small particles. This low solids capture was not a problem since the filtrate was recycled to the fermenter.

[1]J. T. Pfeffer, "Temperature Effects on Anaerobic Fermentation of Domestic Refuse," *Biotechnology and Bioengineering*, Vol. 16, 1974, pp. 771–787.

Table 10.4 Material balance for the refuse separation subsystem

	Dry solids (tons/day)			
	Biodegradable	Combustible	Inert	Water
Refuse feed	225.9	40.8	57.4	75.8
Stockpile	133.5	22.4	23.0	62.2
Incinerator	90.0	15.7	7.7	5.3
Landfill	2.8	2.6	21.7	8.3
Ferrous	0	0	3.8	0
Aluminum	0	0	1.1	0

Final System Configuration and Performance

An extensive evaluation of the refuse separation unit processes was combined with the results of the RefCoM field tests to develop a computer model that can be used to calculate the mass and energy balances and generate cost information on the system. The essential components of the refuse separation system have been presented in Chapter 8 (see Figure 8.22). In addition to the aluminum and ferrous streams, there are three streams generated from the refuse separation processes; a stream rich in organic materials for feed to the fermenter, a combustible oversize for feed to the incinerator and an inorganic stream that goes to the landfill. A computer model of the separation processes has been developed that includes a variety of screen sizes, air knifes, air stoners, and ferrous and aluminum recovery. It does not, however, include the paper separation process, the "wet trommel." Table 10.4 shows the computed splits for a 400-ton/day (2000-ton/week) facility assuming a 5-day operating week for the front end. The combustible fraction represents non-biodegradable organic materials such as plastics, textiles, and rubber. These numbers represent the mass balance only on the separation subsystem. The dewatered cake from the fermenter also goes to the incinerator, and the ash from the incinerator is landfilled.

This scheme represents the baseline processing that would be involved in a complete system. A significant quantity of biodegradable organic components is still being diverted to the incinerator. This material is primarily paper that is part of the coarse screen reject. Recovery of this paper with the installation of a plastic/paper separator ("wet trommel") must be balanced against a reduction in the amount of steam (electricity) that can be produced by the incinerator.

Figure 10.1 is a block diagram of all of the subsystems employed in this integrated system model. Mass balance calculations for the refuse separation processes are based on 5 days/week, while the fermentation process is operated 7 days/week. This example is for 400 tons/day or 2000 tons/week for the 5-day operation. It is easier to consider the total system capacity on a weekly rather than daily basis when doing mass balances.

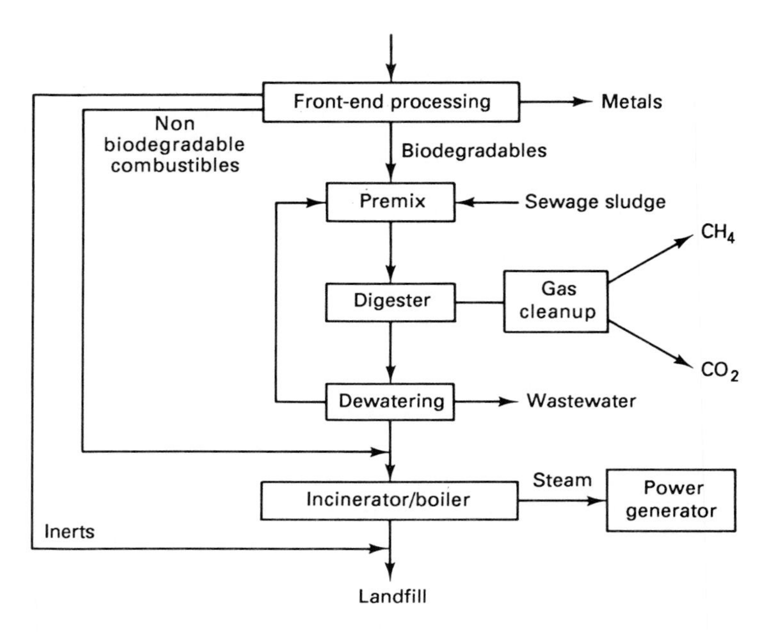

Figure 10.1 RefCoM plant schematic diagram.

The mass balances on the fermenters and the incinerator are shown in Tables 10.5 and 10.6, respectively. Separated refuse is fed from the stockpile to the fermenters daily, which is necessary to maintain a balanced fermentation process. The input streams to the fermenter consist of the sewage sludge, refuse, and recycled filtrate. Fermented slurry and the product gas streams are the output from this process. For this example, the retention time in the reactors is 12 days, resulting in a biodegradable solids reduction of 65%. The methane production is 1 million cubic feet per day, with 933,000 ft^3 of carbon dioxide production. The feed slurry solids concentration is 12%, and the destruction of solids in the reactor results in a reactor slurry solids of 6.8%.

Table 10.5 Mass Balance on the Fermentation Reactors

	Input (tons/day)				Output (tons/day)	
	Sludge	Refuse	Filtrate	Total	Slurry	Gas
Biodegradable	10.0	95.2	10.0	115.2	40.3	74.9
Combustible	0.0	16.0	0.0	16.0	16.0	0.0
Inerts	4.3	16.4	6.9	27.6	27.6	0.0
Water	462.4	44.4	657.7	1164.5	1152.6	11.9
Moisture (%)	97.0	25.8	97.5	88.0	93.2	19.0

Gas production: CO_2, 933,300 scf/day; CH_4, 1,011,000 scf/day

Table 10.6 Mass balance on incineration

	Input (tons/week)			Output (tons/week)	
	Screen	Filter	Total	Ash	Gas
Biodegradable	450.0	169.4	619.4	0.0	619.4
Combustible	78.5	112.0	190.5	0.0	190.5
Inerts	38.5	116.2	154.7	154.7	0.0
Water	26.5	677.6	704.1	0.0	704.1
Moisture (%)	19.1	63.0	42.2		

Gross heat available: 1.1×10^9 Btu/day

The fermented slurry is dewatered and the cake along with the oversize from the coarse disk screen are fed to the incinerator, resulting in a feed rate of 138 tons/day of dry solids (238 tons/day of wet solids). The data in Table 10.6 are expressed in tons per week because of the different operating periods for these processes discussed previously. The 238 tons/day of wet solids fed to the incinerator are reduced to 22.1 tons/day of a dry ash that must be landfilled. The combustion process converts 115.7 tons/day of combustible solids to CO_2 and water and evaporates 100 tons/day of moisture.

The Btu content of the incinerator feed based on 8600 Btu/lb combustible solids is 2.0×10^9 Btu/day. Allowing for heat to evaporate the moisture and heat losses in the incinerator, about 1.1×10^9 Btu/day is available for recovery. Some of this energy is required for process heat (Table 10.7). Heat is required to elevate the temperature of the incoming solids and water in both the refuse and sludge streams. Radiant heat losses from the reactors must be replaced and heat is required for evaporation of water necessary to maintain water vapor in equilibrium with the gas phase at 140°F. When these heating requirements are satisfied, there is about 0.95×10^9 Btu/day available for recovery. There are several options for use of the steam. The magnitude of the electrical power demand makes a strong case for generating electricity for on-site use. With power costs in the range of $0.05 to 0.10

Table 10.7 System heat balance

	Heat (Btu/day)
Heat refuse feed solids	6,622,200
Heat refuse moisture	7,204,200
Heat sludge solids	742,000
Heat sludge moisture	74,945,000
Reactor heat losses	41,064,200
Gas water vapor	24,180,300
Total heat used	154,757,900
Gross heat available	1,100,000,000
Net heat available	945,242,100

Table 10.8 Unit process energy demand

	Energy demand (kWh/yr)
Shredder	356,449
Separation subsystem	291,432
Reactor mixers	2,505,813
Gas processing	13,234,026
Slurry dewatering	183,763
Incinerator	3,176,263
Miscellaneous	1,431,796
	21,178,522

per kilowatt hour, the annual power costs will be in the range of $1,000,000 to $2,000,000. This would buy a lot of capital for the addition of a generating unit to the incinerator.

The electrical demands of several processes are significant. Table 10.8 presents the power requirements for the various unit processes. The refuse separation processes have a line capacity of 50 tons/hour and operate only 8 hours/day, 5 days/week. The reactor mixers, incinerator, and gas processing unit operate continuously, while the balance of the processes operate 16 hours/day, 7 days/week. Gas cleaning and compression is the dominant power user. This produces gas at 975 psig for injection into a gas transmission line. The incinerator and the reactor mixers are the next major consumers of electricity. The shredder is a relative low power user because of the limited size reduction needed.

A mass balance on the total system is given in Table 10.9. These rates are expressed in tons per week. The input stream consists of 2000 tons/week of refuse on an as-received basis and 3336.9 tons/week of sewage sludge that is 97% water. The output streams include the recovered metals and the methane and carbon dioxide. The incinerator stack gases account for 1514 tons/week, about half of which is water from the sludge and refuse. An additional 2858.8 tons/week of water is returned to the wastewater treatment plant. Of the 2000 tons/week of refuse processed, only 331.2 tons/week go to the landfill. This is a reduction in weight of

Table 10.9 Mass balance on total system (tons/week)

	Input		Material	Gas streams			
	Refuse	Sludge	recovered	CH_4/CO_2	Incinerator	Landfill	Sewer
Biodegradable	1130.0	70.0	0.0	524.3	619.4	14.0	42.7
Combustible	204.0	0.0	0.0	0.0	190.5	13.3	0.0
Inert	287.0	30.1	24.5	0.0	0.0	263.2	29.4
Water	379.0	3236.8	0.0	83.3	704.1	41.3	2786.7
	2000.0	3336.9	24.5	607.6	1514.0	331.8	2858.8

disposed material of 83%. However, the volume reduction is even greater. With in-place densities of 1000 lb/yd^3 for refuse in a landfill, the 2000 tons will occupy 4000 yd^3. The in-place density of the residue and ash would be about 2000 lb/yd^3. The volume occupied by the ash is about 330 yd^3, or a volume reduction of 92%. The life of the landfill would be greatly extended.

Economic Evaluation

The following discussion of an economic evaluation of the RefCoM system can be applied to any refuse-processing operation. A number of options are always available in the design of any facility, and the decision regarding which option to select will invariably be based on an economic evaluation. The following example is a synopsis of the evaluation of the RefCoM process as a means for disposal of solid waste while producing a useful form of energy.

The 400-ton/day facility described above, planned to process 400 tons/day on a 5-day/week basis or 2000 tons/week, was used as the base case for this economic evaluation. When the plant is operational, the operating expenses and bond payments must be offset by its revenues. The expenses included here are those for personnel, power, water, equipment replacement, maintenance materials and supplies, landfill disposal, and GS&A expenses. These expenses are inflated each year at 5%.

The plant receives revenues from the sale of methane, carbon dioxide, and recovered metals, the tipping fee for the solid waste, the fee for the sewage sludge disposal, and the interest earned on the bond payment reserve.

Base case. This is the system discussed in Chapter 7. The revenues and expenses for the first year of operation of the base case system are presented in Table 7.2. The base case estimate of the capital requirements and the cash flow for the facility are presented in Tables 7.3 and 7.4 for an equity contribution of 25%.

The cash flow for the 20-year economic life of the system presented in Table 7.4 shows the revenues exclusive of those generated by the tipping fee. The tipping fee is determined by the difference between the bond payment and the operating expenses and other revenues plus the required pretax income. The income from the bond reserve and the expense of the bond payment are constant over the life of the project (with the exception of the return of the bond payment reserve at the end of the project). Expenses and revenues other than methane are assumed to increase with inflation (5%). Methane revenues are inflated according to the Gas Research Institute's gas price projections.[2] The methane revenues increase with inflation as well as resource depletion.

The tipping fee required to cover the difference between the expenses and other revenues presented in Table 7.4 may be determined quite readily. The

[2]Ann Ashby, Gas Research Institute, Washington, DC, 1986.

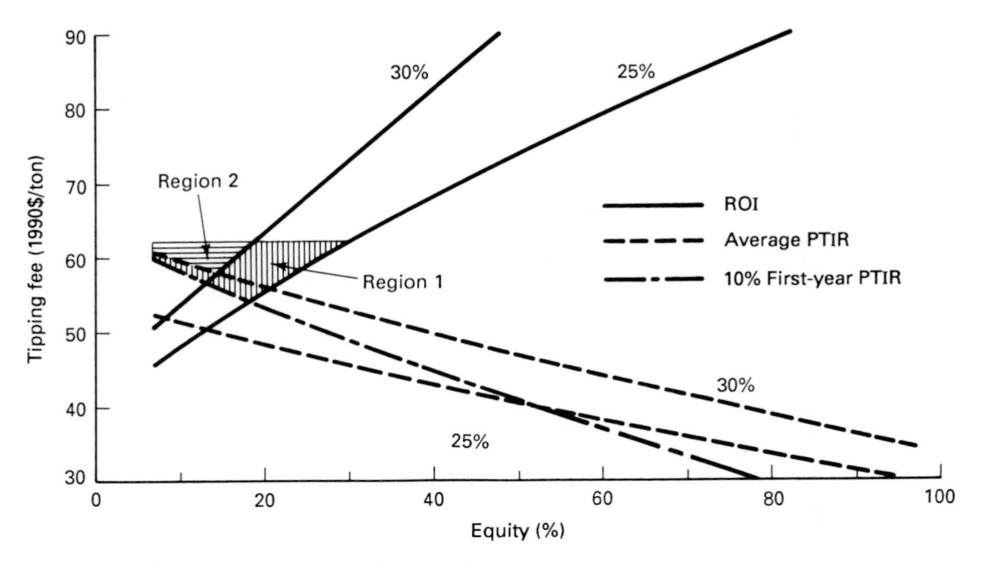

Figure 10.2 Base case economics for the RefCoM process.

additional tipping fee required to provide a profitable operation depends on the measure of acceptable return. The same three measures as discussed in Chapter 7 are investigated here:

1. At least 10% first-year pretax income ratio
2. At least 25 to 30% average pretax income ratio (PTIR)
3. At least 25 to 30% average annual pretax return on invested equity (ROI)

The objective of this exercise is to find the equity contribution yielding the minimum tipping fee that satisfies all three profit measures presented above. Figure 10.2 presents the results of the runs made using the base case cost estimates. The tipping fee for each equity position and profit constraint is calculated and plotted in this figure. Each line plots the 1990 tipping fees that are required to satisfy the foregoing constraints for the range of equity that might be put forward.

The feasible regions shown in Figure 10.2 are equity/tipping fee combinations that satisfy all three constraints. Region 1 satisfies the condition of 25% average return on investment (ROI) and 10% first-year pretax investment ration (PTIR). Region 2 is a subset of region 1 and satisfies the tighter constraints of 30% average ROI and 30% average PTIR. Notice that the requirement for a 10% first-year PTIR exceeds the requirement for a 25% average PTIR in the feasible region. The 10% first-year PTIR rather than the 25% average PTIR is the controlling factor for setting the tipping fee under this scenario. Putting forth 12 to 15% equity gives the lowest tipping fees in feasible regions. Those tipping fees would be in the range of $55 to $57 per ton.

Table 10.10 Composition of the incinerator feed

Constituent	Tons/day
Biodegradable combustibles	89
Nonbiodegradable combustibles	27
Inerts	22
Water	100
	238

Base case with internal energy generation. The base case does not utilize the large amount of heat energy that results from incineration of the fermenter residue and the nonbiodegradable combustible streams. The composition of the stream fed to the incinerator is presented in Table 10.10. Assuming that the biodegradable organic materials have a heating value of 7000 Btu/lb dry solids and the nonbiodegradable combustible solids have one of 14,000 Btu/lb dry solids, a total of 78 million Btu/hr is available. Assuming that it takes 1900 Btu to evaporate 1 lb of water (including furnace efficiencies and the enthalpy of the water vapor), and that the waste heat boiler is 67% efficient, the heat available as steam is estimated at 42,000 lb/hr. About 6600 lb/hr of low-pressure steam is required for digester heating, leaving about 35,000 lb/hr of excess steam. This steam may be sold to other industrial steam users, used to drive in-plant machinery (compressors, mixers, etc.) or used to generate electricity for in-plant use and/or sale. If the latter option is chosen and assuming that the efficiency of a single-stage turbine is about 50%, the potentially available generated power is 5350 kW.

The base case employed an estimate of $5.2 million for the capital, engineering, and construction supervision for the incineration unit. The additional cost associated with power generation (turbine, generator, condenser, etc.) is $8.8 million.

Due to the mode of operation of the plant, there will be different power demands at different times of the day and week. There will also be different rates paid by a power company for on-peak versus off-peak periods. Using rates of $0.02 and $0.016 per kilowatt for on- and off-peak periods, respectively, for the appropriate excess electricity generated throughout the week, a revenue stream of $242,100 per year will be obtained. As a result of the addition of internal generation of electricity, the deviation from the base case is summarized as follows:

1. The addition of $242,100 per year in electricity sales
2. The removal of $1,688,000 per year in power costs
3. The addition of $8,800,000 in capital costs

The results of this case are presented in Figure 10.3. Note that the requirement for a 10% first-year PTIR dominates the requirement for an average PTIR of 25 or 30% and that when combined with the ROI requirement, gives a tipping fee of $44.50 and $46.50 per ton at 25 and 30% ROI, respectively.

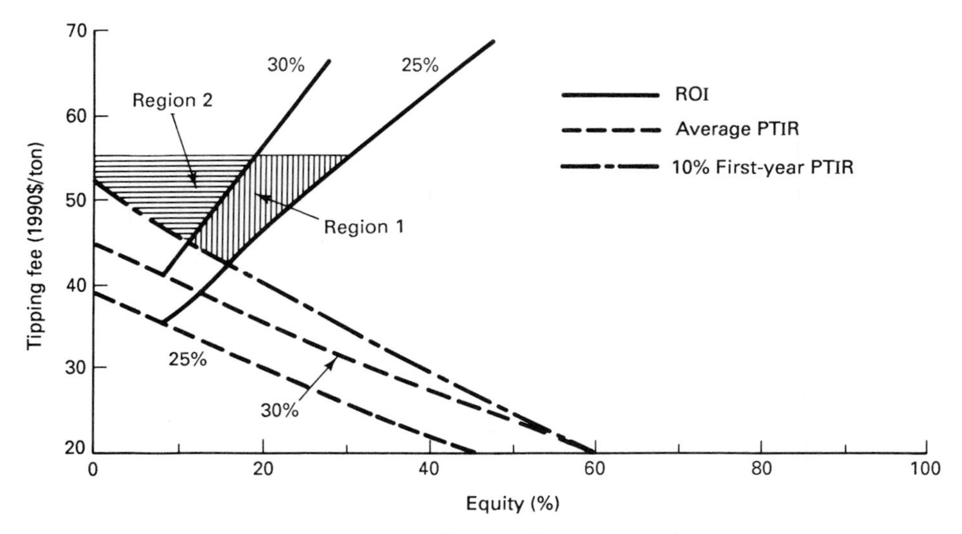

Figure 10.3 RefCoM economics: base case with internal energy generation.

100% public financing. Perhaps private investors would not want to risk their capital even with a 30% ROI and 30% PTIR. What would be the effect of a zero-equity financing arrangement on the economics of a 400-ton/day RefCoM facility? Table 10.11 presents the capital requirements in 1990 dollars for the *base case with and without internal energy generation for zero equity invested.*

Without some equity financing, the bond issue should increase, although not proportional to the amount of equity replaced. In the base case, decreasing the equity from 25% to 0% increased the bond issue from $29,474,000 to $32,600,000 (a 10% increase), while with the base case with internal energy generation the bond issue increased from $39,000,000 to $42,500,000 (a 9% increase). Taking these costs, the other annual expenses, and the revenues other than the refuse tipping fee, a break-even tipping fee can be determined (Table 10.12).

Table 10.11 Capital requirements for zero equity

	Base Case ($1000)	Internal power generation ($1000)
Total construction costs	28,862	37,662
Equity	0	0
Bond capital requirement	28,862	37,662
Bond issue costs	1,154	1,506
Bond payment reserve	0	0
Bond payment during construction	4,618	6,026
Interest earned on reserve	0	0
Interest earned on bond payment	(462)	(645)
Interest earned on construction funds	(1,530)	(1,996)
Bond issue	32,642	42,553

Table 10.12 Tipping fee calculation for zero equity (1990 dollars)

	Other revenue ($1000)	Operating expenses ($1000)	Bond payment ($1000)	Tipping fee ($/ton)
Base case	4156	5937	3306	48.90
Base case with internal energy generation	4400	4249	4301	40.00

Installation of the generating plant substantially increased the capital expenditure and the bond issue. The economic justification rests with the cost of electrical power and the ability to sell the excess power generated. This system has a significant power demand that imposes a major annual operating cost. The economic evaluations are very sensitive to power costs and the market value of the excess power. Even with PURPA, the value of the marketed power is considerably less than the price paid by the consumer. This cost differential drives the economics of the RefCoM facility with internal power generation and use.

As shown in Table 10.12, the break-even tipping fee for the base case with internal energy generation is almost 20% lower than the base case. Assuming (and that may be a big assumption) that the costs used in presenting this analysis are accurate, the larger investment will result in a lower tipping fee. Of course, at this stage of the development of any design, there is a degree of uncertainty associated with any cost information. However, the larger the difference in the computed tip fee, the more confidence one can have in the decision.

STUDY QUESTIONS

1. Describe the biochemical reactions that are required to convert solid waste into a fuel gas containing methane and carbon dioxide.
2. What process environment is necessary to support methane fermentation at high rates?
3. Draw a schematic of a system for the biological production of a fuel gas from solid waste. Identify all of the significant subsystems.
4. What would be typical process parameters for the operation of a fermentation reactor producing methane from refuse?
5. What is the purpose of the refuse separation subsystem? What fraction of the refuse was fed to the fermentation reactor?
6. What was the role of the incineration unit in this system?
7. What are the possible revenue streams for this system?
8. How does the production of electricity by incineration of the residual organic material affect the economic evaluation of this system?

CHAPTER 11

Sanitary Landfill Design and Operation

SITE CONSIDERATIONS

General Requirements

The general requirements for site selection for any regional waste disposal facility have been elucidated in Chapter 3. Section 39.2 of Title X of the Illinois Environmental Protection Act sets the following criteria (these requirements are specific for Illinois sites, but similar requirements can be expected in other states):

1. The facility is necessary to accommodate the waste needs of the area it is intended to serve.
2. The facility is so designed, located, and proposed to be operated that the public health, safety, and welfare will be protected.
3. The facility is located so as to minimize incompatibility with the character of the surrounding area and to minimize the effect on the value of the surrounding property.
4. The facility is located outside the boundary of the 100-year floodplain as determined by the Illinois Department of Transportation, or the site is floodproofed to meet the standards and the requirements of the Illinois Department of Transportation and is approved by the department.
5. The plan of operation for the facility is designed to minimize the danger to the surrounding area from fires, spills, or other operational accidents.

6. The traffic patterns to and from the facility are designed as to minimize the impact on existing traffic flows.
7. If the facility will be treating, storing, or disposing of hazardous waste, an emergency response plan exists for the facility, which includes notification, containment, and evacuation procedures to be used in case of an accidental release.
8. If the facility is to be located in a county where the county board has adopted a solid waste management plan, the facility is consistent with the plan.
9. If the facility will be located within a regulated recharge area, any applicable requirements specified by the board for such areas have been met.

Condition 7 is contained in the legislation, but would not generally apply to new sanitary landfill sites since hazardous wastes would not be accepted along with normal urban refuse. The regulated recharge area referenced in condition 9 are the areas where significant surface water infiltration occurs and recharges the aquifers that are groundwater resources.

These requirements are general in nature and can be used to identify specific site characteristics that are required for a site to be acceptable for a sanitary landfill. The specific characteristics include topographic and geologic factors. The important features are presented in Table 11.1. Steep topography presents a number of problems. Equipment operation and truck access are more difficult when steep slopes are encountered. Because of the exposed soil, erosion problems are more severe as a result of rapid runoff from the steep areas. These problems can be overcome, but at a cost.

Table 11.1 Specific site characteristics for sanitary landfills

	Degree of limitation		
Factor	Severe	Moderate	Minimal
Land slope	>15%	3–15%	<3%
Surface deposits	Clean sand/gravel Heavy organic clay	Sand/gravel with silt	Silty clay
Bedrock depth	<10 ft	10–25 ft	>25 ft
Bedrock type	Fractured limestone	Sandstone	
Groundwater depth	<10 ft	10–25 ft	>25 ft
Distance to:			
Water well	<300 ft	300–1000 ft	>1000 ft
Floodplain	<300 ft	300–1000 ft	>1000 ft
Stream/lake	<1000 ft		>1000 ft
Parks	<1000 ft		>1000 ft
Wetlands	Located within		

Source: Waste Age, June 1986, p. 88.

The geology and hydrogeology of the area are more important and less visible. The availability of sufficient soil suitable for cover material is always a prime consideration. It is equally important to have subsurface conditions that protect any groundwater sources. The soil deposits, both in depth and type, the imperviousness of and distance to the rock strata, and the location of the groundwater table are all important considerations in site selection. The objective is to maximize the distance from the landfill to the groundwater table and to have the maximum depth of impervious material between the fill and the water table. This will reduce the possibility that any leachate that may leak from the fill will contaminate the groundwater.

Soil Properties

The types and quantities of soil available at the site will be significant factors in the cost of operation of a landfill. Soils are needed to provide the moisture barrier in the bottom of the fill and also in the cover. The soil must be able to support the equipment used to transport and place the refuse. When the site is completed, a soil capable of supporting a good vegetative cover is needed.

Soils are generally classified as to the percentage of clay, sand, and silt. There are a number of different soil classifications. Two are shown in Figure 11.1, the U.S. Department of Agricultural Soil Conservation Service (USDA) and the Unified Soil Classification System (USCS). The USDA classification has several more classes than the USCS. Soils are characterized by their particle size as follows:

Clay	<0.005 mm
Silt	0.005–0.05 mm
Sand	0.05–2.0 mm
Gravel	>2.0 mm

Clay is composed of very fine particles, colloidal in nature. These sizes are, in part, responsible for the soil characteristics. However, the presence of different cations and organic compounds can have a major effect on the behavior of the soils, especially clay and to a lesser degree, silt. Clays are plastic and cohesive. The presence of organic compounds will increase the compressibility of clays. It is not sufficient to characterize soils on the basis of particle size alone. The surface properties of the soils are also very different. The ion-exchange capacity is determined by the presence of surface ions that can be exchanged. This exchange can alter the behavior of the soil and change its properties, as well as remove cations from the water passing through the soil.

The soil properties that are important in a sanitary landfill relate to the activities associated with operation of a landfill. The following properties are important.

Permeability. A soil with a low permeability is needed to prevent the passage of water into the fill and the loss of leachate from the landfill. A tight clay is

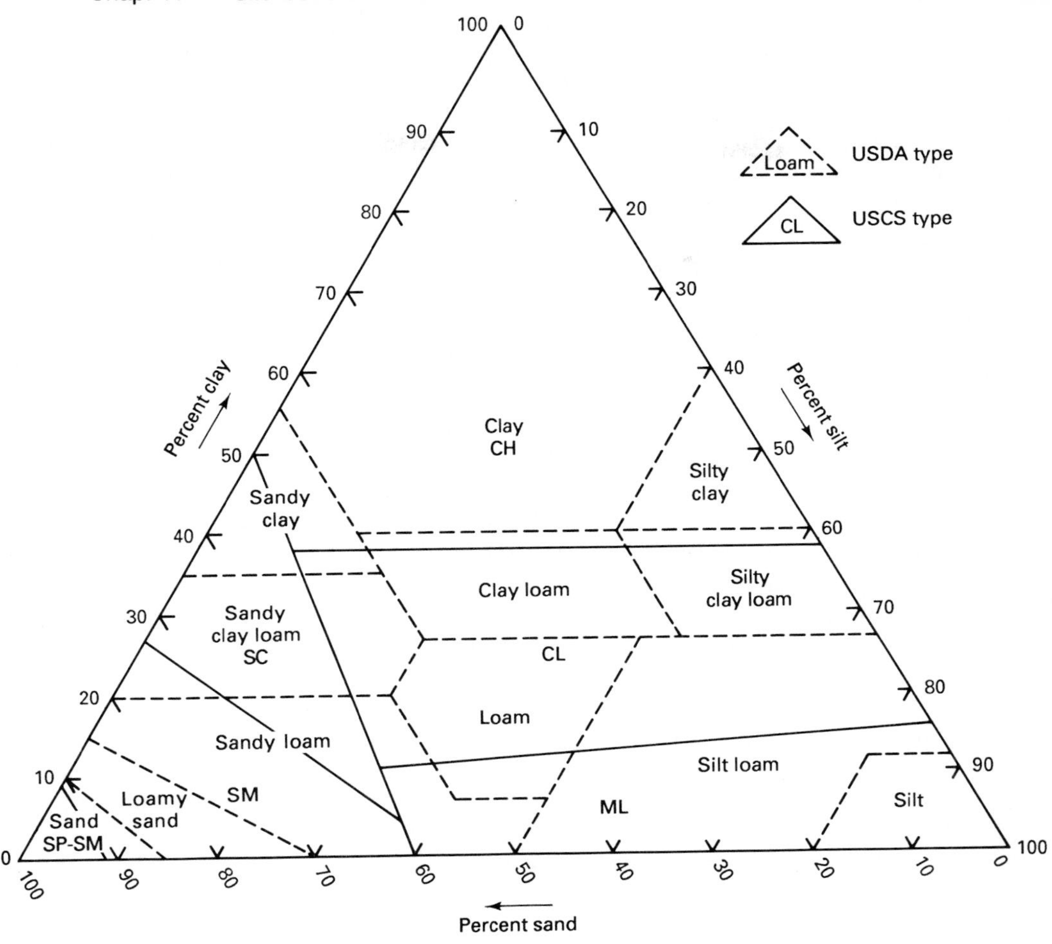

Figure 11.1 Textural classification of soils.

very effective in this role; a silty clay is less effective but still satisfactory. The permeability is a function of the particle size and with larger particles, the particle size distribution. In other cases it is desirable to have a porous soil to control landfill gas. The gas will follow the path of least resistance. Soils that have a large particle size that is relatively uniform will be suited for this role. Uniform gravel or uniform sand is the soil of choice. Such soils also have a role as drainage layers to direct the flow of water within the landfill and cover.

Swelling and cracking. Certain soils are very prone to swelling when they become wet. When these soils dry, they crack, causing a break in the integrity of the soil layer. Clay and silty clays are particularly susceptible to this problem. Clay, especially clay with a high organic content, is plastic and very compressible. When loaded it is subject to heaving. When wet, this type of soil is unsuited for supporting vehicles, especially collection and transfer vehicles.

Support of vegetation. When the landfill is completed, a final cover is placed to isolate the refuse from the environment. It is essential to have a good cover of vegetation to protect the soil from erosion and to serve as a means of dissipating the water that may infiltrate into the top layers of the cover. These soils must be able to absorb and retain significant quantities of water as well as plant nutrients needed for the growth of the vegetative cover. Silt, sandy silt, and to a lesser degree, clay-sandy silt are desirable soils for this purpose. It is important to remember that the cover may be constructed of several layers of soil, each having a specific purpose.

Hydrogeologic Properties of the Site

The selection of a landfill site requires very careful evaluation of the hydrogeology of the area. As indicated in Table 11.1, it is desirable to maximize the distance between the landfill and the groundwater and to have an impermeable stratum between them. Figure 11.2 illustrates the type of conditions that may be encountered. The deep aquifer is a considerable distance below the surface. It is recharged at a distant point. The travel time is long, so the water level will remain relatively constant. It is also protected by an impervious stratum above. Contamination of this aquifer is extremely improbable. The shallow aquifer is more vulnerable; the distance to the landfill is greatly reduced and there is no protective layer. This aquifer has a high risk of contamination if the landfill is not designed to minimize the production of and contain all leachate that may be produced.

This figure also illustrates the importance of site selection relative to the groundwater table. A shallow aquifer is replenished by local precipitation. Conse-

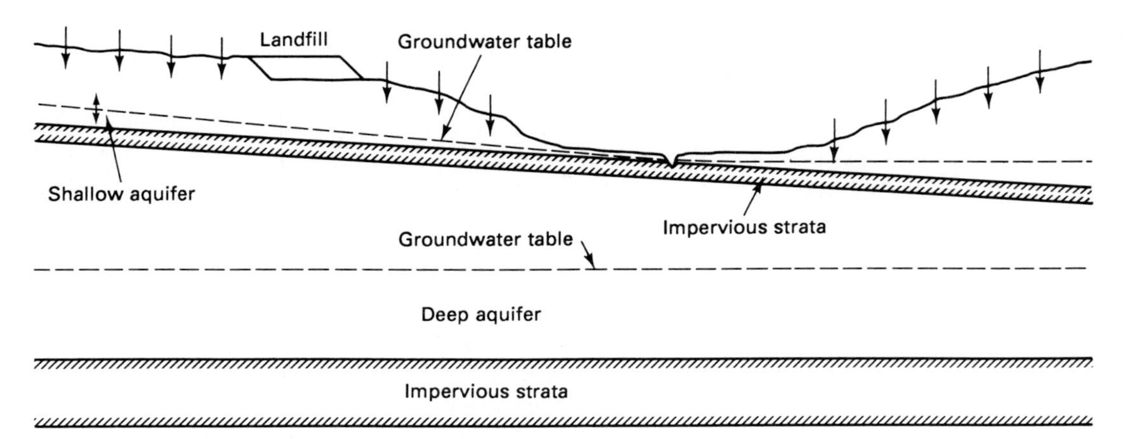

Figure 11.2 Hydrogeological considerations in landfill siting.

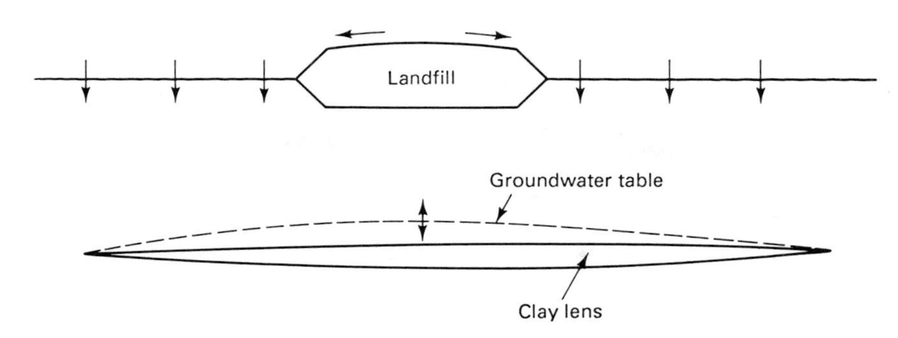

Figure 11.3 Perched groundwater tables.

quently, its level may vary considerably, being near the surface during wet weather. A landfill located on the high ground will not be likely to come in contact with the groundwater. However, if the site is closer to the stream, it is possible for the groundwater to infiltrate the landfill during wet periods. It is very difficult to prevent water from penetrating the bottom of the landfill unless the fill contains a substantial depth of leachate. Lifting pressure backed by many feet of water head can be encountered. Rupture of the lining is a strong possibility and the landfill will become saturated with groundwater. When the water level drops, the contaminated water will also leak from the fill, causing pollution of the shallow aquifer.

Figure 11.3 illustrates a different groundwater problem that is local in nature. An impervious layer of soil such as a clay lens may be 20 to 30 ft below the surface and cover a reasonable area. The precipitation falling in this area will percolate into the ground and form a perched water table. The level of this water table will vary substantially, coming very close to the surface during periods of heavy precipitation. Again, the groundwater may penetrate the landfill, causing saturation of the refuse. When the water level drops, the contaminated water will drain from the fill, causing serious groundwater pollution. These sites are the most difficult to evaluate unless there is a historical record of the groundwater table levels at the specific site. It pays to evaluate any site during wet as well as dry weather conditions.

There usually is a wealth of information available on the hydrogeological conditions within a given area from the state geological surveys or the USDA Soil Conservation Service. They are the repository for the drilling logs from the multitude of water, oil, and gas wells that are drilled. Other drilling logs are also retained by these agencies. This information is used to map the state and can be a valuable source of information in the initial stages of site selection. It will assist in elimination of the sites that have obvious water and soil problems. As potential sites are identified, it will still be necessary to conduct on-site drilling to establish the subsoil conditions as well as the location and variation of the groundwater table.

Each state permitting agency has a required number and spacing for the test drillings. Soil borings on a 100- to 200-ft grid may be necessary to characterize the subsoil adequately. As the borings are being conducted, samples will be tested for moisture content and grain size distribution for use in classifying the soils. Soils that may be considered for liners and cover material will also be tested for permeability, Atterberg limits (effect of moisture on soil properties), moisture content, and moisture–density relationships. The borings should extend to at least 20 ft below the intended bottom of the site. A portion of the borings should terminate below the water table if it is greater than 20 ft below the base of the site. If these borings are located on the periphery of the site, they can be converted into monitoring wells for use after the site becomes operational. The water elevation in these test holes will provide information on the direction of groundwater movement and fluctuations in the water table.

In addition to horizontal movement, groundwater and potential contaminants from the landfill may move vertically within the subsurface formation. Vertical movement can be detected by installing multilevel wells as shown in Figure 11.4. This will identify if the site is in a recharge or discharge area. Recharge areas are identified by the differential level of water in the multilevel test wells, as shown in

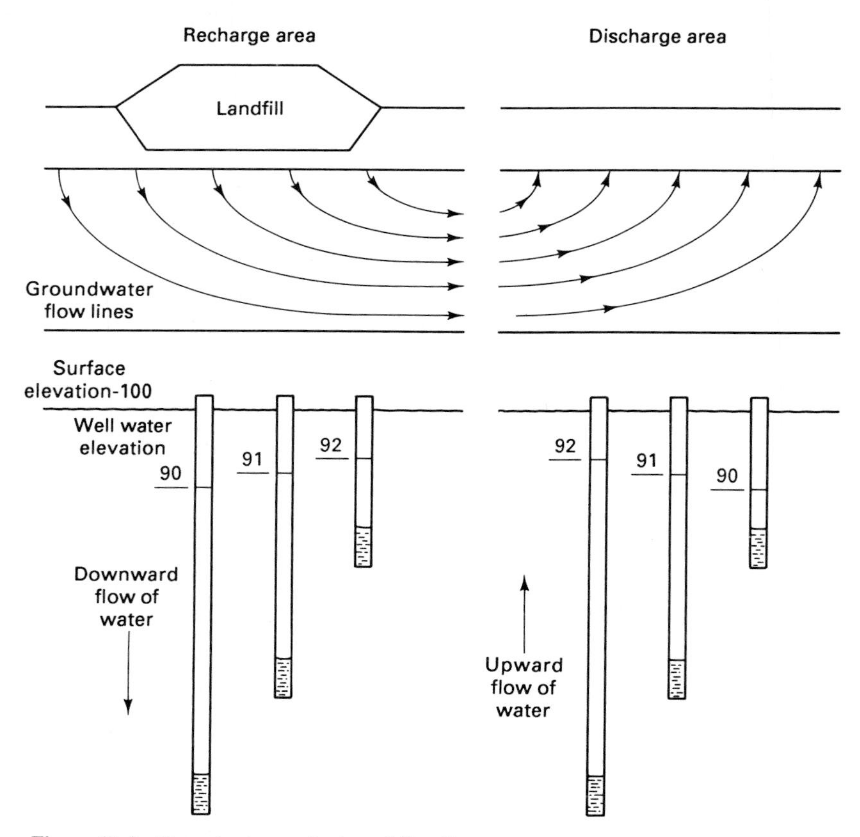

Figure 11.4 Groundwater gradients and flow lines.

this figure. The water level in the deeper wells is lower because of the head loss associated with downward movement of water. The converse is true for a discharge area. Requirements for landfill construction in these areas are more stringent, or may be prohibited, in some states, to provide better protection of the groundwater.

A landfill constructed in a discharge area will require a groundwater management system to prevent the flow of water into the landfill. A well field surrounding the site, or an underdrain system will be necessary to remove the groundwater inflow before it reaches the refuse. Since this is a discharge area, the groundwater will not be susceptible to contamination from the leachate. If, however, the groundwater flow is not controlled, the landfill will be saturated and leachate can escape from the surface of the site and cause surface water contamination.

If the landfill site is in a recharge area, control of surface water is more critical. The percolation of water into the soil will speed the movement of any contamination released from the landfill. It is essential that the surface water infiltration into the landfill itself is minimized, and effective leachate containment and removal systems are installed. Also, collection of the surface water from the surrounding area into a surface drainage system will reduce the infiltration and the rate at which the infiltrated water travels into the groundwater system.

LANDFILL TECHNIQUES

A landfill's basic design parameter is volume. It depends on the area covered, the depth at which the refuse is placed, and the ratio of soil cover to refuse. The air space in this fill site represents the volume available for placing the refuse and cover material. Since the refuse generation rate is measured in tons, an additional parameter that influences the capacity of the landfill is the in-place density of the refuse and soil.

Historically, the site for a landfill was the nearest quarry or hole in the ground. The refuse was used to fill the excavated volume. Little concern was given to the availability of cover material. If soil was available, a final cover of sorts was applied when an area of fill was completed. The current standards and guidelines for the design and operation of sanitary landfills have rendered these sites less desirable. Strict enforcement of the requirement for daily and final cover has increased the operating cost of these sites because of the cost of imported cover material. Also, the need for lining the bottom and sides with a water barrier has increased the site development cost. On-site excavation and placement of soil costs approximately $1.00 to $2.00 per cubic yard. However, if the soil has to be imported from off-site, the costs will be as much as $10.00 per cubic yard, depending on the haul distance. As discussed later, the cost of the cover material can be a significant part of the total operating cost. New sites tend to be on high ground, where the water problem will be minimized and where adequate soil is available for cover. If existing excavations are used, the filling procedure will be the same as used for the area method described below. When cover material was unavailable, alternative landfill techniques have been used in the past. However, the new regulations will probably preclude any system that cannot provide a daily cover.

All landfills have certain commonalities. Appropriate buffer zones and screens are necessary to isolate the landfill from the neighbors. Proper placement of trees and shrubs can greatly reduce the animosity between the neighbors and the site. Also, security fences are needed to prevent unauthorized access to the site, whether by children or for illegal dumping. An all-weather access road from the primary roadways is mandatory. The collection and haul vehicles are essentially highway vehicles. They do not function well on unstable roads. At the site, careful maintenance of the working face access track is required to keep the trucks from becoming stuck. During inclement weather, it may be necessary to have a tractor available to assist the trucks. Of course, the mud that collects on the truck wheels must be spun off before the truck returns to the highway.

At the larger sites, a number of other provisions are required for efficient and safe operation of the facility. Maintaining the equipment used at the site requires a facility properly equipped for the desired on-site maintenance. This may be a complete automotive maintenance shop that can service not only the landfill equipment but also the collection and haul vehicles. Also, it is necessary to have a water supply available for firefighting. When a fire starts in a sanitary landfill, severe air pollution can result if the fire is not extinguished rapidly. A number of techniques are available to handle fires. However, if the fire is in the fill, it may be necessary first to expose the burning material by excavating the fill. Then water or foam can be applied to extinguish the combustion. If the fire zone can be defined, it is also possible to inject an inert gas such as CO_2 to extinguish the fire. Water can be used, but it may add to a leachate management problem.

Trenching Method for Constructing Sanitary Landfills

Figure 11.5 illustrates the more simple technique for constructing a sanitary landfill. In the trench method, the collection truck will dump its load generally from the side of a previously excavated trench that may be 10 to 15 ft deep. A tractor/compactor will spread and compact the refuse by pushing it up the slope of the working face. The tractor makes several passes over the refuse to achieve optimum density. This procedure continues until the end of the day. At that time, cover material is excavated by extending the trench. The volume of cover material excavated creates an initial in-ground volume for the refuse. The additional volume required is obtained by raising the elevation of the top of the site. As each new trench is started, part of the excavated soil is used to construct an aboveground berm on which the initial refuse can be compacted. This berm has a slope of about 1 vertical to 3 horizontal. The balance is stockpiled for use as cover material when the last cell in the trench has been completed.

This placement technique is an inefficient use of land. As can be seen from this figure, there is a considerable volume of soil that is not excavated between the

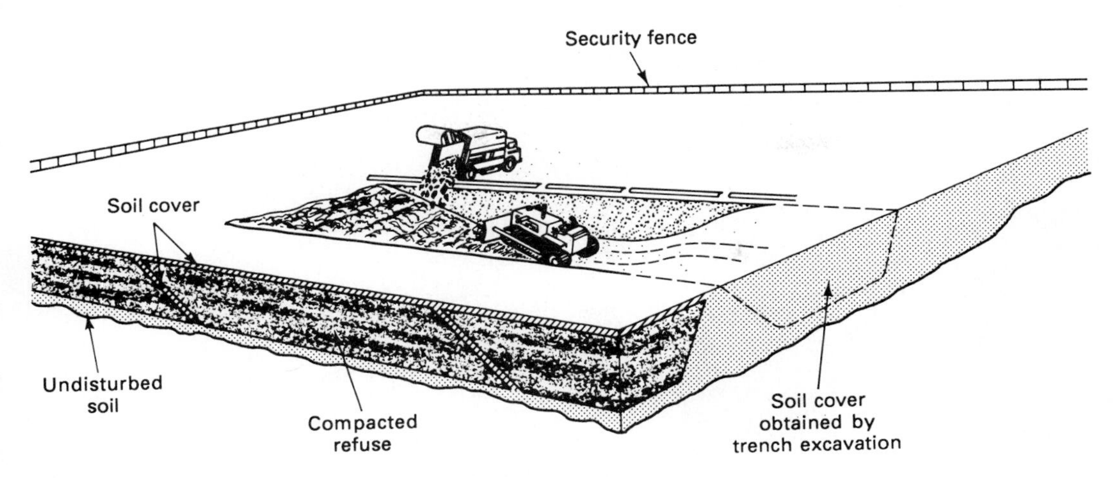

Figure 11.5 Trench method for placing refuse in a sanitary landfill.

trenches. There is very little planning for the site development and management of the on-site soil. However, in a small landfill that has an abundance of area, the inefficiency is offset by the reduced requirements for earth-moving equipment. There is no need to move and stockpile large volumes of soil. It is used as needed for final cover since most of these fills are only one lift. Only one tractor will be needed to spread, compact, and cover the refuse.

At least 3 ft of soil is required as a final cover. Six to 12 in. of soil is used for intermediate cover on the working face, where refuse will be placed the next day. The excavated soil must be of the quality needed for cover: a clay or clay-silt that can be used to form a layer of relatively impervious soil and a silt that can be placed on top of the clay to support a vegetative cover. Evaluation of on-site soils is necessary to ensure the availability of the required type and quantity of soil. Trench construction does not easily permit the construction of liners and leachate collection systems. Therefore, the site selected needs a good layer of impervious subsoil to prevent movement of any leachate. Some planning for the management of the soils is necessary, as it is not uncommon for the operators of these sites to place 20 ft of refuse on top of the needed cover material. A simple, but definite operating plan is required for the site.

Area Method for Constructing a Sanitary Landfill

The area method has a well-developed construction plan (Figure 11.6). It may include a liner and a leachate collection system. Design of such a system requires complete knowledge of the subsoil conditions at the site as discussed above. Is the soil suitable for a liner? Is there sufficient topsoil for use as a final cover? These soils must be excavated and stockpiled for use as they are needed. A portion of the site is excavated to the working depth of the fill, and any required liner and leachate

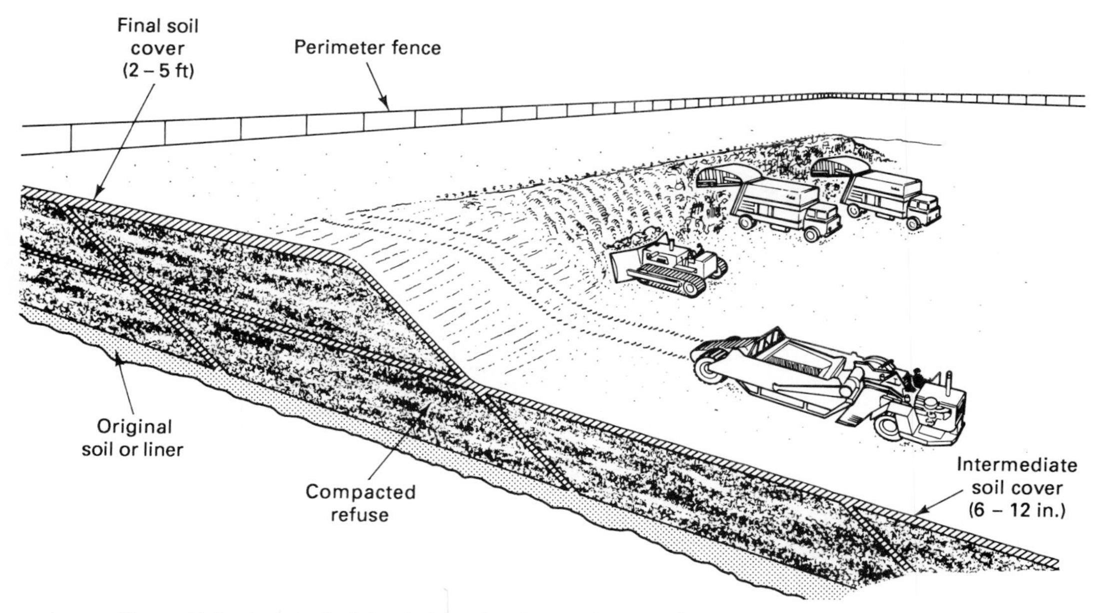

Figure 11.6 Area method for placing refuse in a sanitary landfill.

collection system is installed. Because of the probable volume to be excavated, it is necessary to plan for the use of this soil as cover rather than exporting it.

An embankment with a 1-on-3 slope is constructed by the excavated soil to serve as the initial starting point for the refuse placement. Refuse is dumped at the toe of the working face and spread by compactors in layers of 1 to 2 ft. The spreading and compacting is an important part of the operation of the site. It is desired to achieve a maximum density of the refuse placed in the fill. This layering is continued until the last load is placed. The vertical lift of the refuse will be 10 to 14 ft. Then 6 to 12 in. of cover material is spread over the working face. The top of the intermediate lifts will be covered with at least 1 ft of soil. This cover is sufficient to isolate the refuse for an intermediate time. It is also sufficient to provide a reasonable base for the trucks and equipment when the next lift is placed. If this is the last lift, a final cover will be placed. This may be a specially designed cover to provide an effective control over moisture infiltration into the refuse.

As shown in Figure 11.6, the area method requires a number of pieces of earth-moving equipment. Scrapers are used to excavate the area and stock pile the soil as well as spread cover material on the completed area of the fill. Tractors and compactors are used to spread and compact the refuse as well as the cover material. A road grader may be used to maintain the access road. During dry periods it may be prudent to have a tank wagon to control dust, especially if there are neighbors. Depending on the daily capacity of the site, there may be numerous pieces of equipment involved.

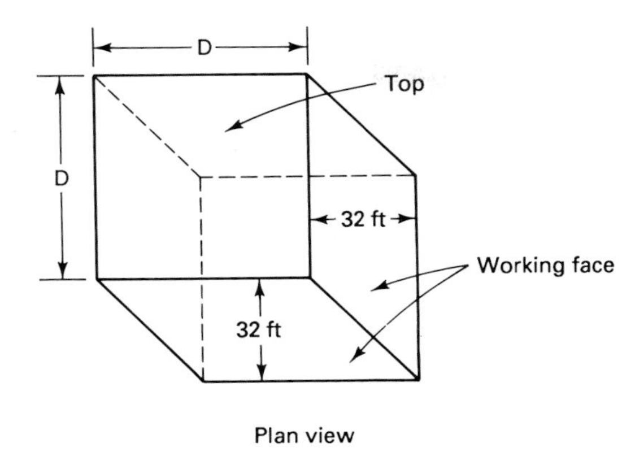

Figure 11.7 Geometric relationships in a landfill cell.

The area method makes the most efficient use of the site. Multiple lifts are used and the height of the fill is limited only by the available cover material, the base area of the site, and any site zoning restrictions. The limits on the side slopes of the fill are generally 1 vertical to 3 horizontal. This limit is imposed primarily by the maximum slope that facilitates the operation of the compaction and earth-moving equipment. Even with a material like solid waste, this slope will result in a stable embankment. If a significant land area is available, one can envision a rather high fill if there are no other limitations imposed.

Cover Material Requirement

The amount of cover material required for a sanitary landfill is a function of the size of the cell. The volume of cover per ton of refuse in the cell will decrease as the cell size increases. This effect can be demonstrated by comparing the surface area-to-volume ratio of different cell sizes. Figure 11.7 is a parallelepiped that represents a typical landfill cell. If the working faces have a 1-on-3 slope, the distance down the working face would be about 32 ft if the vertical depth of the refuse in the cell is 10 ft. One can assume that the top area is a square with dimension D. The refuse volume in any cell will be $10D^2$. The surface area that requires cover material is the area of the top and the two working faces. This area can be computed as

$$\text{surface area} = D^2 + 2(32)D \tag{1}$$

The surface area-to-volume ratios (SA/V) are presented as follows:

D	20 ft	40 ft	80 ft
Volume	4000 ft^3	16,000 ft^3	64,000 ft^3
Surface area	1680 ft^2	4160 ft^2	11,520 ft^2
SA/V	0.42	0.26	0.18

These numbers show that as the size of the cell increases there is a decrease in the ratio of the exposed surface area to the volume of the cell.

This relationship can be used to determine an actual volume of cover material by considering the depth of soil associated with each surface. If there is only one lift to a landfill, the depth of cover material required for the top will be at least 3 ft. If the daily cover on the working faces is 0.5 ft, the following equation can be used to compute the required volume of cover material (CM) and the cover material-to-refuse volume ratio (CM/V):

$$\text{CM} = 3(D^2) + 0.5(2)(32)D \tag{2}$$

D	40 ft	80 ft
CM	6080 ft^3	21,760 ft^3
CM/V	0.38	0.34

The cover material for a single-lift landfill will occupy approximately one-fourth of the total volume of the site.

Assuming an intermediate cover of 1 ft on the top surface and 0.5 ft on the working face, the following equation yields the volume of cover material required an intermediate lift of a multiple-lift fill:

$$\text{CM} = 1(D^2) + 0.5(2)(32)D \tag{3}$$

D	40 ft	80 ft
CM	2880 ft^3	8960 ft^3
CM/V	0.18	0.14

For this cover material scenario, the soil requirements decrease from 15% to 12% of the total cell volume as the cell size increases.

The effect of having multiple lifts on the cover material volume can be seen from the following calculation. If one considers three lifts with 1 ft of intermediate cover and a final cover of 3 ft on the fourth lift, the CM/V can be determined by computing a weighted average as follows:

$$D = 40 \text{ ft}$$
$$\text{CM/V} = \tfrac{1}{4}[3(0.18) + 1(0.38)] = 0.23$$

$$D = 80 \text{ ft}$$
$$\text{CM/V} = \tfrac{1}{4}[3(0.14) + 1(0.34)] = 0.19$$

The cost of the cover material can be estimated by considering the volume associated with the refuse and the amount of cover material used. A site that has received 500,000 tons of refuse will have a compacted refuse volume of approximately 10^6 yd^3. If the CM/V is 0.19, the total fill volume will be $10^6/(1 - 0.19) = 1{,}299{,}000$

yd^3. The cover material volume is 299,000 yd^3. At $1.00 per cubic yard, the cover material will cost

$$\frac{\$299{,}000}{500{,}000 \text{ tons}} = \$0.598 \text{ per ton of refuse}$$

However, if the cost of the cover material is $5.00 per cubic yard, the cost of the cover material alone would be $2.99 per ton of refuse.

The depth of the final cover is an important factor in the cost of the cover material at a specific site. Because a depth of 3 ft or more may be required, it is desirable to have many lifts of refuse covered with the final cover. As the size of the cells and the number of lifts increase, the cost of the cover material per ton of refuse placed will decrease.

Landfill Cover Design

As with all designs, the design of the cover for a landfill is a trade-off—in this case, between the effectiveness of the moisture barrier and the amount of leachate that will have to be collected and treated. It is possible to cap the fill to a point where essentially no water will infiltrate. This is the goal at sites that contain hazardous wastes. Is it worth the cost for a conventional sanitary landfill? This is a major design decision to be made. The regulations for municipal refuse landfills are not specific regarding the design of the cover. As will be seen later, they are definite about reducing the risk of groundwater contamination. The regulations will generally require the collection and treatment of any leachate that is generated unless there are some unusual subsoil conditions.

A modern cover design may contain layers of different soil types, ranging from very tight impervious soils such as clay, to coarse sand and gravel. Each layer has a different function. Figure 11.8 shows the layers that are likely to be considered for a cover design. Each layer has a specific function, and soils with properties that will achieve this function are chosen. Certain functions can be achieved with synthetic materials such as geomembranes, which are moisture barriers, and geotextiles, which can be used for filtration and, to some extent, drainage. This is a relatively new application of these materials and few long-term performance data are available. The following discussion will concentrate on the use of different types of soils to achieve the cover design objectives.

Surface layer. The surface layer has the function of providing a suitable medium for plant growth. Plant growth is essential to protect the surface of the cover from erosion. There will be a slope to the final grade to assist with runoff of precipitation. Without a vegetative cover, the soil will erode, from both wind and water action. It should be a soil with a texture that allows for percolation of water into the soil and that has a good water retention capacity. It should not be an

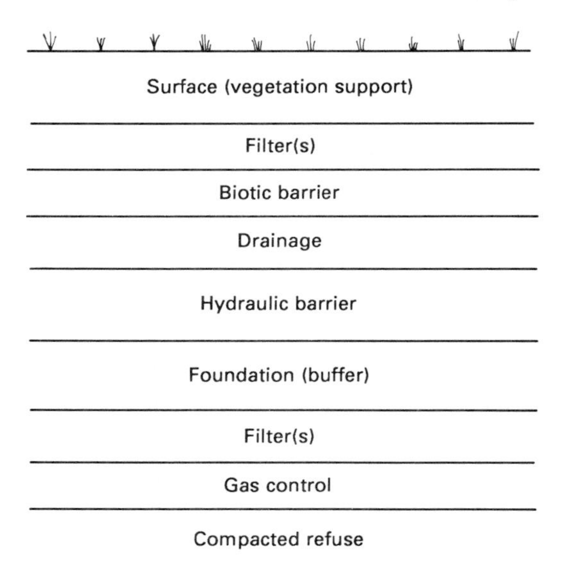

Figure 11.8 Potential layers in a soil cover system.

expansive soil that will swell when wet and crack when dry. Soils that fill this objective are silt, silt-loam, loam, and sandy-loam soils, as defined by Figure 11.1.

The depth of this layer will depend on the type of vegetation and the anticipated role of evapotranspiration in control of the excess moisture. If evapotranspiration is a major factor in the control of excess moisture, this layer should be deep enough to provide a significant soil moisture reservoir. It will need to support deep-rooted plants as well as the more shallow-rooted grasses. Depths of 2 to 3 ft or greater are appropriate. If a drainage layer is installed in the cover, much of the infiltration will be removed by this system. A soil depth of 12 to 18 in. would be appropriate to support a grass cover. This depth of soil with a drainage layer will not contain excessive quantities of water, so it may be necessary to provide for irrigation during the dry periods to establish and maintain the grass cover.

Biotic barrier. The function of the biotic barrier is to prevent penetration of the moisture barrier by either burrowing animals or roots from deep-rooted plants (trees). This layer has been proposed as another safe guard for the integrity of the moisture barrier. It consists of a coarse gravel that will be effective in stopping such animals as woodchucks and prairie dogs from digging into the fill. Such a barrier will not impede root penetration, so maintenance of the surface will be necessary. There is good reason to suggest that the biotic layer is of little value. Mowing of the area to keep down the growth of trees and large plants will solve the root problem. It will also present a habitat that will discourage burrowing animals from establishing residence at the landfill. Judicious use of rodenticide will add any additional control that may be needed.

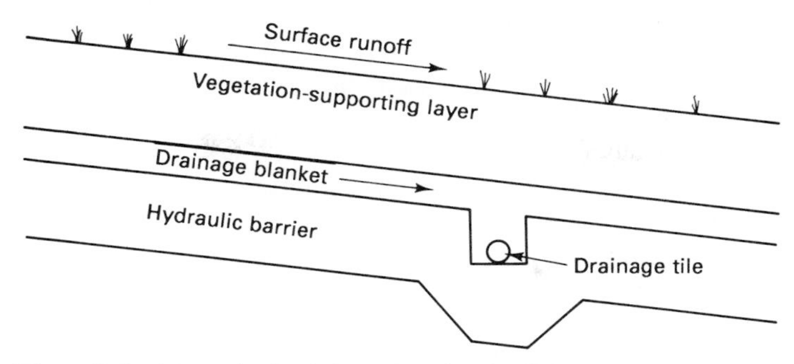

Figure 11.9 Schematic of a drainage layer in a landfill cover.

Drainage layer. This layer will be very important when the landfill is in an area of high precipitation and where evapotranspiration is not an effective mechanism for dissipating the excess water. Its purpose is to remove the water that infiltrates the top layer of the cover. This layer is sloped to a drain line, as shown in Figure 11.9. Water will gradually drain into these pipes and be removed from the cover. The soil used in this layer need only be porous. A coarse relatively uniform sand or fine gravel would be a good candidate material.

The success of this drainage layer depends on maintaining its porosity. As the water from the layers above percolates down through the soil, the fine particles may be transported into the pores of the coarse soil. This is referred to as "piping." These fine particles clog the pores and ruin the open-pore network needed to transport the water. It is advisable to install a "filter" layer above the drainage layer. The filter is a soil of particle size between those of the drainage and surface layers. Ideally, the size gradation will increase in the direction of the drainage layer. It is possible to place more than one layer of filter material, with increasing particle size from top to bottom of the layer. Geotextiles also appear to be suitable for this application.

It is necessary to maintain the slope of the drainage layer as well as the collection pipes. Unfortunately, landfills are noted for settling with time. The additional load of more lifts that are placed in the fill and the vibrations associated with the equipment will cause substantial consolidation of the refuse in the lower levels of the fill. This will occur during the active life of the fill. However, there is a long-term consolidation that is associated with the biodegradation of the organic solids. As these solids are converted to gases and soluble compounds, there is a reduction in the mass of material remaining in the fill. Additional settling will occur as the gas is vented from the site and the soluble solids are removed with the leachate.

If this settling is uniform, it is possible to compensate for it in the design phase. However, uniform settling is not the normal response. The refuse placed in each daily cell may have different characteristics. Some cells may have more moisture than others or may have more nutrients that will increase the microbial activity.

Consequently, the different rates at which the solids are destroyed will influence the settling that can occur. In addition, the components of the refuse placed in the fill may vary from cell to cell. Unless a conscious effort is made to segregate various types of refuse, it is possible to have cells that contain a large fraction of inert material, such as construction and demolition refuse. These cells will behave much differently from the cells containing typical residential refuse.

The drainage system will fail if the differential settling is responsible for the creation of depressions (low spots) in either the drainage layer or the drain pipes. Water from other areas of the cover will concentrate in this depression. This standing water and the associated increase in hydraulic gradient will increase the rate at which water penetrates the soil hydraulic barrier. The concentrated infiltration will cause this portion of the landfill to reach field capacity and start generating leachate. Substantial subsidence will be noted because of the surface depressions that will form, creating surface pools of water. These pools can be eliminated by filling the depression and returning the grade to the original elevation. However, except by excavating to the hydraulic barrier, there is no technique for eliminating the subsurface subsidence. This problem may not be obvious unless there is substantial surface subsidence or very careful attention is paid to the elevations of landfill surface.

There is a strong argument for not installing such a layer in the covers for a sanitary landfill. The added cost of this layer and drain pipes could be used to increase the depth of the hydraulic barrier layer and the surface layer. This will increase the soil reservoir capacity as well as provide additional resistance to the infiltration through the hydraulic barrier. In climates with moderate to light precipitation, this type of cover will control the infiltration to a point where leachate will not be produced. In areas of high precipitation, infiltration will occur and provisions must be made to collect and treat the leachate.

Hydraulic barrier. The hydraulic barrier is instrumental in preventing the infiltration from reaching the refuse. The desired type of soil is a tight clay. Sandy-clay and silty-clay soils are suitable if a clay is not available. The depth of the hydraulic barrier can be adjusted to reflect the permeability of the soil. Two feet of a good clay will substantially eliminate the infiltration if the water is not allowed to pond on the surface and keep the surface layer saturated. If none of these soils is available, relatively impermeable barriers can be created by amending a fine grain soil with bentonite clay, fly ash, or some other soil additive. The cost of importing these additives will be substantial. It is certainly more economical to operate a landfill where these soils are naturally available.

Geomembranes can be used as the hydraulic barrier when suitable soils are not available. They have shortcomings in that they can easily be damaged. A break in the membrane destroys its ability to prevent the water from penetrating the site. It is especially vulnerable in the cover. Significant differential settling can cause elevation differences sufficient to tear the membrane. Also, the potential for mechanical damage during placement and by the refuse material after placement is relatively high unless a significant effort is made to protect the membrane. The

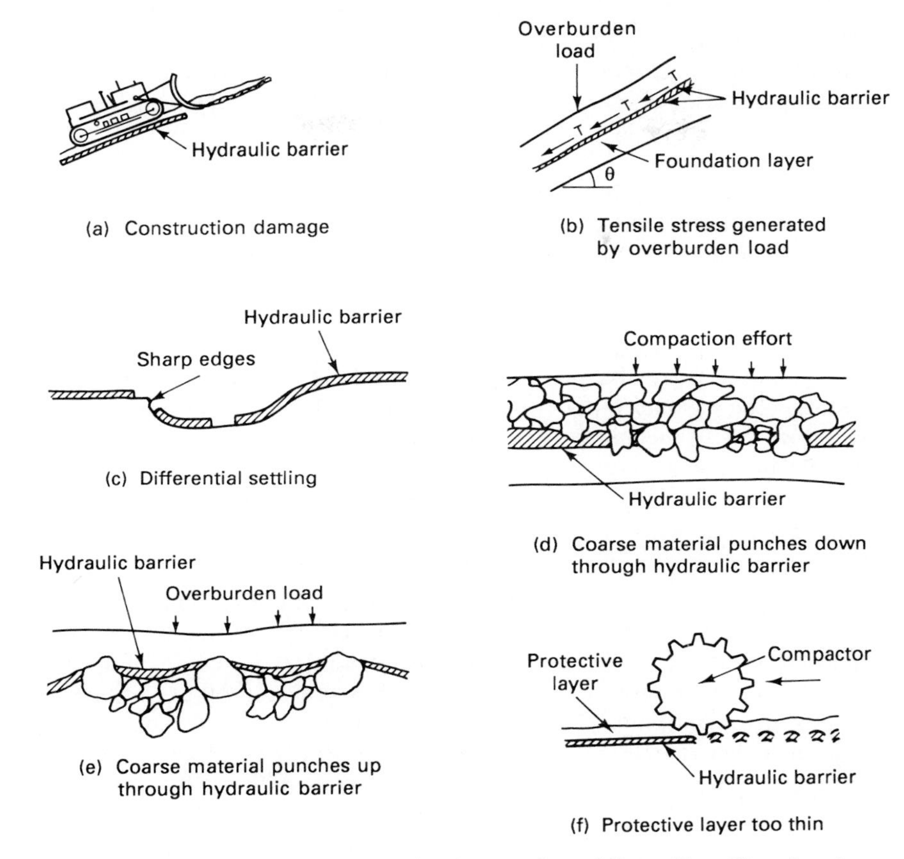

Figure 11.10 Types of mechanically induced geomembrane failures. (From *Waste Age*, Aug. 1987, p. 152.)

typical types of mechanical damage that can occur to the membrane are illustrated in Figure 11.10. Damage may be due to the construction equipment as indicated by Figure 11.10(a) and (f), or due to the manner in which the membrane is placed in the cover as illustrated in the balance of Figure 11.10.

Foundation layer. The foundation layer in Figure 11.8 has special significance when a geomembrane is used. Its role with a soil hydraulic barrier is not clear, and since it is under the geomembrane, it will not be an effective moisture barrier unless this membrane is damaged. It is designed to separate the membrane from the refuse and protect it from damage by the refuse. It can be constructed from any on-site soils that do not contain sharp stones or other objects that would damage the membrane. Another objective of this layer is to provide a "stable" foundation for the membrane, one that will not be subject to subsidence. This is difficult to accomplish because of the long-term biological degradation of the solid wastes and the voids created when the gas is liberated. Subsidence will continue for many years, and frequently at an uneven rate. The differential settling can cause rupture of the membrane and loss of the moisture barrier.

Gas control. Gas will be produced in the landfill as a result of the biodegradation of the organic solids. Soils that prevent infiltration of water will also prevent the gas from escaping. This gas must be released, or it will generate sufficient internal pressure to rupture the restraining membrane or soil moisture barrier. This rupture will provide an avenue for infiltration of water. Consequently, any landfill design requires the installation of a gas collection system. There are several approaches to the design of such a system and they will be discussed later. One such design has a layer of a porous material such as crushed stone placed on top of the refuse. The gas can follow this layer until it reaches a collection well. This will be a rather costly technique for collection of the gas. It probably would not be used except in unusual cases.

Landfill Liner Design

The design of the liner requires many of the same considerations as the design of the cover. The objective is to prevent the movement of the infiltration (leachate) into the groundwater. Figure 11.11 illustrates the components of a liner constructed from soil. The bottom of the landfill is first prepared by excavating to the desired depth. If the bottom soil is a tight clay, it can serve as the liner. If it is a permeable soil, it is then necessary to cover it with a layer of impervious soil that can serve as a moisture barrier. The soil should have a permeability of 10^{-7} cm/sec or less when

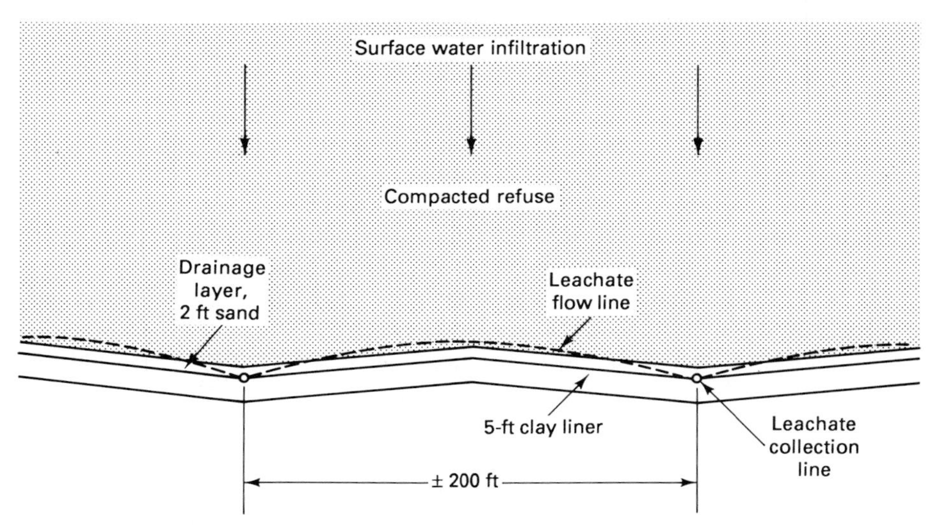

Figure 11.11 Schematic of soil liner for a sanitary landfill.

compacted to its optimum density. Either a clay or an amended sandy or silty clay will meet this criterion. Bentonite can be mixed with the later soils to reduce the permeability to acceptable levels.

However, before the barrier layer is applied, a foundation must be prepared to carry the weight of the imposed load. As mentioned before, a landfill with a depth of 50 ft will have a load of about 2000 lb/ft^2. As the refuse approaches field capacity, this load can double due to the added weight of the water. If an unstable soil is present on the bottom of the fill, it can be deformed as a result of this load, especially if the load is not uniformly applied, and cause a loss of liner integrity. Such soils will have to be excavated prior to the placement of the impervious soil moisture barrier.

The thickness of the soil liner will be dictated by the availability of suitable material. If an impermeable soil is not available, it will be necessary to use on-site material that has a higher proportion of imported bentonite added. The cost of the imported material will be a factor in the final depth of the soil liner. If on-site materials are used, depths up to 5 ft may be used to ensure that leakage will not occur.

The finished grade of the soil moisture barrier is sloped to drain to leachate collection pipes located in the bottom of the fill. These pipes are either plastic or vitrified clay drain tiles. The type of pipe used will depend on the expected load and the ability of the pipe to withstand this load. An additional 2+ ft of a porous sand is placed over the pipes and the moisture barrier. The refuse is then carefully placed on the sand. It is possible to damage the pipes unless care is exercised with the heavy equipment used to place and compact the refuse. The drain tiles are connected to a manifold that conducts the leachate to a wet well for pumping to the leachate treatment and/or disposal system.

When a suitable soil is not available, a synthetic membrane can be used as the moisture barrier. Figure 11.12 illustrates the installation of a geomembrane (synthetic membrane) in a sanitary landfill. The undisturbed soil is graded and compacted to provide a solid foundation for the fill. As with the soil liner, it is necessary to remove any soil pockets that may not be able to carry the superimposed load from the landfill. The graded and compacted base soil is covered with a 12- to 18-in. layer of sand to support the membrane. In an area where it is possible for the groundwater table to rise to the bottom of the fill, an underdrain system may be installed to remove this water and prevent the back pressure from rupturing the liner. This underdrain can also serve for leak detection and may be required under certain conditions.

A variety of synthetic membranes have been proposed for these liners. A 30-mil (0.75-mm) or thicker membrane is usually employed. Each material has different properties. Because of the newness of the use of these liners, there are few historical data on their performance. Many types of polymers have been offered on the market, including a variety of polyethylene polymers, polyvinyl chloride, polyolefin, and rubber fabrics, such as Hypalon, Neoprene, and butyl-cured rubber. Polyethylenes, especially high-density polyethylene (HDPE), have been receiving

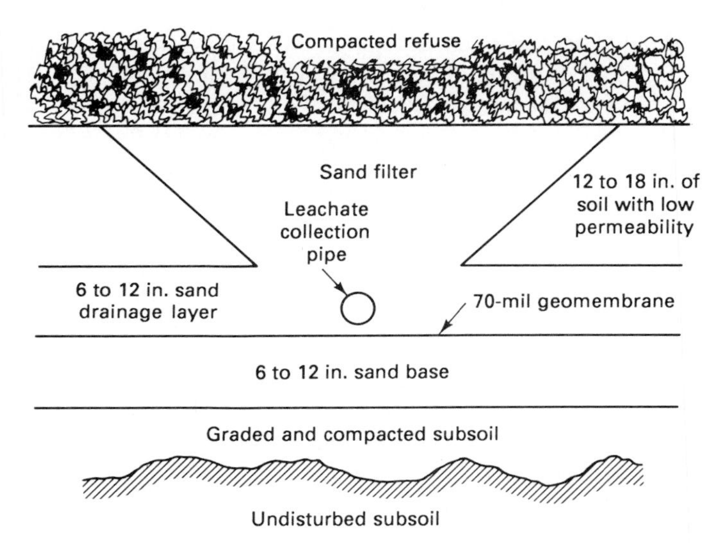

Figure 11.12 Typical installation of a synthetic membrane liner.

the most attention for these applications. They exhibit the properties that are desirable and have a reasonable cost. Properties that are important in the evaluation of these polymers are presented in Table 11.2 together with the recommended testing procedure and recommended minimum values.

After the membrane has been placed and the seams sealed, it is covered with a 6- to 12-in. layer of sand to serve as a protective layer for the membrane. A 12- to 18-in. layer of clay or bentonite amended soil is placed over the sand layer. This soil is compacted to its optimum density. This layer serves as a moisture barrier to direct the leachate to the drainage pipes. The bottom of the fill is sloped toward the drain pipe, which is embedded in the soil layer. The drain pipe also serves to drain any leachate that may accumulate on the membrane.

The complexity of the liner can increase the costs considerably. The more layers placed, the more the construction and material costs. Common excavation and placement of on-site soils will cost about $1.00 to $2.00 per cubic yard. Because of the added care needed in excavating top soil, excavation and placement of on-site top soil will cost between $1.00 and $3.00 per cubic yard. Sand blankets constructed from off-site sources of sand can be expected to cost between $8.00 and $10.00 per cubic yard or more if the haul distance is significant. Any imported soil can be expected to be in the cost range $10.00 per cubic yard or higher. The costs of a synthetic membrane are dependent on a variety of factors ranging from the material type and thickness to the quantity required. In general, the clay liners are less expensive if a good clay is available within a short-haul distance of the site. If imported clay or bentonite admixtures are required, the synthetic membrane liner may be cost-competitive.

Table 11.2 Synthetic liner performance test categories

Test Property	Test Method	Good Minimum Value
Strength category		
Tensile properties	ASTM D638, type IV Dumbbell at 2 in./min	
Tensile strength at yield		2400 psi
Tensile strength at break		4000 psi
Elongation at yield		15%
Elongation at break		700%
Toughness category		
Tear resistance initiation	ASTM D1004 die C	45 lb
Puncture resistance	FTMS 101B method 2031	230 lb
Low temperature brittleness	ASTM D746 procedure B	−94°F
Durability category		
Carbon black percent	ASTM D1603	2.0%
Carbon black dispersion	ASTM D3015	A–1
Accelerated heat aging	ASTM D573, D1349	Negligible strength change after 1 month at 110°C
Chemical resistance category		
Resistance to chemical waste mixtures	EPA method 9090	10% tensile strength change over 120 days
Resistance to pure chemical reagents	ASTM D543	10% tensile strength change over 7 days
Stress cracking resistance category		
Environmental stress crack resistance	ASTM D1693 C condition	1500 hrs

Source: Reprinted with permission from *World Wastes,* 1986 Equipment Catalog. Copyright Communication Channels, Inc., 1986, Atlanta, GA, USA.

ROLE OF MOISTURE IN A SANITARY LANDFILL

The plan for management of the surface and groundwater at a landfill site is an extremely important component of any site development plan. The groundwater problem is generally eliminated during the site selection process. It is not possible to select a site that will not be exposed to precipitation except in some arid regions, such as the desert areas of the southwest and southern California. In fact, most populated areas of the United States are in regions that have significant precipitation. An understanding of the hydrological cycle as it relates to a sanitary landfill is essential in developing a strategy for the management of moisture at a specific site. Because of the effect of climate on this cycle, there is significant variation from site to site in the leachate production potential.

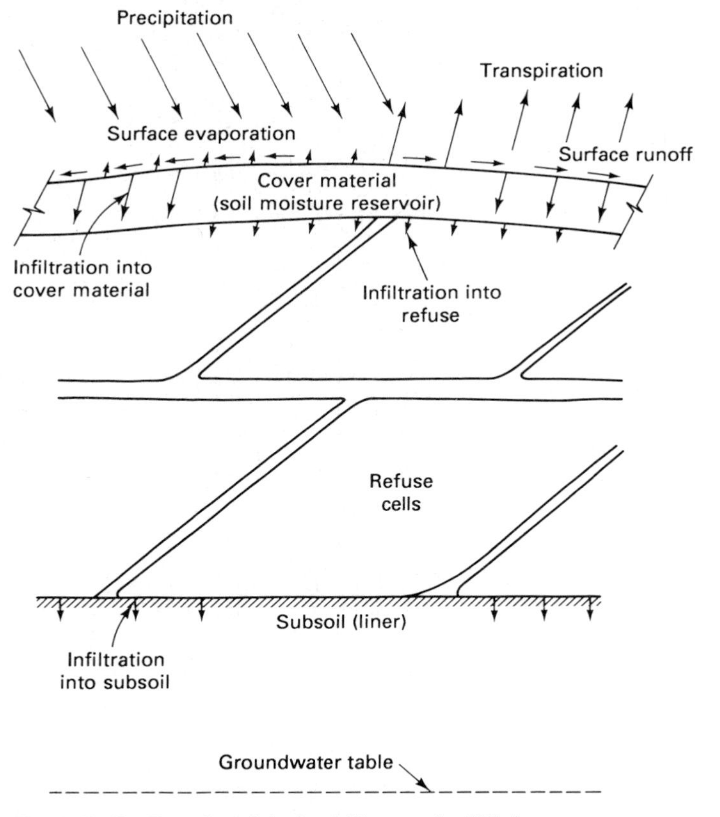

Figure 11.13 Fate of precipitation falling on a landfill site.

Hydrological Cycle for a Sanitary Landfill

Figure 11.13 illustrates the fate of precipitation falling on a landfill site. There are four things that can happen to this precipitation: it can become surface runoff, the water retained on the surface can evaporate, it can infiltrate into the cover material and be extracted by plant transpiration, or it can infiltrate into the refuse, where it may eventually become leachate. Because it is sometimes difficult to separate the surface evaporation from the plant transpiration, these two are frequently combined as one term evapotranspiration.

Infiltration. When there is free water standing on the surface of the landfill, there is an opportunity for this water to percolate into the soil cover. The rate of percolation or infiltration is determined by the characteristics of the soil that is used to cover the site. The water flow can be defined by

$$Q = PIA \tag{4}$$

Table 11.3 Permeability of materials used in landfill construction

Material description	*P* (cm/sec)	*Q* (gpd/ac)
Uniform coarse sand	0.47	4×10^{8}
Uniform fine sand	4.7×10^{-3}	4.4×10^{6}
Silty sand	1.1×10^{-4}	95,700
Uniform silt	5.6×10^{-5}	52,200
Sandy clay	5.6×10^{-6}	5,220
Silty clay	1.0×10^{-6}	957
Clay	1.0×10^{-7}	96
Colloidal clay	1.0×10^{-9}	1

This is the Darcy equation, relating the flow rate (Q) to the permeability of the soil (P), the hydraulic gradient (I), and the surface area (A).

The permeability (hydraulic conductivity) is a laboratory measurement of the rate of flow of water through a column of soil at 60°F and with a hydraulic gradient of 1 ft/ft. The proper selection of the soil that is placed in the cover can have a major effect on the amount of water that can infiltrate into the landfill. Since the soil cover must also absorb some water to maintain a good vegetative cover, more than one type of soil will be required for covering the refuse. The permeability of some common soils is given in Table 11.3.

The significance of the soil permeability is obvious from the amount of water that infiltrates, as shown by the Q values in gallons per day per acre (gpd/ac). A coarse uniform sand passes a lot of water and would be the type of material that would be preferred in the construction of a drainage layer either in a compound cover or in the leachate collection system in the bottom of the fill. Conversely, the amount of water passing through the clay is extremely low. Perhaps a better way to evaluate the clay is to compute the time required for water to penetrate a layer that is 2 ft deep when exposed to a hydraulic gradient of 1 ft/ft. With a P value of 1×10^{-9}, this time is 2000 years. It is not surprising that the goal for the construction of liners and covers for landfills is a permeability of 1×10^{-9}. With a 2-ft depth, a rather extended period is required before any water will pass through the layer. The rate of flow increases if there is a static head of water on the layer, as might be the case with a liner, or a cover if the surface grades are not maintained. A 10-ft static head of water on top of the 2-ft moisture barrier would increase the hydraulic gradient to 6 ft/ft [(10 + 2)/2]. The 2000 years would be reduced to 333 years.

Clearly, the covers and liners are capable of excluding the surface water and retaining water that infiltrates into the fill as long as the integrity of the moisture barrier is maintained. Any significant settling can cause a shift in the soil layer that may cause a break in the barrier. Water can freely pass through this break. It is not difficult to maintain the integrity of the liner if proper attention is paid to the foundation conditions of the underlying soils. A landfill that is 50 to 100 ft high will impose a significant load on the subsoil—2000 to 4000 lb/ft^2. If this is an unstable soil that cannot carry this load, the liner may be expected to fail. Also, the differential settling that occurs in the refuse will make it difficult to maintain the integrity of the cover. Postclosure requirements will call for maintenance of the cover for a period of as long as 30 years.

Evapotranspiration. The water loss associated with the vegetation planted on the final cover of the landfill is often overlooked when evaluating the leachate production potential of a sanitary landfill. A certain amount of the precipitation wets the ground surface as well as the surface of the vegetation. This water evaporates in a short time after the precipitation event. The soil cover serves as a reservoir that captures any of the water that infiltrates the soil. As the water infiltrates, a moisture front moves down through the soil. If the soil layer is deep enough and has sufficient moisture retention capacity, all of the water originating from this precipitation event will be held in the cover. This moisture is the source of water needed for the cover vegetation growth. The water is extracted from the soil and returned to the atmosphere by transpiration. Consequently, little or none of the moisture has an opportunity to reach the refuse.

The quantity of water extracted by the plants depends on the plant species and the climatic conditions. Certain plants have very high water needs and will grow only in wet areas. Other plants have shallow root systems that can extract the moisture from only the top layer of soil. Deep-rooted plants such as trees are not suited to landfills. The roots will not penetrate the refuse because of the adverse environmental conditions: low pH and high salt (dissolved solids) content of the moisture (leachate). Consequently, the selection of plantings for the cover will be important in effectively using this moisture reservoir. The plants should have moderate water needs and a root system that will penetrate the depth of the soil cover.

The transpiration rates are controlled by the factors that affect plant growth. The plants are most active during warm temperatures with good solar radiation. As the temperatures cool, both the evaporation and transpiration rates decrease. There is a potential for significant water loss during the warm summer months. Low humidity and wind will increase the water loss by both mechanisms as long as the water is available. If the cover material is dry, the plants are unable to extract any water. Therefore, excess water must exist in the soil cover before transpiration is effective.

The following information is indicative of the transpiration losses that are possible from various plants.

Trees	6–60 in./yr (cottonwood and willow are at the high end)
Rye grass	18–24 in./yr
Clover	18–24 in./yr
Meadow grass	22–60 in./yr

The various grasses have the ability to extract significant quantities of moisture from the soil. They typically have shallow root zones, and that restricts the transpi-

Table 11.4 Surface runoff potential from land areas

	Runoff (gal/acre)		
Cover (slope)	Sandy loam	Silt loam	Tight clay
Grass			
(0–5%)	16,355 (C = 0.1)	49,000 (C = 0.3)	65,340 (C = 0.4)
(5–10%)	26,136 (C = 0.16)	58,800 (C = 0.36)	89,842 (C = 0.55)
None			
(0–5%)	49,000 (C = 0.16)	81,675 (C = 0.5)	98,000 (C = 0.6)
(5–10%)	63,340 (C = 0.4)	98,000 (C = 0.6)	114,345 (C = 0.7)

Source: Journal of the Water Pollution Control Federation, Vol. 43, No. 10, 1971, pp. 2084–2100.

ration potential. Deeper-rooted broad-leaf plants such as clover are included in the cover vegetation to extract the water from the lower layers of the soil cover.

Surface water runoff. The amount of water that infiltrates is also a function of how long the precipitation remains on the surface of the landfill. If good surface drainage systems are installed, the precipitation is removed from the site rapidly, reducing the time for infiltration to occur. If the soil is impervious and the surface is sloped for drainage, more of the precipitation will be surface runoff. This runoff can be estimated by the rational formula

$$Q = CIA \tag{5}$$

where Q = rate of runoff
C = coefficient of runoff that is determined by the factors above
I = rainfall intensity, in./hr
A = surface area

Table 11.4 shows the total amount of runoff that can be expected for different surface conditions for a rainfall intensity of 1 in./hr for a 6-hr period. The total volume of water associated with this rainfall is 163,550 gallons/acre. The runoff from a tight clay soil with no cover and a slope in excess of 5% would be about 70% of the precipitation. Even a silt-loam soil with a very low slope can be expected to have about a 30% runoff. A small amount of runoff can be very instrumental in the control of the water infiltrating into the refuse. This will be obvious when moisture balance calculations are conducted. The investment in a good surface drainage system is warranted.

Moisture retention capacity. The moisture retention capacity of the refuse and soil must be considered when evaluating the potential for leachate production. Refuse has been found to be able to absorb between 100 and 175 lb of water per 100 lb of refuse.[1] This corresponds to a moisture content between 50 and 64%. The

[1]R. Stone and R. Kahle, *Journal of the Sanitary Engineering Division, ASCE,* Oct. 1972, p. 731.

amount of water retained by the refuse in the landfill will depend on the dry density of the solid waste. Therefore, if the density is 500 lb of dry solids per cubic yard, at 50% moisture the refuse will retain

$$(500 \text{ lb/yd}^3)(1 \text{ yd}^3/27 \text{ ft}^3)(1 \text{ ft}^3/62.4 \text{ lb})(12 \text{ in./ft}) = 3.56 \text{ in./ft}$$

At 800 lb/yd^3 dry density, the water retained by the refuse is

$$(800 \text{ lb/yd}^3)(1 \text{ yd}^3/27 \text{ ft}^3)(1 \text{ ft}^3/62.4 \text{ lb})(12 \text{ in./ft}) = 5.70 \text{ in./ft}$$

This 800 lb/yd^3 represents about the maximum dry weight density that is likely to be experienced in a landfill. It is more common to express the moisture as a volume ratio or in volume per volume units of in./ft. The latter numbers can be related directly to precipitation, which is measured in inches.

The amount of water from infiltration that can be retained by the refuse depends on the moisture content of the refuse when placed in the landfill. A certain amount of water may be incorporated in the refuse due to the precipitation falling while the refuse is placed in the fill. This will add to the moisture content of the refuse. It does not rain every day, so many cells are constructed with no increase in moisture. In fact, significant precipitation is required to exhaust the water retention capacity of the refuse. The following computation shows the amount of additional water that can be absorbed if the refuse has a dry density of 800 lb/yd^3 and an initial moisture content of 25%. The moisture content of the refuse in the fill is assumed to be 50% before the moisture retention capacity of the refuse is exceeded.

$$\text{Initial water} = (800/0.75) - 800 = 267 \text{ lb/yd}^3$$

$$\text{Water absorbed at 50\% moisture} = (800/0.5) - 800 = 800 \text{ lb/yd}^3$$

$$\text{Net water absorbed} = 800 - 267 = 533 \text{ lb/yd}^3$$

$$\text{in./ft} = (533 \text{ lb/yd}^3)(1 \text{ yd}^3/27 \text{ ft}^3)(1 \text{ ft}^3/62.4 \text{ lb})(12 \text{ in./ft}) = 3.80 \text{ in./ft}$$

The moisture retention capacity of soil is also important, especially the soil used in the cover material. There are two characteristics that are important, the field capacity and the permanent wilting percentage. These relationships are shown in Figure 11.14. When precipitation falls on a soil, it will immediately fill the surface pores. Gravity and capillary action will draw the water into the soil. It will continue to fill the pores until the precipitation ceases. A certain depth of soil will be saturated. When the precipitation ceases, the water will continue to be pulled into the soil by gravity. It will reach a certain moisture level that is defined as the field capacity (FC). The field capacity of a soil is determined by subjecting it to a capillary suction head of 100 cm of water. Except for the surface drying, the soil will remain at this moisture level.

If plants are present and the roots extend into the soil of interest, the absorbed moisture will be extracted during plant growth. During growth, the moisture will eventually be reduced to a level at which the plants can no longer extract the water. This is known as the permanent wilting percentage (PWP). The PWP is estimated

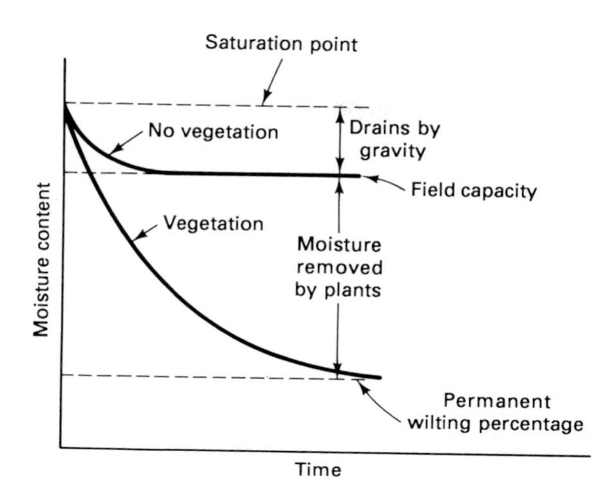

Figure 11.14 Moisture relationships in soils.

by subjecting the soil to an equivalent capillary suction head of 15 atm. The difference between the field capacity and the permanent wilting percentage represents the soil moisture reservoir. When the moisture content is less than the field capacity, any infiltration will be absorbed and later released as transpiration by the plants. If the moisture content exceeds the field capacity, the water will move deeper into the fill. The moisture content will not be reduced below the permanent wilting percentage except by surface drying.

Soils have different moisture retention capacity as evidenced by the data in Table 11.5. The clay, silt, and silt-clay soils will have field capacities in the range 4.0 to 4.5 in./ft. These are the typical soils used in the construction of landfill covers. Soils that are all sand, especially uniform particle sizes, will have a very low field capacity. This is the type of soil that would be used as a drainage layer because of its permeability and inability to hold water.

The soil reservoir capacity is determined by the difference between the two moisture levels. A silt that has a FC = 0.349 volume ratio and a PWP = 0.09 volume ratio is used as an example of this calculation.

$$\text{FC} = (0.349)(12 \text{ in./ft}) = 4.2 \text{ in./ft}$$
$$\text{PWP} = (0.090)(12 \text{ in./ft}) = 1.08 \text{ in./ft}$$
$$\text{Moisture reservoir} = 4.2 \text{ in./ft} - 1.08 \text{ in./ft} = 3.12 \text{ in./ft}$$

Table 11.5 Representative soil field capacities

Soil	Field capacity (in./ft)
Clay	4.5
Silt	4.1
Coarse sand	0.5

The soil has the capacity to store 3.12 in. for each foot of depth. This is the amount of water that can be extracted by the vegetation growing on the cover of the landfill. The time distribution of precipitation relative to this transpiration loss is an important factor in determining the potential of the site for producing leachate. An analysis of this relationship can be conducted to determine the net inflow of precipitation into the refuse.

Moisture Routing

A simplified example of this analysis is shown in Table 11.6. The monthly precipitation and potential evapotranspiration are shown in columns 2 and 3. This information is obtained from records of the area under study. State agencies such as the state water surveys will have historical precipitation data. They may also have evapotranspiration data. The state agricultural service agencies will also be a source of information on the soil–water relationships for specific areas. The difference

Table 11.6 Moisture routing in a sanitary landfill

(1) Month	(2) Precipitation (in.)	(3) Potential evapotranspiration (in.)	(4) Net water (in./month)	(5) CM deficit (in.)	(6) Excess water (Σ in.)
Jan.[a]	3.40	0.5[a]	0	5.10	0
Feb.[a]	2.95	0.5[a]	0	5.10	0
Mar.	4.02	0.62	3.40	1.70	0
Apr.	3.33	1.67	1.66	0.04	0
May	3.53	3.35	0.18	0 (−0.14)	0.14
June	4.07	5.25	−1.18	1.18	0.14
July	4.25	6.10	−1.85	3.03	0.14
Aug.	5.59	5.31	0.28	2.75	0.14
Sept.	3.95	3.74	0.21	2.54	0.14
Oct.	2.91	2.02	0.89	1.65	0.14
Nov.	3.53	0.75	2.78	0 (−1.13)	1.27
Dec.[a]	3.03	0.5[a]	0	0	1.27
Jan.[a]	3.40	0.5[a]	0	0	1.27
Feb.[a]	2.95	0.5[a]	0	0	1.27
Mar.	4.02	0.62	3.40	0	4.67
Apr.	3.33	1.67	1.66	0	6.33
May	3.53	3.35	0.18	0	6.51
June	4.07	5.25	−1.18	1.18	6.51
July	4.25	6.10	−1.85	3.03	6.51
Aug.	5.59	5.31	0.28	2.75	6.51
Sept.	3.95	3.74	0.21	2.54	6.51
Oct.	2.91	2.02	0.89	1.65	6.51
Nov.	3.53	0.75	2.78	0 (−1.13)	7.64
Dec.[a]	3.03	0.5[a]	0	0	7.64

[a]Ground frozen, 100% of precipitation is surface runoff.

between these two numbers represents the amount of water that will be available for withdrawal from storage. When the monthly precipitation exceeds the potential evapotranspiration, the excess water will either become surface runoff or will infiltrate into the soil. When the converse is true, the water is withdrawn from the soil moisture reservoir.

Column 4 is the difference between columns 2 and 3. The difference is treated algebraically to compute the net water produced in in./month. There is a special case associated with the months of December, January, and February. It is assumed that the ground is frozen during these months and no water can percolate into the soil cover. This condition is obviously site specific and may not exist in many parts of the country. The excess water would then be handled in the same manner as with the remaining months. However, the evapotranspiration is temperature dependent and will be considerably reduced during the "winter" months.

It is necessary to define the conditions of the soil as it is placed on the landfill. The initial moisture will be some place between field capacity and permanent wilting percentage. For this example assume the following conditions for the cover material: cover depth = 3.0 ft; field capacity = 4.2 in./ft; permanent wilting percentage = 1.08 in./ft; and initial moisture content = 2.50 in./ft. The following information can be calculated from the foregoing soil conditions:

$$\text{Maximum cover material deficit: } (4.20 - 1.08 \text{ in./ft})(3.0 \text{ ft}) = 9.36 \text{ in.}$$

$$\text{Initial cover material deficit: } (4.20 - 2.50 \text{ in./ft})(3.0 \text{ ft}) = 5.10 \text{ in.}$$

This example assumes that the cover is placed on January 1 and has an initial water deficit of 5.10 in. This is the amount of infiltration that is required to bring the soil in the cover to field capacity. Column 5 indicates the cover material deficit in total inches. Since the soil is frozen during January and February, none of the net water can infiltrate the cover. It either evaporates or becomes surface runoff. However, in March the net 3.4 in. of precipitation will infiltrate the cover and reduce the moisture deficit to 1.7 in. In the following 2 months, an additional 1.66 and 0.18 in. of net precipitation infiltrates the cover. This is sufficient to increase the cover moisture to the field capacity of the soil and have an excess of 0.14 in. of water that will penetrate the refuse.

In the month of June, the water loss by evapotranspiration is greater than the precipitation, so water is removed from the soil reservoir. Consequently, the moisture deficit in the cover material increases to 1.18 in. This continues until August, when there is excess water remaining. The water deficit in the soil decreases to 0 in. in November. At this time, an excess of 1.27 in. of water has reached the refuse. There is no more infiltration in December, when the ground again freezes. As a result of the moisture inflow during the first year, the cover material is at field capacity when the second year starts. The first-year balance is atypical because of the soil moisture level when it was placed. It cannot be used to predict the conditions for the subsequent years. After the first year, the refuse has received 1.27 in. of excess water to add to the refuse moisture content. A wetted front where the refuse has reached field capacity has moved into the refuse. Since this moisture is out of the root zone, it cannot be removed by the vegetation.

The second-year moisture balance represents the normal water accumulation for the precipitation and evapotranspiration conditions. Any significant change in either of these conditions will change the water balance. A wet year will increase the amount of water reaching the refuse, while a dry year will remove all of the moisture from the soil reservoir, allowing for more of the precipitation to be absorbed. In the case presented here, the total net water can be computed by summing columns 2 and 3 except for January, February, and December when the ground is frozen. This is valid as long as the cover material never reaches a maximum deficit.

A continuation of Table 11.6 shows that at the end of 2 years, the excess water reaches 7.64 in. During the second year, 6.37 in. (7.64 − 1.27 = 6.37) of excess water infiltrate into the refuse in the landfill. If the refuse has a moisture-absorbing capacity of 3.80 in./ft, the field capacity moisture front will move into the fill at the rate of 1.68 ft/yr (6.37 in./3.80 in./ft). The time it will take for leachate to be produced will depend on the depth of the refuse and the intermediate cover material. Assume that the refuse is placed in three lifts of 10 ft per lift with 1 ft of intermediate cover between the lifts. It will take 18.5 years [(30 ft/1.68 ft/yr) + (2 ft)(4.20 − 2.50 in./ft)/(6.37 in./yr)] for the leachate to reach the bottom of the landfill. The first year is not included in this calculation and can be ignored since only a limited amount of water reached the refuse. The moisture front will penetrate the refuse only 0.33 ft [1.27 in./(3.8 in./ft)].

The construction of a reasonably impervious cover on the landfill and providing sufficient slope to allow the surface water to drain to a storm drain system can have a significant impact on the amount of water that will infiltrate the refuse. Table 11.7 illustrates the effect if 20% of the precipitation that occurs in the months when the ground is not frozen results in runoff. At the end of the first year, there is an increase in moisture deficit of the soil cover. This suggests that the moisture balance is such that more water is lost by evapotranspiration than infiltrates the cover. During the second year, the moisture content of the cover soil reaches the permanent wilting percentage. At this moisture content, the moisture deficit in the soil is maximum. This cycle will be repeated each year as long as the precipitation remains the same. The refuse will never reach field capacity and the landfill will never produce leachate.

This analysis assumes that the characteristics of the material in the fill remain constant over this time period. This is not necessarily true. The organic material in the fill will be degraded as a result of the microbial action. Much of the mass will be converted to gas and released from the fill. As a result of the imposed load, the density will increase significantly. Also, the loss of the solid material will allow for considerable consolidation of the fill. The moisture-holding capacity of the remaining fill material may increase or decrease, depending on the changes that occur. Most probably it will increase some because of the increased density.

This analysis is based on typical precipitation data for the midwest. In more arid regions, the probability of leachate production from a reasonably designed and

Table 11.7 Effect of runoff on moisture balance in a sanitary landfill

Month	Precipitation-runoff[a] (in.)	Potential evapotranspiration (in.)	Net water (in./month)	CM deficit (in.)	Excess water (Σ in.)
Jan.[b]	0 (3.40)	0.5[b]	0	5.10	0
Feb.[b]	0 (2.95)	0.5[b]	0	5.10	0
Mar.	3.22	0.62	2.60	2.50	0
Apr.	2.66	1.67	0.99	1.49	0
May	2.82	3.35	−0.53	2.02	0
June	3.26	5.25	−1.99	4.01	0
July	3.40	6.10	−2.70	6.71	0
Aug.	4.47	5.31	−0.84	7.55	0
Sept.	3.16	3.74	−0.58	8.13	0
Oct.	2.33	2.02	0.31	8.82	0
Nov.	2.82	0.75	2.07	6.75	0
Dec.[b]	0 (3.03)	0.5[b]	0	6.75	0
Jan.[b]	0 (3.40)	0.5[b]	0	7.75	0
Feb.[b]	0 (2.95)	0.5[b]	0	6.75	0
Mar.	3.22	0.62	2.60	4.05	0
Apr.	2.66	1.67	0.99	3.06	0
May	2.82	3.35	−0.53	3.59	0
June	3.26	5.25	−1.99	5.58	0
July	3.40	6.10	−2.70	8.28	0
Aug.	4.47	5.31	−0.84	9.12	0
Sept.	3.16	3.74	−0.58	9.70 (9.36)	0
Oct.	2.33	2.02	0.31	9.05	0
Nov.	2.82	0.75	2.07	6.98	0
Dec.[b]	0 (3.03)	0.5[b]	0	6.98	0

[a]20% of precipitation becomes runoff.
[b]Ground frozen, 100% of precipitation is surface runoff.

operated landfill is very low. Conversely, areas of high precipitation such as the southeast will experience more problems with leachate and the provisions for leachate collection and treatment represent good design practice.

HELP

A computerized system for conducting the analysis above has been developed for the U.S. EPA by the Waterways Experiment Station of the U.S. Army Corps of Engineers. The Hydrologic Evaluation of Landfill Performance (HELP) program is a quasi-two-dimensional hydrologic model of the water movement across, into, through, and out of a landfill. The model accepts climatologic, soil, and design data and utilizes a solution technique for the effects of surface storage, runoff, infiltration, evapotranspiration, soil moisture storage, and lateral drainage. Figure 11.15 illustrates the landfill system, including various combinations of vegetation, cover soil, waste cells, special drainage layers, and relatively impermeable soil barriers, as well as synthetic membrane covers and liners that are modeled by HELP. The

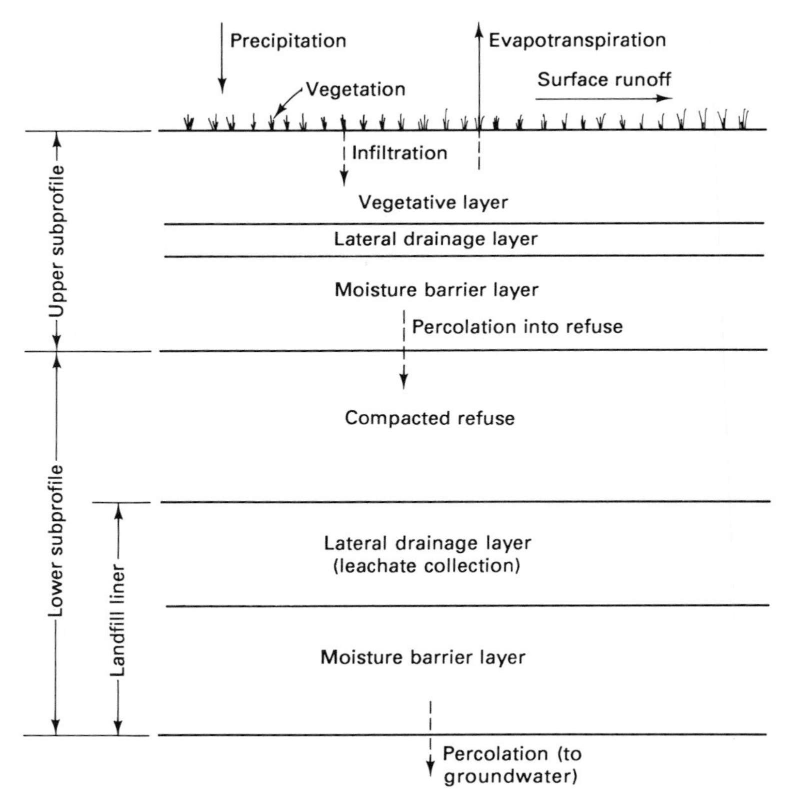

Figure 11.15 Schematic of the components modeled by HELP.

program was developed to facilitate rapid estimation of the amounts of runoff, drainage, and leachate that may be expected from the operation of a wide variety of landfill designs. The model, applicable to open, partially closed, and fully closed sites, is a very useful tool for both designers and permit writers.

The program contains an extensive data base on the characteristics of various soil types, precipitation patterns, evapotranspiration–temperature relationships as well as the necessary mathematical formulations to utilize these data in conducting the analysis of the moisture flow in the landfill. Because of the size of the original program, a minicomputer or possibly one of the newer workstations is required for its use. Version 2.0 can be run on the newer microcomputers.

The transfer of moisture through each layer of a landfill (e.g., the top soil, clay cap, refuse, leachate collection system, and bottom liner) is modeled based on various equations of flow through porous media, evapotranspiration, and runoff. A clear understanding of the limitations and the basic assumptions used by the HELP model is required to interpret the results. The model uses Darcy's equation [equation (4)] for flow through porous media to estimate the infiltration rates through the

Table 11.8 Effect of cover permeability on leachate production

	Existing conditions	Final cover, P (cm/sec)		
		10^{-5}	10^{-6}	10^{-7}
Precipitation (mgd)	0.93	0.93	0.93	0.93
Runoff (mgd)	0.01	0.01	0.10	0.26
Evapotranspiration (mgd)	0.43	0.48	0.60	0.64
Leachate (mgd)	0.47	0.43	0.21	0.01

various layers of the landfill. This equation is valid only when the soil and the refuse moisture content has exceeded the field capacity and free water is present in the pores. In fact, the material should be saturated for this equation to apply. Some of the major limitations are:

1. It may take years for the landfill cap, the refuse, and the landfill liner to approach field capacity. If a final cover is placed on a landfill before the refuse reaches field capacity, the rate of infiltration through the refuse will be greatly reduced and the time required for the refuse to reach field capacity will be greatly extended and may never occur.
2. Leachate may be produced before the entire refuse bed reaches field capacity due to channeling of the water through the refuse.
3. The model does not allow for failure of the moisture barrier. Cracking of the cover can introduce large quantities of water into a limited volume of refuse, resulting in rapid production of leachate.

Even with these limitations, HELP is a valuable tool for analysis of the design of landfills. If the site is maintained so that the system components are functioning, the predictions of this model will be very helpful. An example of the utility of HELP in evaluation of a cover design is an analysis of the leachate production potential of the New York City Fountain Avenue landfill.[2] This 300-acre site operated between 1961 and 1985, receiving up to 3000 tons/day of mixed urban refuse. The calculated quantity of leachate generated under different designs for the final cover is shown in Table 11.8. Existing conditions consisted of 1 ft of intermediate cover of silty sand with a permeability between 10^{-1} and 10^{-3} cm/sec. The fate of the precipitation shown in Table 11.8 results from 18 in. of soil cover with the indicated permeability capped with 6 in. of top soil. The leachate is reduced from about 470,000 gallons/day to only 10,000 gallons/day when the cover has a permeability of 10^{-7} cm/sec.

[2]R. B. Gardner and E. T. Conrad, "The Use of the HELP Model in Evaluating Alternative Leachate Management Plans for Three New York City Landfills," *Proceedings, Waste Tech '86*, National Solid Waste Management Association, Washington, DC, 1986.

STUDY QUESTIONS

1. What are the important topographic and hydrogeologic factors that must be considered when selecting a site for a sanitary landfill?
2. What are the soil classifications according to particle size?
3. What types of soils are suited for the following applications: (a) moisture barrier; (b) drainage barrier; (c) vehicle support; (d) final cover material?
4. What is the importance of permeability when considering soils for landfill applications?
5. What is the objective in the siting of landfills relative to the groundwater table?
6. What is the difference between a groundwater recharge and discharge area relative to landfill sites?
7. Explain how a trench landfill is constructed.
8. Explain the advantage of an area method over a trench method for the construction of a sanitary landfill.
9. What supporting activities/facilities must be included in the development of a landfill site?
10. How does the size of the cell and the number of lifts in a landfill affect the percentage of the fill that is occupied by cover material?
11. What are the primary objectives of the final cover on a sanitary landfill? What types of soils are best suited to meet these objectives?
12. What are some of the problems associated with using a drainage layer in a landfill cover to remove excess filtration?
13. What types of materials can be used as hydraulic barriers in both landfill covers and landfill liners?
14. Explain the possible fate of precipitation that falls on the surface of a landfill.
15. What factors determine the rate of infiltration into the soil?
16. What is the field capacity? What is the permanent wilting percentage?
17. A soil has a field capacity of 0.35 volume ratio and a permanent wilting percentage of 0.08. What is the moisture retention capacity?
18. What is the relationship between climate and evapotranspiration?
19. A refuse with an in-place density of 600 lb of dry solids per cubic yard reaches field capacity when the moisture content reaches 60%. How much water expressed in in./ft of refuse can the landfill hold if the moisture content of the refuse is 30% when it is placed in the landfill?
20. If a sanitary landfill is located in central Illinois, estimate the time required for leachate to be produced under the following conditions:
    (a) Three lifts, 12 ft of refuse per lift, 1 ft of intermediate cover, 3 ft of final cover.

Soil	Refuse
FC, 4.1 in./ft	In-place density, 600 lb/yd^3 dry solids
PWP, 1.0 in./ft	FC at 55% moisture
Initial moisture, 2.5 in./ft	Initial moisture, 25%

(b) Cover is flat, so no surface runoff occurs except when ground is frozen, at which point 100% runoff occurs.

Month	Precipitation (in./month)	Potential evapotranspiration (in./month)
Jan.[a]	2.2	0.7
Feb.[a]	1.9	0.9
Mar.	3.3	1.0
Apr.	3.7	2.3
May	4.1	3.4
June	4.2	4.4
July	3.5	4.6
Aug.	3.1	4.6
Sept.	3.3	3.8
Oct.	2.9	2.2
Nov.	2.7	0.9
Dec.	2.2	0.8

[a]Ground is frozen, so all precipitation will become runoff.

21. Using the same data in Question 20, calculate the time required to produce leachate if the cover is sloped to allow 20% runoff of precipitation when ground is not frozen, at which time 100% runoff occurs.
22. What is HELP?

CHAPTER 12

Control of Leachate and Gas from Sanitary Landfills

LEACHATE CONTROL

The control of leachate can be accomplished in two ways: (1) prevent the precipitation from entering the refuse by an appropriate cover, or (2) collect and treat the leachate produced by the site. If the moisture conditions are such that the formation of leachate can be prevented without an expensive cover, this is the preferred technique for control. However, if excessive precipitation exists and leachate production is likely, it is necessary to provide for its collection and treatment. It is still appropriate to design a cover that will reduce the inflow of water. The treatment of leachate is costly, as will be shown later.

Leachate Collection System

The leachate collection requires the establishment of an underdrain system as illustrated in Figure 11.5. A collector line connects the individual drain lines to a wet well located outside the active fill area. All drain lines are designed for gravity flow, but since only a liquid is being transported, the slopes on these pipes can be very flat, but a slope to the collector line is necessary. A major problem with the flat slopes is the potential subsidence that may occur if the base of the fill is not properly constructed to support the imposed load of the refuse and cover material. If this occurs, the leachate will remain in the fill and have the potential of leaking from the site.

A lift station is required to pump the leachate to the surface for treatment or storage prior to transporting to a disposal site. The size of the wet well must be sufficient to store enough leachate to keep the pump in the lift station from cycling too frequently. The pump size will be dictated by the predicted leachate production rate. A 50-gal/min pump would require a wet well capacity of at least 1000 gallons. Any less volume would require the pump to cycle too frequently. If on-site treatment is employed, the treatment system should have sufficient surge capacity to accommodate the flow rates that occur when the pump operates.

If the projected volume of leachate is low and the decision is made to haul it to a treatment facility, on-site storage is required. Tank trucks with volumes of about 5000 to 8000 gallons can be used to haul the leachate. The size of truck will be determined by the road conditions and allowable weights. It is necessary to have sufficient surface storage capacity to be able to fill the tanker perhaps several times. The haul costs may be more attractive if several loads can be hauled during each cycle. A surface storage volume equal to several tanker volumes may be desirable. A high-capacity transfer pump will be needed to fill the tanker in a reasonable time.

The surface storage tank will have to be freeze resistant in colder climates. The tank can be insulated if aboveground installation is selected, or it can be covered with soil to provide the necessary insulation. The leachate pumped from the landfill will be at moderate temperatures of perhaps 50 to 60°F, so all that is necessary is sufficient insulation to prevent loss of this temperature. The corrosiveness of the leachate will require a material that is corrosion resistant. A concrete or carbon steel tank with an epoxy coating on the interior or fiber-reinforced plastic tank will be satisfactory. Local costs will determine the selection.

Leachate Properties

Before effective processes for treatment of leachate from a land disposal facility can be identified, the characteristics of the leachate must be defined. The composition of leachate from a landfill receiving only residential refuse will be substantially different from the leachate originating from a site that has been receiving industrial solvents. A complete characterization of the leachate is necessary before treatment is attempted. Contaminants may range from simple short-chain organic acids that exert a biochemical oxygen demand (BOD), to high total dissolved solids, to soluble metals (cadmium, iron, lead, mercury, zinc, etc.), to volatile organic compounds, to toxic organic compounds such as chlorinated hydrocarbons, and so on.

The quality of leachate is determined by several factors, including the age of the fill, the biological activity in the fill, the infiltration rate, and the material placed in the fill. Biological decomposition of the organic material will significantly change the chemical environment in the fill. Many solid materials may become soluble as a result of this change and the biological transformation of the organic solids that occurs. As this biological activity continues, the organic material is converted into gases that may migrate from the fill site. The longer the period of biological stabilization, the more stable the organic solids become, resulting in less material being available for leaching. High rates of infiltration will extract the contaminants from the fill at a much faster rate as well as reduce the concentration of contaminants in the leachate.

The contamination can be classified in general categories according to the type of treatment processes that are required for their removal. Biodegradable organic compounds (BOD) represent those compounds that can be stabilized by aerobic or anaerobic biological processes. This represents a wide variety of organic substances, toxic and nontoxic. Nonpolar compounds that can be adsorbed on activated carbon represent the majority of the remaining organic compounds. Some organic contaminants may be in such dilute concentrations that neither biological nor activated carbon treatment processes are practical. If the organic compound is volatile, removal by air stripping may be possible. If not, it may be necessary to employ a strong chemical oxidant such as chlorine, ozone, or peroxide. Finally, the inorganic ions, such as heavy metals, must be removed by chemical precipitation or ion exchange. As a contaminant category, total dissolved solids is not as binding as the above and is usually not a consideration in leachate treatment.

Biological transformations. The sanitary landfill contains a complex microbial ecosystem. In general, the initial moisture content of the refuse is low and the biological activity is substantially inhibited. However, there are pockets of moisture that will support a vigorous population of microorganisms. Therefore, long before the landfill reaches field capacity and starts to produce leachate, a substantial number of biological transformations has occurred. After the initial oxygen that is trapped in the refuse voids is depleted, the aerobic biological stabilization rapidly turns anaerobic. The anaerobic biological transformations discussed in Chapter 10 are the same as those occurring in a sanitary landfill.

The chemical environment changes as a result of this anaerobic decomposition. The acidity produced by the formation of the organic acids and the carbonic acid can decrease the pH to levels of less than 5.0 during the active acid fermentation phase. The acid pH will substantially alter the solubility of metals as illustrated in Table 12.1. Of the metals shown, only ferric iron (Fe^{3+}) and mercury (Hg^{2+}) are insoluble in pure water at moderately acid pH levels. The other metal ions become very soluble at the lower pH.

There is also a significant change in the oxidation-reduction potential (ORP) of the environment. As the ORP becomes negative, the various oxidized species become reduced. The impact of the ORP change on some metal ion solubility in water is also illustrated in Table 12.1. For example, the solubility of iron at pH of 7.0 increases from 1.2×10^{-13} mg/L in the ferric form to 1.2×10^{4} mg/L in the ferrous form. This table is concerned only with the solubility of these ions in pure water since the insoluble form is the hydroxide species. The introduction of other anions can significantly alter the solubilities of metal ions. Table 12.2 shows but a few examples of the solubility of metal salts in water. The presence of the carbonate ion and the sulfide ion significantly alter the solubility of many of the metal cations.

Table 12.1 Metal ion solubility[a] in pure water as a function of pH

Metal ion (mg/L)	pH				
	4.0	5.0	6.0	7.0	8.0
Cd	V.S.	V.S.	V.S.	1200	12
Cu	V.S.	V.S.	14.0	0.14	1.4×10^{-3}
Fe^{2+}	V.S.	V.S.	V.S.	1.2×10^{4}	120
Fe^{3+}	1.2×10^{-4}	1.2×10^{-7}	1.2×10^{-1}	1.2×10^{-13}	1.2×10^{-15}
Pb	V.S.	V.S.	V.S.	2.5×10^{4}	250
Hg	6.0×10^{-1}	6.0×10^{-3}	6.0×10^{-5}	6.0×10^{-7}	6.0×10^{-9}
Ni	V.S.	V.S.	V.S.	6.0×10^{3}	60
Zn	V.S.	V.S.	V.S.	130	1.3

Source: Data from A. J. Bard, *Chemical Equilibrium*, Harper & Row, New York, 1966.
[a]V.S., very soluble

The carbonate and sulfide ions are not normally present in the surface water that infiltrates into the landfill. The carbonate ion forms from the carbon dioxide that dissolves in the water. However, the pH must be above 7.0 before the carbonate ion becomes a significant factor in the solubility of the metal cations. The sulfide ion is also a result of the biological activity. In an anaerobic environment, any sulfate ions present in the leachate can be reduced by sulfate-reducing bacteria. The sulfide produced will react with the metal cations to form insoluble metal sulfide precipitates. Unfortunately, the quantity of sulfur present in most refuse is limited and the available sulfide is soon reacted with the large supply of metal ions. Soluble metals will appear in the acidified leachate.

Also, metal solubility is affected by a variety of other factors, some of which are of biological origin. For example, it has been shown that certain bacteria have the ability to convert some metals into an organic form. This was demonstrated with mercury, when it was shown that the methylmercury found in natural waters resulted from the biological transformation of deposits of mercury metal into the

Table 12.2 Solubility of metal salts in water

Compound	mg/L
CdS	1.1×10^{-9}
CuS	1.6×10^{-13}
$FeCO_3$	0.33
FeS	1.4×10^{-4}
$PbCO_e$	0.07
PbS	1.7×10^{-15}
HgS	2.0×10^{-21}
NiS	5.9×10^{-8}
ZnS	1.0×10^{-7}

Source: A. J. Bard, *Chemical Equilibrium*, Harper & Row, New York, 1966.

mobile methylmercury. Organic compounds that are found in the leachate may also form a complex with the metal cation that is very soluble and mobile. The form of the metal is an important consideration in deciding on appropriate treatment processes.

The biological activity is also responsible for a major change in the form of the solid waste. Much of the solid organic material is biodegradable and the microorganisms are able to convert the insoluble polymers into soluble monomers. Cellulose, a common constituent in paper, yard wastes, and food wastes is a good example of this change. A hydrolysis reaction converts the insoluble cellulose fibers into glucose. Not only is the glucose soluble, but it is also a prime substrate for the fermentative bacteria that are responsible for the formation of organic acids.

The soluble organic material released by the biological action in the fill joins the soluble metals and other ions released by the acid pH to produce the contamination that can be expected in leachate from any sanitary landfill. The concentration of the soluble organic material (BOD and COD) will be determined by the amount of additional stabilization that occurs from methane fermentation. Because of the active methane production that may occur in an established older fill, the leachate will be substantially different from a young fill. The BOD will be significantly lower, the pH will be nearer to neutral (7.0), and the concentration of metal ions will be lower. The dissolved solids and the nonbiodegradable organic compounds, as measured by the difference between the BOD and the COD, will probably remain at high concentrations.

In addition to the metals made soluble by the acid and reduced conditions and the products of the biological fermentation, leachate may contain a variety of other contaminants. These contaminants will depend on the types of materials that were placed in the fill during its life. One may find everything from small quantities of household pesticides to industrial chemicals. This depends on the degree of control exercised over the types of material received at the landfill. Before any attempt can be made to develop a treatment train for leachate from an existing sanitary landfill, it will be essential to know the composition of the leachate.

Leachate composition. The literature contains numerous reports on the characteristics of leachate from sanitary landfills, and each site is unique. Table 2.4 lists the range of values reported for the different contaminants. Table 12.3 presents the average composition of leachate collected each year over a 4-year period from one site. These data show several characteristics of leachate that are important in devising a treatment system. The high BOD values suggest that much of the organic material is biodegradable and can be removed with a biological process. There is, however, a significant amount of biologically inert material as indicated by the difference between the BOD and COD concentrations. The ultimate BOD or COD of the biodegradable components can be estimated at 1.5 times the BOD reported in Table 12.3.

The data in this table illustrate the variability in the yearly average of the concentrations of the various contaminants. The variability between each sample was even more pronounced. Substantial variations occurred in the BOD and COD

Table 12.3 Landfill leachate characteristics

	Concentration (mg/L except pH)			
Contaminant	Year 1	Year 2	Year 3	Year 4
BOD	4,460	13,000	11,359	10,907
COD	11,210	20,032	21,836	18,533
Dissolved solids	11,190	14,154	13,181	13,029
pH	7.1	6.6	7.3	6.9
Alkalinity ($CaCO_3$)	5,685	5,620	4,830	5,404
Hardness ($CaCO_3$)	5,116	4,986	3,135	4,652
Calcium	651	894	725	818
Magnesium	652	454	250	453
Phosphate	3	3	3	3
Kjeldahl-N	1,660	760	611	984
Sulfate	114	683	428	462
Chloride	4,816	4,395	3,101	4,240
Sodium	1,177	1,386	1,457	1,354
Potassium	969	950	968	961
Cadmium	0.04	0.09	0.10	0.09
Chromium	0.16	0.43	0.22	0.28
Copper	0.44	0.39	0.32	0.39
Iron	245	378	176	312
Nickel	0.53	1.98	1.27	1.55
Lead	0.52	0.81	0.45	0.67
Zinc	8.70	31	11	21
Mercury	0.007	0.005	0.011	0.007

Source: J. D. Keenan, R. L. Steiner, and A. A. Fungaroli, "Chemical–Physical Leachate Treatment," *Journal of the Environmental Engineering Division, ASCE,* Vol. 109, No. 6, 1984, pp. 1371–1384.

from the first year. Also, the concentration of some of the metal ions changed substantially as did the hardness and the total dissolved solids. Treatment of wastewater with this variability is difficult unless provisions are made to cope with such variation. A surface storage basin should be considered to equalize the variation in volume and concentration that will likely occur.

Any treatment train will probably include processes for removal of biodegradable and nonbiodegradable organic components. The required reduction in the BOD and COD concentration will be determined by the point of discharge. BOD levels of 200 mg/L will generally be accepted for discharge to a sanitary sewer. For discharge to a stream or lake, the BOD may have to be reduced to 30 mg/L or less. Standards for COD concentrations have not been defined for most discharges. The concern is more for the specific organic compounds contributing to the COD. However, a surface water discharge permit for an effluent containing 4000 to 5000 mg/L COD would probably not be granted. Discharge of a high COD leachate to a sanitary sewer is an option.

The metal ions are of particular concern because of the very strict discharge standards, especially for such metals as cadmium, lead, and mercury. Table 12.4

Table 12.4 State of Illinois water quality standards

	Standard (μg/L)	
Metal	Acute	Chronic
Arsenic	360	190
Cadmium	$e\{1.128[\ln(\text{hardness})] - 2.918\}$	$e\{0.7852[\ln(\text{hardness})] - 3.490\}$
Chromium (total)	16	11
Copper	$e\{0.9422[\ln(\text{hardness})] - 1.464\}$	$e\{0.8545[\ln(\text{hardness})] - 1.455\}$
Iron (total)	1000	
Lead	$e\{1.273[\ln(\text{hardness})] - 1.46\}$	
Mercury	0.5	
Nickel	1000	
Zinc	1000	

Source: Illinois Environmental Register 366, Illinois Pollution Control Board, Chicago, Sept. 1, 1988.

lists the Illinois water quality standards for metal cations commonly found in leachate. Where chronic standards are listed, they refer to the concentration outside of the mixing zone in the receiving water. Otherwise, the acute standard applies at all points in the receiving water. Because these metals are removed in wastewater treatment plants and become part of the sludge, most municipalities and sanitary districts are imposing pretreatment regulations that require metal removal prior to discharge to the sanitary sewer system. This metal-laden sludge may have to be treated as a hazardous waste. Consequently, all leachate will probably require treatment to remove metals prior to discharge to a natural water body or sanitary sewer.

The remaining contaminants shown in Table 12.3 may or may not be required to be removed prior to discharge. The nitrogen and phosphorus levels are very low for the biological treatment of the leachate. Phosphorus will definitely have to be added as a nutrient to obtain an acceptable population of microorganisms. Depending on the biological process, nitrogen may also be limiting. The other cations and anions per se do not present a problem. The composite of these ions as represented by the dissolved solids is in excess of the 1000 mg/L Illinois water quality standards. The hardness, as represented by the calcium and magnesium ions, and the alkalinity can be removed by lime precipitation. However, the chloride salts of sodium and potassium require a more expensive process such as ion exchange or reverse osmosis. If discharge is to a sanitary sewer, the dilution with the sanitary sewage will generally eliminate the need for total dissolved solids removal.

Finally, one must be concerned with the organic compounds that are on the list of priority pollutants. These compounds may fall in one of the foregoing categories and be removed as part of the BOD or COD. Those compounds that are biodegradable will be metabolized in the biological process. Nonbiodegradable organic compounds could be expected to adsorb to the biomass or to the activated carbon that may be used for COD removal. It is possible that some of these compounds may be air stripped in an aerobic biological process. The allowable concentrations of these materials in the effluent are so low that extra care is required to ensure that the required efficiency is obtained. If not, additional treatment will be required to achieve the specified standard.

Leachate Treatment Systems

Reduction of the high BOD common in the leachate from landfills, especially relatively young sites, is accomplished with either an aerobic or anaerobic biological process. In any biological process, the concentration of metals is of concern. The zinc concentration in the leachate shown in Table 12.3 is very high, a level that would be toxic to microorganisms in the biological process. The other metal ions are at marginal levels and may not be toxic. The fate of these metal ions in the biological process must be ascertained.

The pH of either an aerobic or anaerobic system must be 7.0 or above for successful operation. If the acidity is caused by organic acids, very little base will be required to obtain the neutral pH. The acids are metabolized by the organisms, consuming the hydrogen ion that causes the acidity. As the pH increases, the solubility of the metal ions decreases (see Table 12.1) and iron (ferric) and mercury in particular become very insoluble. The presence of alkalinity, the carbonate ion, also precipitates some of the metal ions (see Table 12.2). A careful analysis of the solubility of the metals known to be present in the leachate must be made to determine if pretreatment of the leachate is necessary before attempting to treat it with biological processes.

Anaerobic treatment processes. Any time the BOD concentration is in excess of 1000 to 2000 mg/L, an anaerobic biological process is the likely choice for the initial treatment step. The sulfate present in the leachate is reduced to sulfide in this process. This sulfide plus additional sulfide that may be present in the leachate provide an important anion for the removal of several metal cations (see Table 12.2). The heavy metal removal efficiency in the anaerobic biological process is generally high as long as sufficient sulfur is present to precipitate these metals. If sulfur is inadequate, it can be added to the anaerobic process as a sulfide salt.

A variety of anaerobic processes is available; ranging from a conventional CSTR (completely stirred tank reactor) with or without sludge recycle to a packed-bed anaerobic reactor to a fluidized-bed reactor. Figure 12.1 is a schematic of these reactor types. The CSTR in Figure 12.1(a) is an enclosed gastight tank with a mixing system to keep the contents from stratifying. Because of the relatively low rate of the anaerobic fermentation, this process is normally heated to 35°C. When a CSTR has solids separation and recycle added, it is termed an anaerobic contact process, as shown in Figure 12.1(b). The solids recycle increases the solids retention time (SRT) to a level greater than the hydraulic retention time (HRT). The efficiency of BOD reduction is a function of the biological SRT; consequently, this process can achieve a higher reduction of contamination at a short HRT without the need for heating. The fixed-film reactor [Figure 12.1(c)] can be a packed-bed reactor or a fluidized-bed reactor.

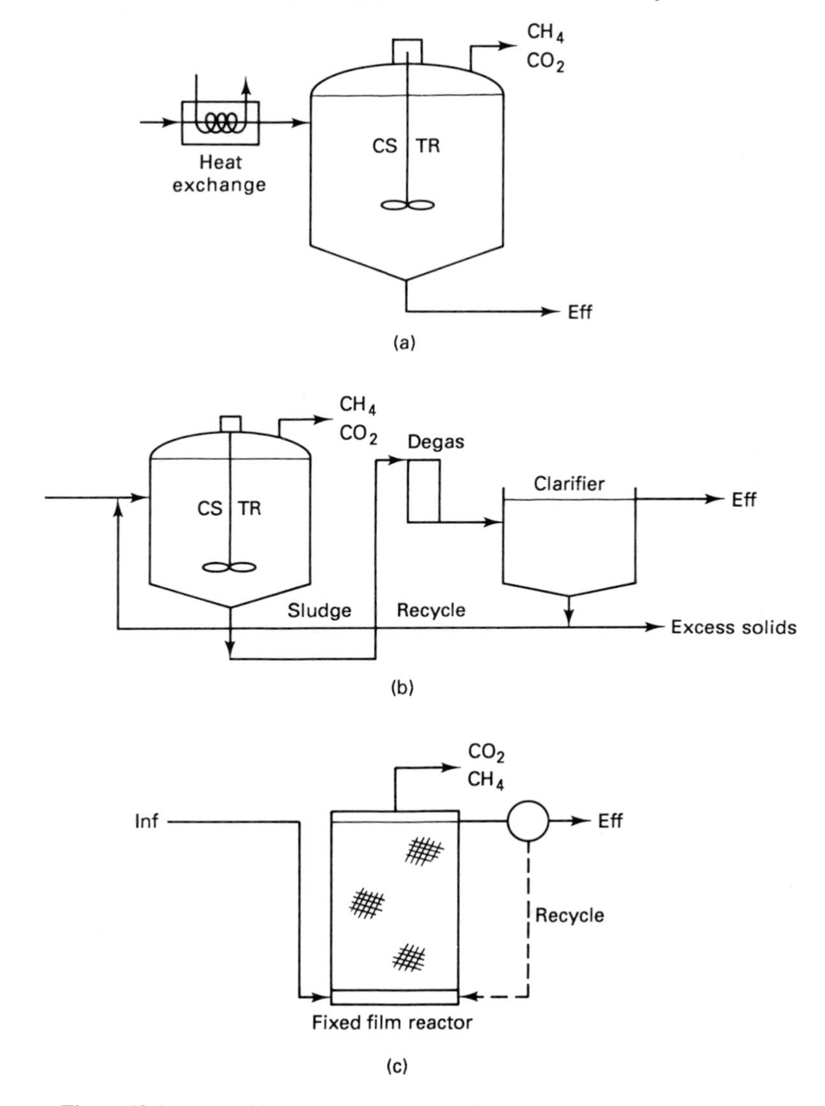

Figure 12.1 Anaerobic processes to consider for treating leachate.

The anaerobic biofilm (fixed-film) reactor offers several advantages as a process for treating leachate. As evidenced by the packed-bed or fluidized-bed reactors, the biofilm reactor is the most effective means of achieving a SRT that is not a direct function of the hydraulic retention time. The reactor contains a packing that has a large specific surface area. The microorganisms attach to the column packing medium. With the proper design considerations, extremely long SRTs can be obtained. Because of this very long SRT, the anaerobic biofilm reactor is capable of producing much better treatment efficiencies than a CSTR or an anaerobic contact process. Since anaerobic methane fermentation is the active biological process, power costs for oxygen transfer are nonexistent.

The anaerobic biofilm reactor is generally operated in an upflow mode. Without effluent recycle the column operates as a plug-flow reactor. If the waste strength is high, process failure can occur because of the multistep nature of the methane fermentation process. This problem is easily solved with effluent recycle. At high recycle rates, the flow regime in the reactor will approach a CSTR. In the fluidized-bed reactor, a bed of small particles is expanded with the aid of effluent recycle. The small particles provide a high specific surface area for a thin dense microbial growth. This permits a high concentration of biomass within the reactor, and because the biofilm is thin, substrate diffusion limitations are minimized. The metal precipitation that occurs in these reactors can pose a severe operating problem. Sludge will accumulate in the voids and metal salts will precipitate on the packing. An operational procedure is necessary to remove the sludge and clean the precipitates from the packing. The fluidization velocity in the fluidized-bed reactor prevents the accumulation of solids. It is also possible to remove a portion of the media from the reactor periodically for cleaning. The cleaned media can be returned to the reactor for reuse.

Many types and strengths of wastewater have been treated with fluidized-bed anaerobic reactors. The data in Table 12.5 show the performance of packed-bed and fluidized-bed reactors for the treatment of still bottoms from an alcohol production plant. The BOD of the stillage is composed of many of the same fermentation products as leachate. The major difference between the two treatment systems was the specific surface area of the media; the packed-bed system was 187 m^{-1} (613

Table 12.5 Performance of anaerobic packed-bed and fluidized-bed reactors in the treatment of high-strength wastewater

	Packed-bed			Fluidized-bed		
Parameter	Run 1	Run 2	Run 3	Run 1	Run 2	Run 3
Retention						
Time (days)	1.8	0.9	0.9	1.8	0.92	0.23
Influent COD	9900	11,500	26,000	48,500	44,200	27,900
Effluent COD	380	910	5,764	900	800	3,400
lb COD						
ft^3-day	0.34	0.81	1.8	1.7	4.6	11.5
% COD						
removed	96	92	78	98	98	88

Source: P. Muhs, *Treatment of an Ethanol Fermentation-Distillation Wastewater with an Anaerobic Fluidized-Bed and Anaerobic Fixed-Film Filters,* unpublished report, Department of Civil Engineering, University of Illinois, Urbana, IL, 1985.

ft^{-1}) and the fluidized-bed system was 4400 m^{-1} (13,114 ft^{-1}). The data in this table clearly illustrate the greater capacity of the fluidized-bed reactor in treating wastewater containing a high concentration of biodegradable organic compounds.

The use of granular activated carbon (GAC) in lieu of an inert material as the medium in an anaerobic fluidized-bed reactor significantly improves treatment efficiencies, especially when the leachate contains nonbiodegradable components. The GAC medium has superior biofilm attachment properties and provides a longer SRT within the reactor. The adsorption capacity of the GAC provides a buffer to smooth fluctuations in feed strength as seen during shock loads to the treatment system. Wastewater containing a high concentration of phenolic compounds plus a variety of other compounds produced in the thermal gasification of coal have been treated successfully by this process. The wastewater was treated in a two-stage system consisting of a GAC fluidized-bed anaerobic reactor followed by an aerobic polishing process. The aerobic process was necessary to oxidize the high level of ammonia and the cyanide and thiocyanate present in the wastewater. It will generally be true that an aerobic process will be required after anaerobic treatment to provide final polishing of the effluent prior to discharge to other than a sanitary sewer system.

A summary of the steady-state performance of the two-stage system when treating this wastewater is presented in Table 12.6. The only organic compounds that were identified in the GAC anaerobic reactor effluent were the five phenolic compounds. However, none of these compounds were detected in the effluent from the aerobic treatment system. Steady-state total and filtered COD removal efficiencies in the GAC reactor were 94 and 95%, while the corresponding removal efficiencies of these constituents across the total treatment system were in excess of 98%. In addition to reducing the COD in the anaerobic reactor effluent, the aerobic

Table 12.6 Steady-state performance of the two-stage treatment system

	Concentration (mg/L)		
Parameter	Influent	Reactor effluent	Final effluent[a]
Total COD	14,658	907	272
Filtered COD		753	208
Cyanide	275	275	30
Thiocyanate	218	218	36
Ammonia-N	1,244	1,244	59
Phenol	1,957	0.05	BDL
o-Cresol	663	8.75	BDL
m- and *p*-Cresol	1,473	17.09	BDL
2,4-Dimethylphenol	67	0.05	BDL
3,5- and 2,3-Dimethylphenol	233	0.30	BDL

Source: J. T. Pfeffer and M. T. Suidan, "Anaerobic–Aerobic Process for Treating Coal Gasification Wastewater," *Proceedings of the Industrial Waste Symposium, 58 Annual Conference*, Water Pollution Control Federation, Alexandria, VA, 1985, pp. 235–249.

[a]BDL, below detection limits of gas chromatography.

polishing system was responsible for removal of ammonia, cyanide, and thiocyanate. The removal efficiencies for these three constituents were 95, 89, and 83%, respectively.

The success of this process in treating wastewater containing toxic compounds is controlled by the ability of the GAC to adsorb the toxic substances. During start-up, significant concentrations of toxic compounds in the wastewater will inhibit the growth of the microorganisms necessary to degrade the COD. This is true even if the toxic compounds are biodegradable. Adsorption of the biodegradable toxic compounds allows the microorganism population to develop to the level required to metabolize these substances at the rate at which they are being fed to the system. This adsorption capacity of the GAC serves as a sink that can adsorb surges of toxic organic compounds and prevent toxicity from occurring. The excess biodegradable material is held until the soluble concentration of these organic compounds is reduced. Desorption can then occur and this adsorbed COD can be metabolized by the biofilm.

The nonbiodegradable adsorbable organic compounds must be adsorbed by the GAC as they are introduced into the reactor. Adsorption is the only mechanism for removal of this material. Adequate adsorption capacity must be available to keep the toxic compounds below their threshold level. As the carbon becomes exhausted, the soluble concentration of these toxic compounds begins to increase. This increase begins to inhibit the biofilm, causing an increase in the concentration of the biodegradable organic compounds. Since many of the biodegradable compounds can be adsorbed, the capacity of the carbon is consumed even more rapidly. As a result, process failure is rapid. This condition clearly shows the importance of maintaining the capacity of the carbon by replacing spent carbon with fresh carbon. This replacement rate is controlled by the rate of application of the nonbiodegradable COD and not the rate of application of total COD. The higher the percentage of biodegradable COD, the lower the rate of carbon replacement.

The efficiency of the biofilm in metabolizing the biodegradable COD is controlled by the SRT of the biological process. Since the biofilm is attached to the carbon, the SRT is determined by the replacement rate of the carbon. If 1% of the carbon is replaced per day, the SRT of the reactor would be 100 days. A 5% per day replacement rate would yield a 20-day SRT. As discussed above, the carbon replacement rate is controlled by the rate at which nonbiodegradable COD is applied to the reactor and the adsorptive capacity of the carbon. The COD application rate is a function of the concentration of nonbiodegradable COD in the influent and the hydraulic retention time of the reactor. As the nonbiodegradable COD loading rate increases, the SRT will decrease.

The presence of certain organic compounds may complicate treatment. If leachate is contaminated with such organic compounds as PCBs or dioxin, treatment processes specific for these materials will be required. The presence of compounds such as organic solvents (benzene, chloroform, methyl chloride, etc.) needs special consideration. Many of these compounds are nonpolar, having a limited solubility in water. In general, such compounds have an affinity for activated carbon and should show a high removal efficiency in the GAC reactor. However, the standards imposed may be impractical to meet when employing the high COD loading commensurate with efficient operation of the anaerobic GAC fluidized-bed reactor. Additional treatment for these compounds may be necessary. They are generally quite volatile and could be expected to be stripped in the aeration basin of an aerobic process. Therefore, it may be possible to remove these residual organic materials from the leachate by air stripping. However, in some air quality districts there are regulations for the control of volatile organic compound (VOC) emissions from wastewater treatment processes.

The variability of leachate requires characterization of each site regarding the BOD, COD, BOD/COD ratio, metals, and organic compounds that are on the priority pollutant list. An anaerobic fluidized-bed GAC reactor offers an efficient system for removing both the biodegradable (BOD) and nonbiodegradable (COD) organic materials. Metals will also be precipitated as metal sulfides if adequate sulfur is available in the leachate or added to the reactor as a sulfide salt. Periodic replacement of a portion of the carbon is necessary to maintain the adsorption capacity of the carbon. If the carbon becomes saturated, the process will fail rapidly. The fluidized-bed system will prevent the accumulation of sludge in the reactor and the carbon replacement will remove the metal precipitates that form on the carbon surface. Carbon adsorption will remove many of the organic priority pollutants, but because of the very low levels permitted in the effluent, subsequent treatment may be required.

Aerobic treatment processes. There are a variety of aerobic treatment processes that one may consider for the treatment of leachate. The biological processes have the ability to stabilize only the BOD. Nonbiodegradable material is removed only incidentally by limited adsorption to the biomass and other solids that are generated in the process. Aerobic processing can render many of the metals insoluble due to the neutral pH, the presence of carbonate ions, and the oxidation state resulting from the aeration. The insoluble metals are incorporated into the biomass and removed along with the sludge that is generated. Therefore, a solids removal mechanism is desirable for the treatment of leachate.

When selecting an aerobic treatment process, one should remember that the volume of leachate produced per day is generally small. Also, the staff support for the operation of this process will be limited unless a significant labor cost is acceptable. Consequently, the simpler the process, the more suitable it will be. One of the most simple processing systems is a multistage lagoon system. Figure 12.2 is a representation of a two-stage lagoon system that can be designed to treat raw leachate or to serve as a polishing unit for an anaerobic process. The first stage is an aerated lagoon designed to have a rapid rate of BOD stabilization. Because the leachate will have a high BOD, it is necessary to transfer oxygen to the lagoon artificially to keep it from becoming anaerobic and creating significant odor problems. The second stage is a stabilization pond that provides additional stabilization and also provides a solids removal mechanism. The solids will deposit on the bot-

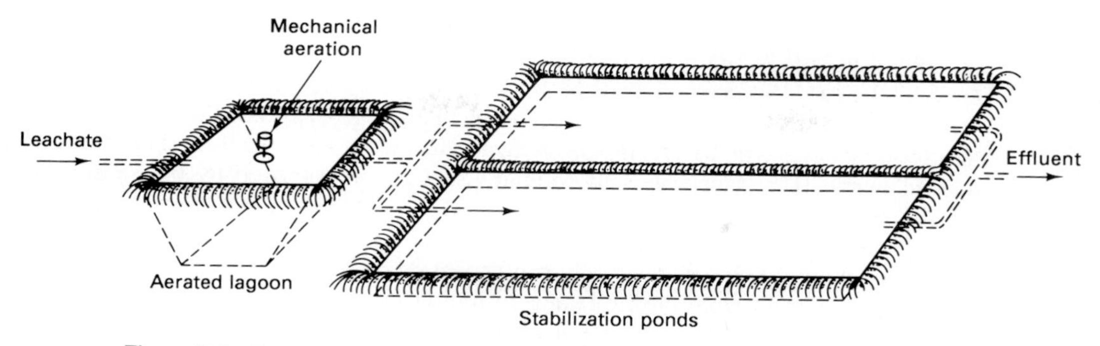

Figure 12.2 Two-stage lagoon system for leachate treatment.

tom of the second-stage ponds. By constructing more than one cell, a cell can be allowed to dewater and dry so that the deposited solids can periodically be removed.

The first-stage lagoon can be analyzed as a CSTR. The BOD stabilization rate is first-order, equal to kS, where S is the BOD in the basin and k is the rate constant in $(\text{days})^{-1}$. A mass balance on the CSTR will yield

$$\frac{S}{S_O} = \frac{1}{1 + k\theta} \tag{1}$$

From this equation and a value for $k = 0.55 \text{ day}^{-1}$, the BOD removal efficiency of the aerated lagoon can be calculated. For example, if the hydraulic retention time (θ) is 15 days, the BOD removal efficiency at 20°C will be 89.2%. The value of k is temperature dependent and a "rule of thumb" is that it changes by a factor of 2 for each 10°C change in temperature. At 10°C, the value of k would be approximately 0.275 day^{-1}. The BOD removal efficiency will be reduced to 80.5% for a 15-day retention time. If the influent BOD (S_o) is large, say 5000 mg/L, the effluent BOD (S) will be high. At 20°C, there will still be 490 mg/L of BOD in the effluent [5000(1 − 0.892)].

The aeration in the aerated lagoon is accomplished with either a surface mechanical aerator or a diffused air system. The surface aerator is preferred because of the ease of installation. These units can be mounted on pontoons and floated in the lagoon. This is especially attractive when the lagoons are constructed earthen basins. Depending on the soil conditions, it may be necessary to line these lagoons to prevent the leachate from infiltrating into the soil. The power required for the aeration can be estimated from the BOD stabilization. Each pound of BOD_5 will require approximately 1.5 lb of oxygen for stabilization. A typical surface aerator will deliver about 2 lb of oxygen for each horsepower-hour of energy used. This relationship will permit sizing of the aerator and calculating the energy costs.

The first-stage lagoon is relatively deep, 12 to 15 ft, depending on the volume required. It does not rely on any static air–water surface interactions. Mechanical energy is used to assist in the transfer of the oxygen from the air. Conversely, the second-stage lagoon relies heavily on the static air–water surface for oxygen transfer. This lagoon will be shallow, 3 to 4 ft deep. Because of the various mechanisms active in this stage, there is no model that adequately describes the BOD reduction. Solids separation will be an important factor in the BOD reduction. The biomass formed in the first stage has a substantial BOD that is removed by the settling in the second stage.

The retention time is one design parameter used to size the lagoon. With a retention time of about 60 days, one can expect an additional 90% reduction of the BOD in the effluent from the first stage. An alternative design is based on a BOD_5 loading rate. A typical rate that will achieve about 90% BOD reduction is 15 lb BOD_5/ac-day. It is relatively easy to reduce the leachate BOD to a level that is acceptable for discharge to most sanitary sewer systems. However, for discharge to a receiving water, a third-stage stabilization pond with an additional retention time of 60 days will probably be required for the lagoon system of treatment. In many areas of the country where the evaporation rate from a free water surface exceeds the precipitation rate, the retention time in the three-stage system may be sufficient so that all of the water is lost by evaporation and there is no effluent.

Expected Life for Biological Activity in a Landfill

One issue that is frequently underestimated when considering the postclosure liability of a landfill and the associated leachate treatment or disposal is the length of time leachate will be produced. As discussed previously, it may be years after closure before the fill reaches field capacity and leachate is produced. The environment that develops in the fill can be expected to retard the biological activity and prolong the release of soluble material. The acid production phase essentially "pickles" the refuse by reducing the pH to inhibitory levels. As these acids are leached, additional organic polymers are hydrolyzed and fermented to more acids. Methane fermentation will also slowly reduce the acidity, allowing for more biodegradation. This continues until all of the biodegradable material is either leached from the site or converted to methane and carbon dioxide.

The time required for this to happen can be very long. The following example can be used to illustrate how long the leachate production may continue. The conditions chosen will actually increase the rate of stabilization of the fill. Assume that the BOD of the leachate is 20,000 mg/L at the start of leachate production. Since there is a gradual reduction in the concentration of BOD with time, one can assume an average BOD of 5000 mg/L over the period of biological activity in the fill. Also, the rate of infiltration will be important in determining the mass of organic material leached. Assume that 24 in./yr of precipitation infiltrates into the site. That will produce 0.653 million gallons (MG) per acre per year of leachate. The rate of BOD leaching is calculated as follows:

$$(0.653 \text{ MG/ac-yr}) (8.34 \text{ lb/gal}) (5000 \text{ ppm}) = 27{,}250 \text{ lb/ac-yr}$$

$$= 13.63 \text{ tons/ac-yr}$$

Now the tons of refuse per acre of fill can be computed given the depth of refuse (assume 30 ft) and the density (assume 500 lb/yd^3 dry solids):

$$\frac{\text{tons of refuse}}{\text{acre}} = (30 \text{ ac-ft})\left(1610 \frac{\text{yd}^3}{\text{ac-ft}}\right)\left(0.25 \frac{\text{ton}}{\text{yd}^3}\right) = 12{,}075 \frac{\text{tons}}{\text{ac}}$$

The question now is to determine the quantity of biodegradable solids in this refuse and their fate. The following assumptions are made:

1. 75% of the dry solids is organic material.
2. 25% of the organic solids is nonbiodegradable.
3. 65% of the organic solids is converted to methane and carbon dioxide.
4. 10% of the organic solids is leached as BOD_5.

$$\text{Leached organic solids} = (12{,}075 \text{ tons/ac}) (0.75) (0.10) = 906 \text{ tons/ac}$$

If the ratio of BOD_5 to leached organic material is 0.75, the BOD_5 available per acre for leaching will be

$$\text{leachable BOD}_5 = (906 \text{ tons/ac}) (0.75) = 679 \text{ tons/ac}$$

The time required for all of the BOD_5 to be leached from the landfill is

$$(679 \text{ tons/ac}) (1 \text{ ac-yr}/13.63 \text{ tons}) = 50 \text{ years}$$

This extended time period has significant implications on the long-term costs of landfill closure. The current requirement for sufficient capital to fund postclosure activities for only 30 years after closure may be totally inadequate. If a municipality owns the facility, it will have to continue funding the postclosure operating costs as long as needed. The funds can be obtained from the general tax revenue or from surcharges on the current refuse disposal system. If a private corporation owns the land, the funds for this long-term operation will be paid out of corporate profits, or if the company is no longer in business, it will become a public cost. The message is to avoid as much as possible the production of leachate.

Discharge of Leachate to Publicly Owned Treatment Works

One option for leachate disposal is to discharge it to a publicly owned treatment works (POTW). This option may be available to landfills that are located in an area that has a large POTW. However, in a smaller community in which the landfill receives all of the community refuse, the impact of the leachate on the POTW can be devastating. The following example illustrates the impact of the direct discharge of leachate to the POTW handling the wastewater from the population that produces the refuse. If one assumes that the landfill has an operating life of 20 years and, as discussed above, the site has 12,075 tons of refuse per acre, the population served can be computed:

$$\text{population served} = \left(12{,}075\ \frac{\text{tons}}{\text{ac}}\right)\left(\frac{1}{20\ \text{yr}}\right)\left(0.6\ \frac{\text{ton refuse}}{\text{person-yr}}\right) = 1006$$

The volume of wastewater produced by this population is 36.7 million gallons per year [(1006)(100 gal/person-day)(365 day/yr) = 36,700,000]. The calculated volume of leachate is 650,000 gal/ac-yr if the infiltration is 24 in./year. The BOD_5 of the influent to the POTW during the initial period of leachate production is calculated by taking a weighted average of the BOD of the leachate (say 10,000 mg/L) and raw sewage (150 mg/L).

$$\text{BOD}\left(1 + \frac{0.65}{36.7}\right) = (150\ \text{mg/L})\ (1) + (10{,}000\ \text{mg/L})\left(\frac{0.65}{36.7}\right)$$

$$\text{BOD}_5 = 320\ \text{mg/L}$$

The contaminants in the leachate are primarily soluble compounds, so the BOD can be expected to pass on to the secondary treatment processes. The influent BOD to the secondary portion of the POTW is typically between 100 and 150 mg/L. With the leachate addition the load on the secondary processes will increase by a factor of 2 to 3. Since the design of this portion of a POTW is based on the BOD loading rate, the secondary process capacity must be doubled or tripled to accept the increased load.

There are numerous other assumptions that can be made for the example above with an equal number of results. However, as long as the quantity of leachate is significant, the impact on the POTW will be significant. A leachate BOD of 10,000 or even 5000 mg/L will increase the BOD of the influent to the POTW to a level that will overload the secondary treatment processes. The volume flow rate of the leachate must be considerably less than 1% of the wastewater flow if this overload is to be prevented.

Most POTWs require pretreatment of any discharge to a sanitary sewer to a level comparable to normal domestic sewage. Under certain conditions, higher BOD concentrations will be accepted, but there is a significant charge for the discharge. The charge is intended to recover the cost of treating the wastewater. The sewer use charge varies for different POTWs. An example is the charge imposed in 1990 by the Urbana and Champaign Sanitary District.[1] The charge based on the volume of sewage was $1.04 per 1000 gallons. In addition, a charge of $104.00 per 1000 lb of BOD is levied. This sewer use charge can substantially increase the operating cost of the landfill.

[1]Ordinance 481, Urbana and Champaign Sanitary District, Urbana, IL, 1990.

In the example discussed above, 27,250 lb/ac-yr of BOD was discharged from the landfill. This calculates to an annual cost for the BOD load of $2834 per acre-year. With a flow of 650,000 gallons/year, the volume charge will be an additional $676 per acre-year. The total cost is $3500 per acre-year. To be meaningful, this annual cost must be related to the 12,075 tons of refuse (dry solids) placed in each acre of the fill. The annual cost for treating the leachate for this example is $0.29 per ton-year of dry solids. How long will the landfill be generating the leachate: 20 years; 50 years? If 20 years, the cost of treating the leachate will be $5.80 per ton of refuse dry solids. If 50 years, the cost will be $14.50 per ton of refuse dry solids. These are significant sums that must be considered in the pricing of the tip fee at the site. All too often, the postclosure costs are grossly underestimated.

FATE OF THE LEACHATE CONTAMINANTS IN THE SUBSOILS

The effect of the leachate on the groundwater after it leaves the landfill is determined by the characteristics of the underlying geological formations. There is a time factor relating to the time of travel (TOT) to a compliance point, such as a water well or monitoring well. Depending on the porosity of the substrata, the TOT can be a few days to a few centuries. While the leachate is passing through these formations, there is a possibility for removal of the contaminants by a variety of biological, chemical, and physical processes. Again, these processes will depend on the characteristics of the formations.

The Effects of Soil Porosity and Hydraulic Gradient on TOT

The Darcy equation can be used to define the flow of water in a saturated soil:

$$Q = PIA \tag{2}$$

where Q = flow rate
P = permeability of the soil
I = hydraulic gradient
A = cross-sectional area of the flow path

Since Q/A is a velocity term, this equation can be expressed as follows, where the velocity of flow (V) in a saturated soil is defined by the soil permeability and the hydraulic gradient:

$$V = PI \tag{3}$$

If a landfill is fortunate to have a thick clay deposit under it that may have a permeability of 10^{-7} cm/sec, the flow velocity will be

$$V = \left(1\,\frac{\text{ft}}{\text{ft}}\right)\left(10^{-7}\,\frac{\text{cm}}{\text{sec}}\right)\left(\frac{60\text{ sec}}{\text{min}}\right)\left(\frac{1440\text{ min}}{\text{day}}\right)\left(\frac{365\text{ days}}{\text{yr}}\right)\left(\frac{1\text{ ft}}{30.5\text{ cm}}\right) = 0.1\,\frac{\text{ft}}{\text{yr}}$$

If water is not ponded on top of this soil, the hydraulic gradient will be 1 ft/ft. It will take 10 years for the leachate to pass through 1 ft of this soil. Unfortunately, there may be layers of more porous material mixed with the clay. Of course, the leachate will flow through the porous material at a much higher rate. The value of placing a 5-ft clay liner in a landfill is obvious. If a leachate collection system is installed, there will not be any ponded water, so the hydraulic gradient will be 1 ft/ft. It will be 50 years before any leachate will pass through a clay layer with this permeability. As discussed below, one can expect significant improvement in the quality of this leachate as it passes through the clay.

The horizontal movement of the leachate through the clay will be almost nonexistent. The hydraulic gradient is very low, so the driving force is minimal. However, if the permeability of the soil is high, the horizontal movement of the leachate can be substantial. For example, a uniform fine sand may have a permeability of 10^{-3} cm/sec. Even with a flat hydraulic gradient of, say, 53 ft/mile, the distance the leachate can move is significant. The same calculation as above will show that the contamination can move about 52 ft/year. Since these landfills are likely to produce leachate for several decades, it is conceivable that the contamination could move 2000 ft or more even with such a flat hydraulic gradient.

In-Situ Biological Activity

The soil is generally well inoculated with bacteria and other microorganisms. The presence of biodegradable material in the leachate will increase this population significantly. The increasing microbial population rapidly depletes any oxygen that may have been present in the soil. Anaerobic fermentation is established. The oxidation-reduction potential is lowered, reducing the oxidized elements that are present in the leachate. A particularly important step is the conversion of sulfate into sulfide. As discussed earlier, many metals are insoluble in the presence of sulfides. This side effect of biological stabilization can be very effective in immobilizing heavy metals, causing them to be held in the soil matrix as insoluble precipitates. These precipitates, as well as the biomass growth, can also plug the soil pores and greatly reduce the permeability.

This biological activity will occur in all soils, pervious and impervious. Even though a sandy soil may allow for the rapid movement of the leachate, only the nonbiodegradable materials are generally found any distance from the landfill. Unfortunately, these substances are frequently the toxic materials that have been deposited in the landfill. If the concentration of these toxic substances is sufficiently high, it is possible to inhibit the biological activity in the soil. Then the biodegradable materials may be found at significant distances from the site.

This biological activity does change the quality of the groundwater. The reduced sulfur may be present as hydrogen sulfide, rotten-egg gas. Some naturally occurring metals such as iron and manganese may become soluble under the reduced conditions. The carbon dioxide produced will depress the pH, which will increase the solubility of calcium carbonate, thereby increasing the hardness of the water. These changes degrade the quality of the groundwater but do not render it unsafe. Technology exists and is frequently used to remove these contaminants, which also occur naturally.

Cation-Exchange Capacity

For aqueous concentrations of metals (cations) in the same order of magnitude as those of competing ions, adsorptive interactions are generally described as ion-exchange reactions. This is a reversible process by which cations and anions are exchanged between the solid and liquid phases. Cation ion exchange is generally the more important when considering the attenuation of ions contained in the leachate. The fractions of soil that are the seats for cation exchange are the organic and mineral particles having effective diameters less than 20 μm. This includes a portion of the silt, all of the clay particles, as well as colloidal organic matter.

Positively charged cations are attracted to surfaces that are negatively charged. The negative charge on the organic particles arises from the ionization of the carboxyl (COOH) and hydroxyl (OH) functional groups present in the organic particle. The charge on the clay arises from two sources: ionization of the hydroxyl groups attached to the silica atoms at the broken edges of the tetrahedral planes, and by isomorphous substitution. The latter is the substitution of "proxy" atoms for the silica and aluminum found in the crystalline structure. Examples are Al^{3+} or Fe^{3+} for Si^{4+} or Mg^{2+} for Al^{3+}. This causes a fairly uniform net negative charge to develop on the clay particles.

Mineral clays in soils are of two general classes. The 2:1 clays are composed of layers consisting of a sheet of alumina interposed between two silica sheets. Montmorillonite, beidellite and vermiculite are 2:1 clays. The 1:1 clays are composed of layers that contain one silica sheet and one alumina sheet. The charge on 2:1 clays results primarily from isomorphous substitution, while the lesser charge on the 1:1 clays is a result of the ionization of the hydroxyl groups at the broken edges of the plates.

The negative charge that develops on the organic and mineral colloids is neutralized by cations attracted to the surface of these colloids. Numerous cations, including calcium, magnesium, potassium, sodium, ammonium, aluminum, iron, and hydrogen, are adsorbed to the charged sites with varying degrees of tenacity, depending on the cation charge and hydrated radii. As a general rule, di- and trivalent cations are more strongly bound to the colloid than are monovalent cations. Also, the greater the degree of hydration of an ion, the less tightly it will be held.

The cation-exchange capacity (CEC) of a soil will be affected by the nature and the amount of mineral and organic colloid present. Soils with large amounts of clay and organic matter will have a much higher CEC than that of sandy soils. Also, soils with 2:1 clays will have a higher CEC than that of soils with 1:1 clays. Typical values for CEC are 100 to 200 mEq/100 g for organic colloids, 40 to 80 mEq/100 g for 2:1 clays, and 10 to 20 mEq/100 g for 1:1 clays. The pH also affects the CEC of all colloids. For example, a soil humus with a CEC of 150 to 300 mEq/100 g at a pH of 7.0 will see this value drop to 10 to 20 at a pH of 4.0. At these lower pH levels, the H^+ concentration is sufficiently high to compete for the exchange sites. These values for the CEC indicate the quantity of metal that can be removed by soils. If the CEC for iron is 100 mEq/100 g, the soil would be able to remove about 2.8 g of iron per 100 g of soil. Considering the mass of soil contained in a 5-ft cover, one can calculate the amount of metal retained in the liner.

Adsorption Properties of Soils

Many soils exhibit a significant capacity for surface adsorption of organic materials from water passing through the soil matrix. The organic carbon content of the soil has been observed to be a common parameter that can be used to define the adsorptive capacity of soils for specific organic compounds.[2] This relationship can be expressed as a partition coefficient (K_{oc}):

$$K_{oc} = \frac{\mu\text{g chemical/g organic carbon}}{\mu\text{g chemical/g water}} \tag{4}$$

The organic matter in most soils is intimately bound to clay as a clay–metal–organic complex. As a result, there are two major types of adsorbing surfaces available to a chemical: clay-organic and clay (mineral). The relative contribution of the organic and inorganic surface areas depends on the extent to which the clay is coated with organic matter. Comparative studies between known clay minerals and organic soils suggest that most, but not all, organic molecules have a greater affinity for organic surfaces than mineral surfaces. However, the influence of clay on organic chemical adsorption can be significant, especially in soils where the organic matter is below 1%.

Table 12.7 presents empirical regression equations determined for some classes of organic chemicals. These values for K_{oc} are correlated with the solubility

Table 12.7 Regression Equations for the Estimation of K_{oc}

Equation	Chemical classes
$\log K_{oc} = -0.557 \log S + 4.277$ (S in μmol/L)	Chlorinated hydrocarbons
$\log K_{oc} = -0.55 \log S + 3.64$ (S in mg/L)	Variety of pesticides
$\log K_{oc} = -0.54 \log S + 0.44$ (S in mole fraction)	Aromatic and polynuclear aromatics

Source: James Dragun, *Hazardous Materials Control,* Sept./Oct. 1988, p. 24.

[2]James Dragun, *Hazardous Materials Control,* Sept./Oct. 1988, p. 24.

of the organic material in water. Very strong correlations have also been found between K_{oc} and the octanol–water coefficient (K_{ow}).

The importance of this mechanism in control of the typical quantities of toxic organic compounds that may be observed in the leachate from a landfill receiving only residential refuse can be seen with the following calculation. For example, 1,1,2-trichloroethane (TCE), a chlorinated hydrocarbon has a solubility of 4.5 g/L or 33,700 μmol/L. From the equation in Table 12.7 for chlorinated hydrocarbons, the K_{oc} is calculated to be 57.0. If the concentration of TCE in the leachate is 1 mg/L, the TCE retained per gram of organic carbon in the soil can be computed from equation (4) to be 57.0 μg TCE/g organic carbon. A typical organic-clay soil may contain as much as 5% by weight of organic carbon. The quantity of TCE adsorbed by this soil would be 2.85 μg/g of soil. A soil with a specific gravity of 2.0 will contain 124.8 lb of soil per ft^3. The total amount of TCE adsorbed by this soil will be 0.16 g/ft^3. Consider a liner that is 5 ft deep; how much leachate will need to pass through this liner to exhaust its capacity to adsorb TCE if the TCE is at a concentration of 1 mg/L?

$$\text{Leachate volume} = \frac{(5\ \text{ft}^3/\text{ft}^2)\ (0.16\ \text{g/ft}^3)}{10^{-3}\ \text{g/L}} = 800\ \text{L/ft}^2$$

This is equivalent to infiltration of 340 in. of precipitation through the landfill.

Consider the time required to obtain this much leachate if the cover prevented 90% of the precipitation from infiltrating the landfill and the leachate collection system was 90% effective in capturing the leachate. Under the assumptions above, an area receiving about 31 in./yr of precipitation would have 0.3 in./yr of leachate escaping the landfill. It would require about 1000 years for enough leachate to pass through the landfill to exhaust the adsorptive capacity of the clay liner.

The scenario above is subject to many assumptions, but the same conclusion is obtained regardless. A properly designed, constructed, and maintained landfill will provide almost a certain guarantee that groundwater pollution will not occur as long as the refuse is restricted to the normal urban solid waste, and wastes from toxic and hazardous materials generators are not accepted.

LANDFILL GAS CONTROL/RECOVERY

The gas produced by the decomposition of the refuse placed in a landfill has a bad history. There are several recorded instances of explosions in structures built on or near a landfill. Several of these explosions have resulted in deaths. In addition to this hazard, the gases have been associated with intense odor problems. Release of hydrogen sulfide plus a variety of decomposition products such as volatile amines create severe odor problems for neighbors of a landfill. The sanitary landfill has been identified as a significant source of hydrocarbon and carbon dioxide emissions contributing to the concern for local and global air quality. Consequently, control of the gases generated by the landfills is receiving more attention. Both the federal and state regulatory agencies are increasing the requirements for landfill gas control.

There are two potential sources of gases from the landfill: gases produced by the decomposition of the organic solids in the refuse, and volatilization of volatile organic compounds (VOCs) such as organic solvents that have been placed in the landfill. The former will continue to be a problem, while the latter will be greatly reduced as a result of the ban on the disposal of hazardous wastes in sanitary landfills. Existing sites may continue to generate VOCs if they were accepted into the site. There will always be a very limited amount of VOCs, due to the household chemicals that are part of the urban waste stream.

When the refuse is placed in the fill under conditions that support biological activity, the oxygen present in the refuse voids is rapidly used by the aerobic microorganisms to metabolize the biodegradable solids. Carbon dioxide is produced by this reaction. In a short time, the gas composition is changed by replacing the oxygen with carbon dioxide so that the gas is about 20% carbon dioxide and 80% nitrogen. The initial anaerobic decomposition that develops when the oxygen is exhausted is fermentative, producing additional carbon dioxide. The added volume of CO_2 increases the internal pressure of the fill and displaces the nitrogen by forcing it from the fill. The gas will be composed primarily of CO_2, with a small percentage of hydrogen that results from the acetogenic reactions. This transient gas composition is eventually replaced with an equilibrium composition that develops as the methane fermentation becomes active. Since the majority of the solid waste is carbohydrate, the gas composition will approach 50% CH_4 and 50% CO_2. This is the gas composition that is considered when evaluating the control/recovery of landfill gas. Contamination with nitrogen and oxygen may occur in a site that is not well contained. The landfill will "breathe" as a result of the changes in atmospheric pressures. During high-pressure periods, air will flow into the fill as the void volume in the fill is compressed. During low pressure, the converse is true. In a fill that has a significant gas production rate and is reasonably tight, this contamination will not be noticeable.

Gas Migration in Soil

Normally, the gas will move from the fill by molecular diffusion through the soil cover and the adjacent soil. If the internal pressure of the fill is above atmospheric but not sufficient to rupture the cover, the pressure differential will increase the rate of diffusion by increasing the partial pressure of the CH_4 and CO_2. In all cases the gas will move along the path of least resistance. If the soil in the liner is very tight, the gas will not move through it easily. If a tight cover exists that has a significant resistance to gas diffusion, the internal gas pressure may increase sufficiently to lift the soil or synthetic membrane, causing a rupture to occur. When this occurs, the gas is flowing through a free opening and is driven by a pressure differential only. Molecular diffusion no longer applies.

Fick's law defines the molecular diffusion as a function of the concentration gradient and a property of the gas defined as diffusivity. Gas concentrations are expressed as lb-moles per cubic foot. The gradient includes a length term to define the distance over which the concentration changes. The definition of this length has a significant impact on the manner in which gas diffuses from the site, and results in two different equations for gas diffusion. The cover has a fixed dimension and the concentration of CH_4 and CO_2 above the fill will remain constant at the typical values for the atmosphere. Therefore, the following equation can be used to define the rate of gas diffusion through the cover material:[3]

$$q = \frac{(C_{\text{fill}} - C_{\text{air}})(D_p)}{L} \tag{5}$$

where

q = net velocity or transfer rate, lb-mol/ft²-day

C_{fill} = lb-mol/ft³ of gas in the fill

C_{air} = lb-mol/ft³ of gas in the atmosphere

L = depth of landfill soil cover, ft

D_p = diffusivity of gas in soil, ft²/day

The diffusivity of gas through the soil (D_p) is a function of the soil porosity (p = ratio of the volume of pores to total soil volume) and the gas diffusivity in free space (D_0):

$$D_p = 0.66pD_0 \tag{6}$$

The gas diffusion through the bottom and the sides of the landfill is modeled as a point source since the gas diffuses radially in all directions except through the cover. Also, the concentration of the gas in the soil pores changes with time until it reaches the composition of the gas in the fill. Therefore, the concentration gradient is not constant, but time dependent. The following equation describes the rate of gas diffusion horizontally and downward from the fill:

$$q = C_0\left(\frac{D_p}{\pi t}\right)^{1/2} \tag{7}$$

This equation assumes that similar soils surround the fill. C_0 is the concentration of the gas at the soil–refuse interface. As the value of t increases, the diffusion rate decreases as the square root of time. Examination of these equations will show that assuming the same porosity of soil, the gas diffusion will be much greater through the cover.

[3]*In-Situ Investigation of Movements of Gas Produced from Decomposing Refuse,* California Water Quality Control Board Publication 35, 1967.

Gas Production Potential

The gas production potential can be determined from the quantity of solid waste placed in the landfill. If the biodegradable fraction of the refuse is assumed to be primarily cellulose, the stoichiometric volume of gas generated per pound of solids destroyed is 13.3 scf. Using the example presented on page 273, there are 12,075 tons of dry solids per acre in a 30-ft-deep landfill. If 65% of these solids are converted to CH_4 and CO_2, the maximum gas that can be generated by the methane fermentation process is

$$(12{,}075 \text{ tons})\ (2000 \text{ lb/ton})\ (0.65)\ (13.3 \text{ scf/lb}) = 209 \times 10^6 \text{ scf/ac}$$

The unknown in this analysis is the rate at which the gas will be produced. There are many factors that determine this rate. Refuse is known to be deficient in nitrogen and phosphorus. These nutrients are essential to the growth of the microorganisms necessary for the fermentation reactions. A low moisture content will also retard the bacterial activity. Whenever the moisture content is less than 75%, biological activity is a function of the moisture level. A low temperature will also reduce the rate of activity. While the average environmental conditions in the fill may be detrimental to a high rate of bacterial activity, there may be pockets of the fill (garbage and grass clippings) where the moisture and nutrient levels are satisfactory for good bacterial growth. As the refuse in these pockets is stabilized, the moisture and the nutrients become partially available for biological activity in more of the refuse. It is not practical to attempt to add the required nutrients and moisture. It has been proposed that sewage sludge be deposited in the landfill to enhance the growth conditions for the fermentative bacteria. Many landfills do accept sewage sludge in various states of dewatering. There are some implications of accepting the extra water that can be expected to exacerbate the leachate production.

There is little good information on the rate at which gas is generated under the typical conditions encountered in a landfill. Numerous laboratory lysimeter studies have indicated that the refuse could be stabilized in a matter of months. These are ideal conditions that are never encountered in a full-scale landfill. The rates of stabilization observed in the field have been much slower. A large-scale test facility was constructed at the Mountain View, California, landfill site to obtain data on the gas production rates from refuse that was placed in the test cells with each cell receiving different treatments. No discernible difference in the treatments was observed, due in part to faulty monitoring systems. The baseline conditions did produce some indication of the production rates that could be expected. The first-year production rate was low but increased substantially for the next 5 years. The rate then decreased to a lower rate, which might be expected for an extended period (10 to 20 years). These production rates were as follows:

Year 1	0.20 scf/lb dry solids
Year 2–6	0.50 scf/lb dry solids
Year 7–?	0.25 scf/lb dry solids

Table 12.8 Gas production potential

	Production [1000 scf/day (Mcf/day)] for operating year:										
Year	1	2	3	4	5	6	7	8	9	10	ΣMcf/day
1	171										171
2	427	171									598
3	|	427	171								1025
4	|	|	427	171							1452
5	|	|	|	427	171						1879
6	|	|	|	|	427	171					2306
7	214	|	|	|	|	427	171				2520
8	|	214	|	|	|	|	427	171			2734
9	|	|	214	|	|	|	|	427	171		2948
10	|	|	|	214	|	|	|	|	427	171	3162
11	|	|	|	|	214	|	|	|	|	427	3162
12	|	|	|	|	|	214	|	|	|	|	3162
13	|	|	|	|	|	|	214	|	|	|	3162
14	|	|	|	|	|	|	|	214	|	|	3162
15	|	|	|	|	|	|	|	|	214	|	3162
16	|	|	|	|	|	|	|	|	|	214	3162
17	0	|	|	|	|	|	|	|	|	|	2948
18		0	|	|	|	|	|	|	|	|	2734
19			0	|	|	|	|	|	|	|	2520
20				0	|	|	|	|	|	|	2306

These rates can be used to estimate the gas production potential for a landfill. Assume that sections of the fill are completed annually and a gas recovery system is installed when each section is completed. Also assume that the fill receives 800 tons/day (600 tons of dry solids) of refuse. With the figures above, Table 12.8 can be constructed. It is assumed that the fill has a 10-year operating life. The gas produced by each section of the fill as it is completed annually is listed in the columns. The total gas production from the fill each year after the start is computed by summing the rows. After some time in the fill, the refuse will cease to produce gas, so the total gas production will peak when the initially filled material ceases to produce measurable quantities of gas.

Assuming that 75% of the solids are organic solids and that 75% of these solids are biodegradable, the theoretical gas production for each year's worth of refuse placed in the fill would be 2.3×10^9 scf. The total gas produced in each column can be determined by summing the daily gas production for each of the years indicated. If this is done for year 1, the total is 1.6×10^9 scf. This represents about 70% of the theoretical gas production in this period. Additional gas may be produced for a number of years, but the production rate will be substantially lower than the rates used to construct Table 12.9. It has been observed in the field that the gas will continue to be produced for many years. The active life of the fill for gas production is difficult to predict because of the many factors that determine the rate at which the gas is produced. By monitoring the gas production during the first

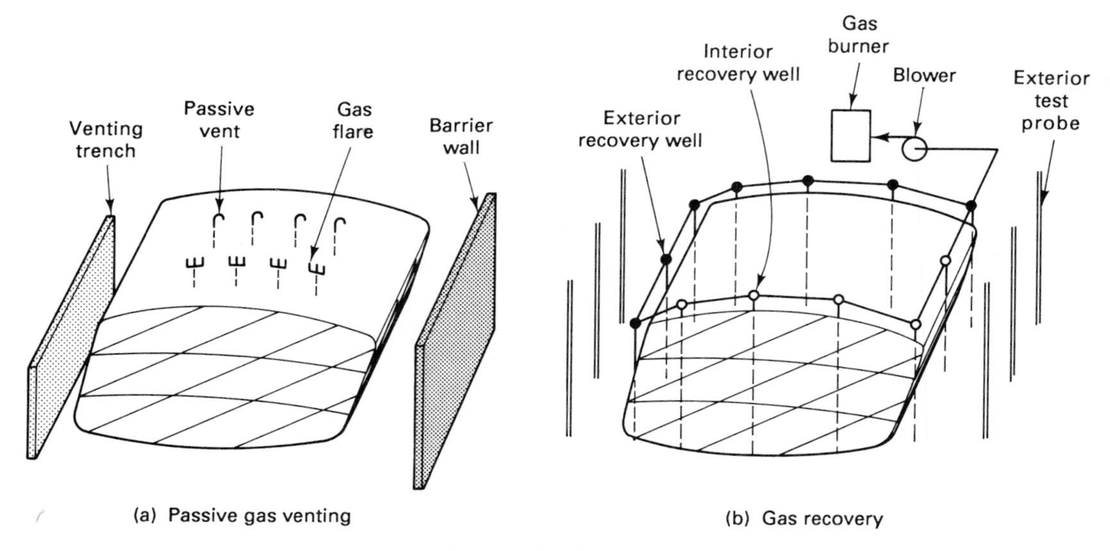

Figure 12.3 Schemes for the control of gas migration.

few years and comparing it with the theoretical gas production, an estimate can be made of how long gas production can be expected.

Gas Recovery/Migration Control

The requirements for control of gas migration and, in some areas, the discharge of hydrocarbons (CH_4) to the atmosphere generally require installation of some type of landfill gas collection system. Once the collection system is installed, it may be economically desirable to use the gas as an energy source. The installation of the gas control system will account for a significant portion of the cost. The additional cost for gas utilization is easily recovered from the sale of the energy.

Figure 12.3(a) illustrates a passive system for control of gas migration. The fill is surrounded by either an impervious barrier of a tight clay or a slurry trench, or it has a synthetic membrane liner that prevents the flow of gas from the fill. Another option is to install a venting trench around the periphery of the fill. This trench is filled with crushed stone that allows the gas moving horizontally to escape by diffusing up through the trench to the atmosphere. In addition, wells or vents may be installed on the surface of the fill to allow the gas to escape. These wells may be vented to the atmosphere or to a gas flare that burns the gas as it is liberated. Local permit conditions will dictate the system to be installed. Figure 12.3(b) illustrates the installation of gas recovery wells on the periphery of the fill. The gas is collected by a header pipe and transported to a gas utilization system.

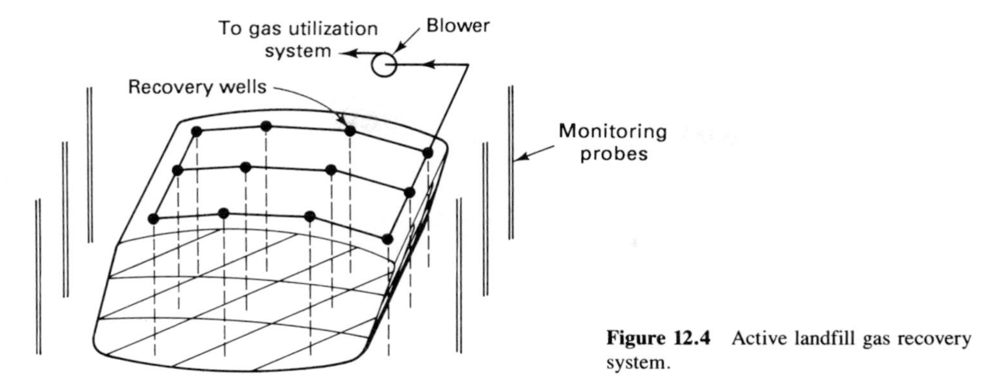

Figure 12.4 Active landfill gas recovery system.

If gas recovery is the prime objective, additional wells will be installed in the fill to assist in more efficient recovery of the gas. Figure 12.4 illustrates the general approach to the installation of the gas recovery wells. The well extends to about 80% of the fill depth. Spacing will vary, depending on the construction of the fill. If a tight soil is used for the intermediate and daily cover, it will be necessary for each cell to be penetrated by a well. Spacings will typically be on a grid with distances between wells varying from about 100 to 200 ft. The individual wells are collected to a header pipe that conveys the gas to a central point for processing.

Figure 12.5 illustrates an approach used by the Los Angeles County Sanitation Districts that is a combination of wells and collection trenches. As the landfill lifts are completed, a trench is constructed in the refuse directly underneath the soil

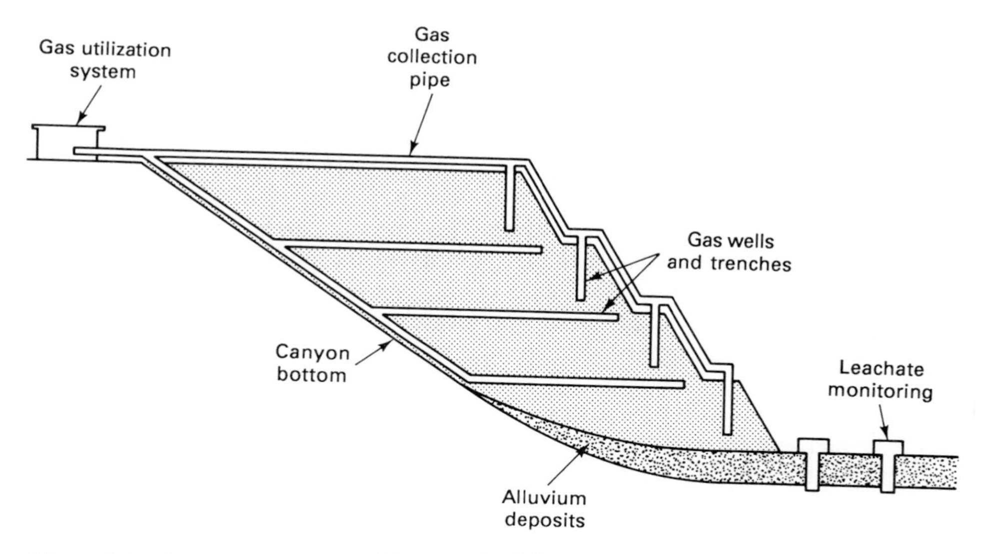

Figure 12.5 Gas recovery system used in canyon landfills.

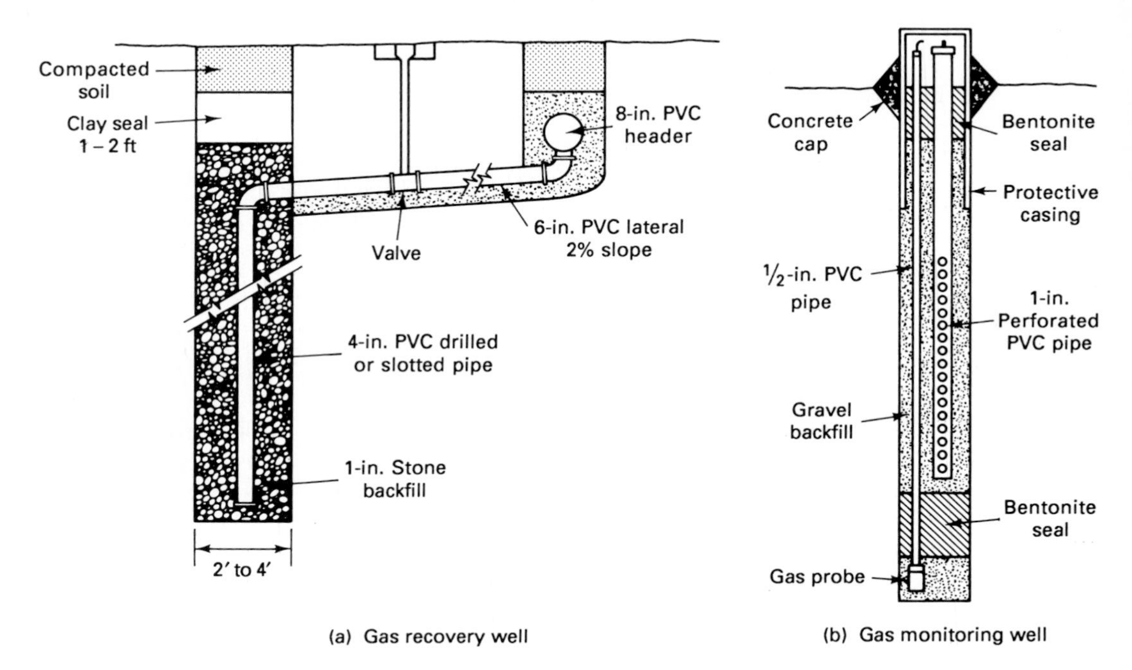

Figure 12.6 Gas monitoring and recovery wells.

cover. This trench is filled with crushed stone and connected to a collector pipe at the back of the landfill. Since these are canyon fills, construction of these collector pipes and the gas recovery system is possible before the fill is completed. It is possible to initiate gas migration control and gas recovery before the site is closed. When the site is closed, gas recovery wells can be added on the surface as needed to ensure complete coverage of the landfill with a gas removal unit.

The construction of the gas recovery wells has not been standardized. Each gas recovery designer has a typical design. Figure 12.6(a) illustrates such a design for installation of peripheral wells. This concept can be applied to wells located on the fill. A major concern is the integrity of the piping. Settling, especially differential settling, can cause failure of the pipe unless sufficient flexible connections are provided. Some designers will locate the collector pipes above the fill surface. Construction is simplified since it is not necessary to excavate the refuse to place the pipes. Also, stresses due to differential settling can be managed. Protection from freezing will be necessary in cold climates. The gas released from the fill will be wet and the condensation that will occur in the exposed pipes can freeze.

The gas monitoring probe illustrated in Figure 12.6(b) is installed in the soil adjacent to the landfill. Its purpose is to determine if methane is migrating from the site. The probe can be a simple perforated PVC pipe that is embedded in a bore hole backfilled with gravel or coarse sand. The top of the well is sealed with a tight clay cap or concrete. Internal pressure as well as methane concentrations can be measured with this probe. A more complicated probe can sample from different levels by separating the bore hole into zones. Each zone has a sampling pipe open to the gas environment in that zone. The $\frac{1}{2}$-in. pipe in Figure 12.6(b) shows how such a multilevel monitoring well can be constructed.

The gas extraction rate must be carefully monitored to prevent drawing a vacuum on the landfill. If this occurs, air will be drawn into the fill. This reduces the heating value of the gas and the air contamination may prevent use as a pipeline quality gas. The gas processing systems are designed to separate methane from carbon dioxide, not from oxygen and nitrogen. The changing atmospheric pressures will significantly alter the volume of gas that can be extracted. If the barometer is rising, the gas volume in the fill is decreased by the pressure increase. The extraction rate will decrease accordingly. A falling barometer will result in an increase in the extraction rate. Control of the extraction rate is achieved by monitoring the pressure differential between the fill and the atmosphere. Also, oxygen monitors can detect the presence of air in the landfill gas, indicating the need to reduce the extraction rate.

Gas Utilization

The typical composition of the gas extracted from a landfill is presented in Table 12.9. The heating value of this gas is 476 Btu/scf. There are a number of other gases that may be found in this gas, depending on the materials received at the site. Older fills may have received industrial wastes that contained VOCs. It is common for these compounds to be found in the landfill gas. New fills will not experience this problem due to the prohibition on the acceptance of hazardous wastes. The household solvents and related substances will not be in sufficient quantity to be detected in the gas.

The heating value of the gas is sufficient to use directly as boiler fuel or as fuel for a gas turbine or internal combustion engine. Combustion of the gas at the high temperatures of a boiler or turbine will oxidize the various organic contaminants present in the gas. The steam generated by the boiler would be used for process heat by an industry. For the gas to be used for this purpose, the industry must be relatively close to the landfill. The cost of transporting the gas any significant distance is prohibitive. An alternative use is to generate electricity with a gas turbine. A steam boiler can also be added to the gas turbine to generate low-pressure

Table 12.9 Typical landfill gas composition

Constituent	Volume (%)
Methane	47.4
Carbon dioxide	47.0
Nitrogen	3.7
Oxygen	0.8
Miscellaneous	1.1

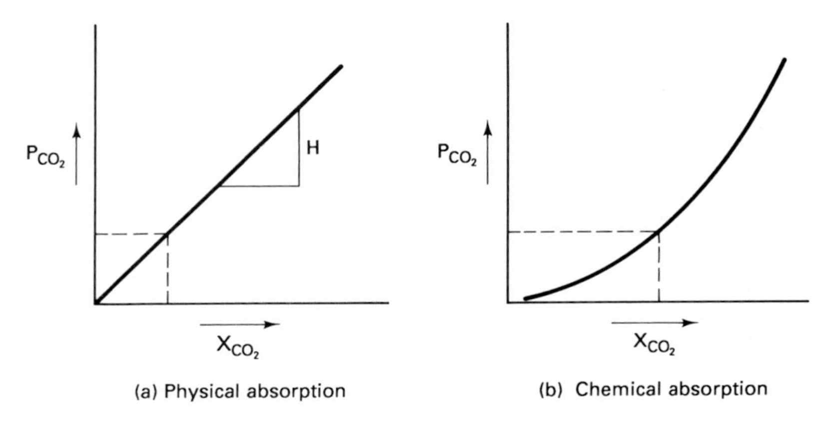

Figure 12.7 Absorption mechanisms for removal of CO_2.

steam if a market exists. Marketing the electricity is guaranteed under PURPA, but the price may not be very attractive. If the power can be used during peak demand periods, the electric utility may be interested in paying a higher price. The success of marketing the energy will definitely be a factor in the economic viability of gas recovery.

Some gas purification is required even for on-site use as a fuel. The gas will be wet, so a drying system will be needed. This is typically done with a chiller. The gas temperature is lowered to decrease the moisture content to an acceptable level. After it is dried, the gas is reheated with the rejected heat from the chiller. This step will eliminate the formation of condensation in the gas lines and the associated corrosion. Because the CO_2 concentration is high in the gas, the pH of the condensate can be 5.0 or less. This will be corrosive to many metals.

This condensate is a unique problem for landfills that have VOCs present in the gas. Because some of the VOCs are considered a hazardous waste, their presence in the condensate makes it a hazardous waste. Some states prohibit return of this condensate to the landfill. The new federal guidelines prohibit return to the landfill unless a liner exists and the leachate is collected and treated.

The separation of carbon dioxide from the methane can be accomplished with several different processes. To date, there are three different techniques to consider: physical absorption, chemical absorption, and membrane systems. In physical absorption, a solvent is used to absorb one component preferentially. Henry's law determines the effectiveness of the solvent used for the separation. As shown in Figure 12.7(a), there is a linear relationship between the partial pressure of CO_2 and the mole fraction in solution. To obtain a very low value for P_{CO_2}, it is necessary to maintain a low concentration of CO_2 in the solvent (X_{CO_2}). The landfill gas is passed into an absorption column, generally under pressure, and contacted with the solvent. The CO_2 is absorbed by the solvent. The purified gas is then piped to the end use. The solvent is regenerated and recycled back through the absorber. Some commercial processes and the solvent employed are the following:

Fluor process	propylene carbonate
Rectisol process	methanol
Selexol process	dimethyl ether of polyethylene glycol

Figure 12.7(b) shows the relationship between P_{CO_2} and the mole fraction of CO_2 in the solvent when chemical absorption is involved. There is a reversible chemical reaction between the CO_2 and the solvent. As illustrated with this curve, a low P_{CO_2} can be obtained with a high mole fraction of CO_2 in the solvent. These processes have been used extensively in natural gas processing. After passing through a pressurized absorption column, the solvent is regenerated for recycle back to the absorption column. The purified gas is sent to its end use. Monoethanol amine, diethanol amine, and glycol amine are solvents commonly used for chemical absorption.

The solvents used typically absorb all acid gases. This includes such gases as hydrogen sulfide. When regenerating the solvent, the CO_2 is normally vented to the atmosphere. Consequently, the presence of hydrogen sulfide can create significant odor problems. The removal of this gas before the CO_2 is removed is necessary to eliminate this odor problem. A common process used for this purpose is an "iron sponge." This is a column packed with a medium coated with ferric oxide. The removal mechanism involves a chemical reaction between ferric oxide and hydrogen sulfide. The ferric ion is reduced to ferrous ion, which reacts with the sulfide to form ferrous sulfide on the surface of the packing material.

Figure 12.8 is a schematic of an absorption tower showing how the process operates. After passing through the "iron sponge" for hydrogen sulfide removal, the gas is introduced into a pressurized absorption tower and contacted with a "semilean" solvent. This solvent has been only partially regenerated, but due to the pressure differential, has capacity to remove some CO_2 from the gas coming from the landfill. The partially purified gas passes up the column, where it is contacted with regenerated "lean" solvent for removal of the remaining CO_2. The flow of the gas in the absorption tower is countercurrent to the solvent. The gas with the lowest concentration of CO_2 is contacted with the solvent having the lowest concentration of CO_2.

The solvent exits the absorption tower to a flash chamber, where the pressure is released. Because of the pressure drop, the solvent is supersaturated with CO_2. The gas is released from solution and discharged to the atmosphere. The "semilean" solvent is pressurized and returned to the tower to absorb more CO_2. A portion of the semilean solvent is sent to the solvent regeneration column. After heating the solvent, it is passed through a stripping tower, where the CO_2 is air stripped. The regenerated or "lean" solvent is returned to the absorption tower.

Semipermeable membranes have been applied successfully to the removal of CO_2 from landfill gas. Membranes are thin barriers that allow preferential passage of certain molecules through the membrane pores while retaining other molecules. Membranes have been developed that allow CO_2, H_2S, and H_2O to pass while retaining the CH_4 molecule. These membranes are formed either as flat sheets or as

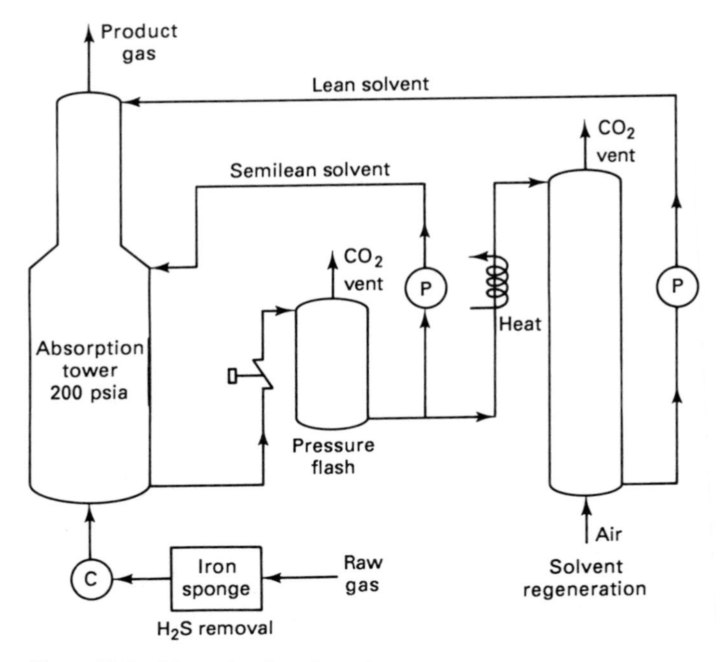

Figure 12.8 Schematic of an absorption process.

hollow fibers. The flat sheets are packaged into "spiral-wound" modules to increase packing efficiency and to withstand the internal pressures that develop. Gas membrane separation is pressure driven with pressures up to 2000 psi. The higher the pressure, the more efficient the separation. The hollow fibers are packaged as "bundles" so that a large surface area can be available in a small volume. The fibers have a very small diameter that keeps the wall stress low when the system is operated at high pressures.

The cost of gas purification by membrane processes can be significantly affected by the design of the process. In some applications, a single-stage membrane is adequate, and consequently, the capital cost for gas purification is low. A single stage will result in low product recovery. A more expensive design may be justified to achieve a higher product recovery of a valuable product. Improved product recovery is achieved through the use of a multistage system with recycle as shown in Figure 12.9. The methane content of stream D will be 30 to 35%, depending on the desired methane concentration in the product gas, stream B. This stream is compressed and passed through a second-stage membrane. The product gas, stream E, is recycled and mixed with stream A. The recycle along with the raw gas is passed through the first stage to produce a product that is 98% methane.

In addition to the added capital cost for the second-stage membrane, the recycle stream (D) has to be compressed to the operating pressure of the membranes, which is usually around 350 psi. The volume of the recycle stream can be equal to

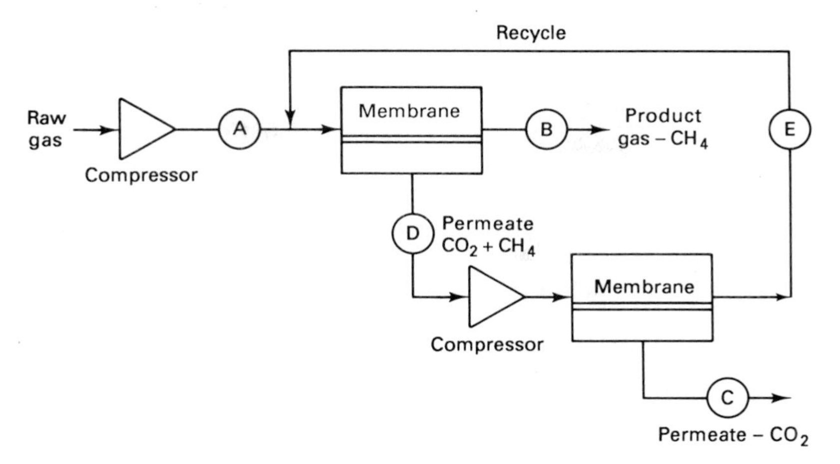

Figure 12.9 Process flow diagram for membrane separation of CO_2.

or greater than the raw gas stream. The compression costs are not insignificant. Consequently, serious consideration should be given to the economic justification for purification of the landfill gas. Purification to pipeline standards will cost approximately \$1.00 to \$1.50 per 1000 ft^3. Compression to a pipeline pressure near 1000 psi can increase the cost by an additional \$1.00 per 1000 ft^3.

FINAL LAND USE CONSIDERATIONS

In addition to the postclosure requirements in the federal guidelines, some plan needs to be developed for productive use of the site. In general, only open space uses are suitable. There are five major problems that must be addressed when considering final use of a landfill site: fill settlement, both total and differential; allowable foundation loads; corrosive nature of fill material; explosive hazards associated with the landfill gas; and excavation of the decomposing refuse in the fill.

The foundation problems with landfills has been discussed in the literature.[4] The maximum supportive foundation load is only 500 to 800 lb/ft^2. Any structure will have to be supported by piles driven to the original base of the fill. The settling that occurs with time will present continued maintenance problems. As the surface subsides, it will separate from the base of the structure, exposing the underside of the floor system. The utility lines will be stressed if not broken by the load of the subsiding cover and related material. Recognizing that settlement up to 30% of the fill depth may occur, the maintenance requirements may be significant.

The settling experienced in a landfill is influenced by a number of factors, some external to the fill. Most of the natural settling will occur within 5 years after

[4]"Foundation Problems in Sanitary Landfills," *Journal of the Sanitary Engineering Division, ASCE,* Feb. 1968, p. 103.

the refuse is placed. Additional settling may result from activities associated with the final use of the landfill. A superimposed load associated with a building or additional fill material can cause additional consolidation of a fill that was considered stable. The consolidation can be increased by the vibrations associated with equipment used in construction, by traffic, and so on. If water seeps into the fill, it can increase the consolidation by changing the physical nature of the fill material as well as increasing the rate of biological decomposition of the solids. Differential settling is also experienced. This creates very difficult problems in maintaining the integrity of the landfill cover as well as any pipes, and so on, that are placed in the cover system.

The refuse and the compounds that are produced by biodegradation create a very corrosive environment. The pH may be acidic, sometimes as low as 4.5. The acidity is a result of the CO_2 and the organic acids generated by the biological degradation of the solid wastes. The corrosivity of a soil is determined by its ability to conduct electricity. The higher the resistance, the lower the ion concentration and the less the corrosivity. The U.S. Bureau of Standards classifies a soil as highly corrosive if the electrical resistivity is less than 2000 ohms/cm^3. Typical electrical resistivity measurements for landfill material is 800 to 1000 ohms/cm^3. Because of this property, any material placed into the refuse portion of the fill must be corrosion resistant.

The most appropriate use for completed landfill site is open space and recreational areas. Parks, playgrounds, ballparks, and so on, are ideally suited for installation on these sites. Other successful uses include parking lots, open industrial storage areas, and runways for light aircraft. In general, the land use must be initially planned to be compatible with any applicable zoning regulations. During the active life of the landfill, a variance to the local zoning code may be needed. This can generally be obtained as long as the ultimate use of the site fits the planned land use. It may be true that the ultimate use of the site is years in the future, but it may be a selling point that can be used to generate local support. Many residents are amenable to having usable open space in the neighborhood, especially in congested areas.

SUBTITLE D REGULATIONS

The 1984 revision to the Resource Recovery and Conservation Act required U.S. EPA to rewrite Subtitle D rules to address the needs for design and operation standards for sanitary landfills. This revision was published in the *Federal Register,* August 30, 1988, with a request for comments by the public. These comments were considered in any revisions made prior to the rules ''going final.'' These rules have had some significant impacts on both existing and new urban solid waste landfills. The Subtitle D regulations set forth revised minimum criteria that include:

1. Location standards
2. Facility design and operation criteria

3. Closure and postclosure care requirements
4. Financial assurances standards
5. Groundwater monitoring and corrective action standards

The following presents the significant changes and additions that are important in the siting, design, and operation of urban solid waste landfills.

Subpart B: Siting Limitations

Certain locations have been identified as incompatible with the operation of a sanitary landfill. Some of these location restrictions previously existed (airports and floodplains), while the others are new.

Airports. Landfills are known to attract birds that may pose a hazard to the operation of aircraft. Any landfill closer than 10,000 ft of an airport used by turbojet aircraft and 5000 ft of one used by piston-type aircraft will have to demonstrate that they do not pose a bird hazard to aircraft.

Floodplains. Any landfill located in a 100-year floodplain will have to be designed so as not to restrict the flood flow, reduce the temporary water storage capacity of the floodplain, or result in washout of solid waste so as to pose a hazard to human health and the environment.

Wetlands. New landfills will not be able to locate in wetlands unless the following demonstrations are made:

1. No practical alternative with less environmental risk exists.
2. Violations of other state and local laws will not occur.
3. The unit would not cause or contribute to significant degradation of the wetland.
4. Appropriate and practicable steps have been taken to minimize potential adverse impacts.
5. Sufficient information to make the determination is available.

Fault areas. New landfill units cannot be sited within 200 ft of a fault line that has had a displacement in Holocene time (past 9000 years).

Seismic impact zones. New landfill units located within a seismic impact zone will have to demonstrate that all contaminant structures (liners, leachate collection systems, and surface water control structures) are designed to resist the maximum horizontal acceleration in lithified materials (liquid or loose materials consolidated into solid rock) for the site.

Unstable areas. Landfill units located in unstable areas must demonstrate that the design ensures stability of structural components. The unstable areas include areas that are landslide prone, in karst geology susceptible to sinkhole formation, and undermined areas where subsurface mining has taken place. Existing facilities that cannot demonstrate the stability of the structural components will be required to close within 5 years of the regulation's effective date.

Subpart C: Design and Operational Changes

The new Subtitle D rules impose several new general design and operating criteria for sanitary landfill receiving urban solid waste. The following rules identify specific criteria that in the past were considered to be good design practices but not specific design requirements.

Hazardous waste exclusions. Landfill operators will be required to develop a program to detect and prevent the disposal of regulated hazardous wastes. The program will include at a minimum:

1. Random inspection of incoming loads
2. Inspecting suspicious loads
3. Maintaining records of inspections
4. Training personnel to locate hazardous waste in the general refuse stream
5. Enacting reporting procedures when such wastes are found

Cover material. Landfills will be required to cover disposed waste with a suitable cover material daily or more frequently to control disease vectors, fires, odor, blowing litter, and scavenging.

Disease vector control. Landfills will have to prevent or control on-site populations of disease vectors using techniques that are appropriate for the protection of human health and the environment.

Control of explosive gases. Routine monitoring for explosive gases in facility structures and at the facility property boundary will be required. The allowable level will be 25% of the lower explosive limit (LEL) for the facility structures and the LEL at the facility property boundary. If a landfill is found to exceed these levels, gas control systems will be required to eliminate the problem.

Air criteria. Landfill sites will be required to meet applicable requirements under a State Air Pollution Control Implementation Plan pursuant to Section 110 of the Clean Air Act. Open burning of refuse is prohibited.

Access. Public access to landfills must be controlled to protect public health and the environment and to prevent unauthorized vehicular traffic and illegal disposal of wastes.

Run-on/run-off control systems. Storm water control systems that divert water from the adjacent areas and carry the run-off from the landfill must be designed, constructed, and maintained for each landfill. The design capacity should be sufficient to handle the water from a 24-hr, 25-yr storm event.

Surface water requirements. Landfills will be required not to:

1. Cause a discharge of pollutants into waters violating any requirements of the Clean Water Act, including the National Pollution Discharge Elimination System
2. Cause a nonpoint pollution of water that violates an approved water quality management plan under the Clean Water Act

Liquid restrictions. Landfills will be prohibited from accepting bulk and noncontainerized liquids, and containers holding free liquids unless these liquids are household or septic wastes. Also, recirculation of leachate or landfill gas condensate will be prohibited unless the landfill has the proper leachate collection and liner system to prevent the release of liquids.

Record-keeping requirements. Owners and operators of sanitary landfills will be required to record and retain environmental monitoring testing results, or analytical data; inspection records, training procedures, and notifications; and closure and postclosure plans.

Subpart C: Closure and Postclosure

This subpart has significant changes regarding how the site is managed after its useful life is expended. The long-term consequences will require careful planning for the finances needed for this period.

Closure criteria. Owners and operators of landfills will be required to develop a closure plan for performing closure in a manner that minimizes the need for further maintenance and minimizes the probability for release of leachate.

Postclosure care. There are two phases of postclosure care; the first is set by U.S. EPA and is for a minimum period of 30 years. The second phase is for a period to be set by the state. The EPA specifications for the first phase are:

1. Maintaining the effectiveness and integrity of the final cover
2. Maintaining and operating the leachate collection system

3. Maintaining the groundwater monitoring system and monitoring the groundwater quality
4. Maintaining and operating the gas monitoring system

In addition, the EPA will require the state to mandate that the second phase include a provision for groundwater and gas monitoring.

Financial assurance criteria. It will be required that owners and operators of landfills demonstrate that they have the financial ability to pay for the post-closure and any required remedial activities. The exception to this requirement is state and federal agencies for which the debt would be that of the state or federal government. States will be allowed to specify the financial assurances mechanisms that would be acceptable in meeting this obligation.

Subpart D: Risk Assessment

Instead of the federal proscription of specific design requirement (i.e., liner design or leachate collection and treatment), the states have flexibility in determining the allowable risk levels and the point of compliance within the federally mandated guidelines. As specified by the EPA, these levels will be for a minimum excess lifetime cancer risk level within the 10^{-4} to 10^{-7} range (i.e., one cancer incidence to 10,000 exposures to one in 10,000,000 exposures). This proposal has received a storm of protest from environmental organizations. They want the risk reduced to essentially zero.

In establishing the design goal, a state will be required to consider at least the following factors:

1. Hydrogeological characteristics of the site area
2. Climatic factors
3. Volume and physical characteristics of the leachate
4. Proximity of groundwater users
5. Groundwater quality

The EPA is not specific as to how the states are to determine compliance. They do provide a guidance document to accompany the final regulations on methods that will adequately determine compliance. There are three options for the states to consider: (1) a risk-based algorithm, (2) a categorical approach, and (3) an empirical method.

These requirements are for new facilities. Existing landfills will be required to be equipped at closure with a final cover that is designed to prevent infiltration through the cover and into the refuse. Existing units are not required to be retrofitted with liners and leachate collection systems.

Risk-based algorithm. The extensive data base that the EPA has developed has provided them with sufficient information that they are willing to propose a mathematical relationship that characterizes the site's potential for groundwater contamination. The algorithm represents the level of contamination as the excess lifetime cancer risk associated with human consumption of groundwater at the facility's compliance point. This algorithm is

$$R = 4.5 \times 10^{-4}\left(\frac{Q_R}{Q_A}\right)e^{(\mathrm{TOT})\,(-0.209)} \tag{8}$$

where R = lifetime risk posed by groundwater consumption at the compliance point
Q_R = predicted leachate release rate to the uppermost aquifer, m^3/yr
Q_A = groundwater flow rate in the uppermost aquifer, m^3/yr.
TOT = leachate time of travel in the aquifer from the unit boundary to the compliance point (TOT = 0 for a unit boundary equal to the compliance point)

Categorical approach. The EPA's guidance for using the categorical approach will consist of a 2 × 2 matrix based on annual precipitation (*P*) and time of travel (TOT) through the unsaturated zone from the bottom of the landfill to the aquifer. Precipitation at the site is available from published rainfall records or it may be measured at the site. Calculation of the TOT can be obtained through:

1. Calculation methods prescribed for determining vulnerable hydrogeology under RCRA's hazardous waste section
2. Calculations of Darcy's law
3. Calculations based on the linear velocity of water
4. A wetting front approach for unsaturated soil

The location category for each landfill unit can be determined as follows:

Category I: P is less than or equal to 40 in./year; TOT is equal to or greater than the unit's active life.

Category II: P is greater than 40 in./year; TOT is equal to or greater than the unit's active life.

Category III: P is less than or equal to 40 in./year; TOT is less than the unit's active life.

Category IV: P is greater than 40 in./year; TOT is less than the unit's active life.

A minimum TOT of 20 years has been specified by EPA to preclude the siting of short-duration projects in relatively poor locations.

The landfill category can then be used to determine the need for liners and leachate collection systems.

Leachate Collection Systems

Landfills in Categories I and III are not required to install a leachate collection system unless it is determined that leachate will accumulate to a depth of 1 ft or greater.

Landfills in Categories II and IV must have a leachate collection system that will prevent the leachate from accumulating to a depth greater than 1 ft over the unit's bottom.

Liner Criteria. All units must have a liner or overburden, or a combination of both, that will prevent the migration of leachate to an aquifer during a time period equal to 20 years.

Categories I and II. A natural base consisting of the vadose zone will be allowed unless a liner is required to meet the performance standard above or is needed for a leachate collection system.

Categories III and IV. Install earthen or synthetic liners or modify the existing subbase such that in combination with the overburden, the composite TOT value meets the above performance standard.

Under the categorical approach, a final cover system that prevents infiltration of surface waters into the waste after closure is required. States will have the opportunity to determine the requirements for the final cover system, as long as the design assures compliance with the health-based performance standard in the design criteria in this section.

Empirical method. This approach allows owner and operators of existing facilities or those planning to build new units in similar locations to use the groundwater monitoring results from the existing units to determine the design. If the monitoring results do not exceed the design goals, the design of the new unit would not have to be any more complicated than the existing unit. For this method to be allowed, the following conditions must be met:

1. The new unit must have sufficiently similar location and waste characteristics.
2. The existing unit must have operated groundwater monitoring wells over a period long enough to allow leachate generation, release, and migration through the unsaturated and saturated zones to the wells. (This requirement essentially negates the use of this option since the objective of a good design is to prevent the formation and migration of leachate. If this design objective is achieved, the leachate will never reach the monitoring wells, and consequently, the design is not acceptable.)

3. The groundwater monitoring parameters must address the phase I parameters, and phase II parameters if triggered.
4. The monitoring data must be supplemented with appropriate fate and transport modeling of hazardous components over a time period equal to the postclosure care period.

Subpart E: Groundwater Monitoring

The groundwater monitoring and corrective action requirements are similar to those being used for RCRA hazardous waste sites. These requirements will apply to all new and existing landfills unless it can be demonstrated that there is no potential for leachate migration to the uppermost aquifer during the active life of the landfill, including the postclosure care period. The monitoring system required for all sites must be able to detect the presence of leakage from the site. The number, spacing, depth, and construction of the wells will be based on site-specific information. System design will include a thorough characterization of the aquifer thickness, groundwater flow rates, and directions. Identification of the saturated and unsaturated geologic units overlying the uppermost aquifer, including thickness, sequencing, hydraulic conductivity (permeability), and porosities, is required. The monitoring wells are installed at the compliance boundary used for the design.

Under this monitoring program, EPA requires the states to establish groundwater trigger levels that are protective of human health and the environment for some 234 parameters. The trigger levels will be the maximum contaminant levels (MCLs) promulgated under the Safe Drinking Water Act, state-established health risk levels where MCLs do not exist, or background levels. There are two phases incorporated in the monitoring program, with the second phase increasing in scope and detail.

Phase I. Monitoring consists of semiannual sampling of wells during operation and closure, and at a state-specified frequency during postclosure care for:

1. 15 indicator parameters such as sodium, chemical oxygen demand, chloride, and iron
2. 9 heavy metals
3. 46 volatile organic compounds

If a statistically significant increase or decrease occurs over background in two or more indicator parameters, or one or more heavy metals or volatile organic compounds, phase II of the monitoring program would be triggered.

Phase II. Monitoring in this phase consists of analyzing the groundwater for any of the 234 parameters detected as significantly different above background concentrations at the compliance point. The minimum sampling frequency for these parameters is quarterly during the active life of the fill and closure period, and a state-specified frequency during the postclosure care period. If the increase in any of these parameters is substantiated, the owner or operator will be required to proceed to corrective actions. Corrective action requirements would mandate the assessment of available corrective measures, selection of a corrective measure, establishment of a groundwater protection standard, and implementation of the corrective action program. The goals of the corrective action program are to protect human health, attain the groundwater protection standard, and control further releases into the environment that may pose a threat to human health and the environment.

These regulations have a significant impact on the design of new landfills. One can expect the costs of this disposal method to increase. The cost is a long way from equaling that for incineration, but it will increase the avoided cost associated with recycling. The impact on existing landfills will not be as severe, but the closure requirements and postclosure care may be a major liability for an operator that does not have the capital or cash flow needed for the postclosure care, particularly the monitoring requirements. Any required corrective actions can be expected to bankrupt the smaller operators. The tab for site remediation will fall to the state or federal government.

STUDY QUESTIONS

1. How does anaerobic and aerobic biological activity affect the solubility of the common metal contaminants found in leachate?
2. What biological transformations occur in the landfill that are responsible for the contaminants found in leachate?
3. What processes are appropriate for the removal of biodegradable organic material? Nonbiodegradable organic material?
4. Draw a schematic of an anaerobic biological treatment process for treating leachate with a BOD_5 = 2500 mg/L and a COD = 6000 mg/L if the effluent is discharged to a municipal sewer.
5. Draw a schematic of an aerobic biological treatment process for treating leachate with a BOD_5 = 2500 mg/L and a COD = 3000 mg/L if the effluent is discharged to a municipal sewer.
6. What role can GAC play in leachate treatment systems?
7. If a leachate is produced at the rate of 20,000 gpd and has a BOD_5 of 4500 mg/L, size an aerated lagoon and stabilization pond to treat the leachate for discharge to a municipal sewer. Compute aerator horsepower requirements. State any assumptions.
8. What is the potential impact of the discharge of untreated leachate to a municipal wastewater treatment system?
9. How much added cost can leachate treatment add to the tip fee for a sanitary landfill?

10. A landfill is leaking leachate to a groundwater aquifer. If the aquifer has a permeability coefficient of $10^{-3.5}$ cm/sec, how long will it take for the contamination to reach a water well located 1 mile from the fill if the hydraulic gradient is 0.5%?
11. What effect does in-situ biological activity have on the leachate as it passes through the subsoil?
12. What type of soils have the greatest capacity to limit the movement of di- and trivalent metal ions? What is the mechanism?
13. An organic clay has been found to have a CEC for cadmium of 75 mEq/100 g. If the cadmium content in the leachate is 0.1 mg/L, how much leachate will have to pass through a cubic meter of this soil before the CEC is exhausted?
14. What is the fate of soluble organic material in the leachate when it is in contact with soils that have a high organic carbon content?
15. What is K_{oc}?
16. Naphthalene, a polynuclear aromatic ($C_{10}H_8$), has a solubility of 34.4 mg/L. How much leachate can be passed through a cubic meter of a soil before breakthrough of the naphthalene? The soil is 3% organic carbon and the leachate contains 2.5 mg/L of naphthalene.
17. What is the mechanism for gas production in a sanitary landfill? What is the origin of volatile organic compounds that are sometimes found in this gas?
18. What determines the migration path for landfill gas?
19. Why is gas control important?
20. What would be the effect of a cover containing a gastight membrane with no gas recovery wells installed at the site?
21. What factors in the landfilled refuse determine the potential gas production rate?
22. How can gas be recovered from a landfill? What is the quality of this gas? What is its value?
23. Identify some processes for gas purification. What contaminants are removed by each process?
24. What characteristics of the landfill limit the final use that can be made of the site?
25. What are the siting limitations for sanitary landfills that were promulgated under Subtitle D of RCRA?
26. What significant design and operational changes have been established under Subtitle D?
27. What is the impact of the postclosure requirements on the cost of operating a sanitary landfill?
28. What design objectives for sanitary landfills are being specified by U.S. EPA?
29. What is the difference between the risk-based algorithm and the categorical approach for determining if a liner and leachate collection system is required?
30. What is the implication of the requirement for phase II groundwater monitoring?

Index

A Stranger's Mirror

ALSO BY MARILYN HACKER

Diaspo/Renga (with Deema K. Shehabi)

Unauthorized Voices: Essays on Poets and Poetry

Names

Essays on Departure: New and Selected Poems 1980–2005

First Cities

Desesperanto

Squares and Courtyards

Winter Numbers

Selected Poems 1965–1990

The Hang-Glider's Daughter

Going Back to the River

Love, Death, and the Changing of the Seasons

Assumptions

Taking Notice

Separations

Presentation Piece

MARILYN HACKER

A Stranger's Mirror

New and Selected Poems, 1994–2014

W. W. NORTON & COMPANY

New York • London

Printed in the United States of America
First Edition

For information about permission to reproduce selections from this book,
write to Permissions, W. W. Norton & Company, Inc.,
500 Fifth Avenue, New York, NY 10110

For information about special discounts for bulk purchases, please contact W. W. Norton
Special Sales at specialsales@wwnorton.com or 800-233-4830

Manufacturing by Courier Westford
Book design by Ellen Cipriano
Production manager: Louise Parasmo

Library of Congress Cataloging-in-Publication Data

Hacker, Marilyn, 1942–
[Poems. Selections]
A stranger's mirror : new and selected poems 1994/2014 / Marilyn Hacker. — First edition.
pages ; cm
ISBN 978-0-393-24464-9 (hardcover)
I. Title.
PS3558.A28A6 2015
811'.54—dc23
2014037031

W. W. Norton & Company, Inc.
500 Fifth Avenue, New York, N.Y. 10110
www.wwnorton.com

W. W. Norton & Company Ltd.
Castle House, 75/76 Wells Street, London W1T 3QT

1 2 3 4 5 6 7 8 9 0

for Linda Gardiner

and for

Mimi Khalvati

CONTENTS

NEW POEMS

FROM *WINTER NUMBERS*

FROM *SQUARES AND COURTYARDS*

FROM *DESESPERANTO*

SOME TRANSLATIONS

FROM *NAMES*

ACKNOWLEDGMENTS

Winter Numbers received the Lenore Marshall Award of the Academy of American Poets and a Lambda Literary Award, both in 1995.

Squares and Courtyards received the Audre Lorde Award of the Publishing Triangle in 2001.

The new poems have appeared, sometimes in different versions, in *American Poet*, *The Baffler*, *Banipal* (UK), *The Cortland Review* (online), *Critical Muslim* (UK), *FIELD*, *The Kenyon Review*, *Little Star*, *Mslexia* (UK), *New Letters*, *Plume* (online), *PN Review* (UK), *POEM* (UK), *Poetry Daily*, *Poetry Review* (UK), poets.org, *Prairie Schooner*, *Upstreet*, *The Warwick Review* (UK), and *The Wolf* (UK).

The three poems on Fado themes appeared in *Saudade: An Anthology of Fado Poetry*, edited by Mimi Khalvati, published by the Calouste Gulbenkian Foundation (UK) in 2010.

"Ghazal: A Woman" won a Pushcart Prize for 2014 and appeared in the 2014 *Pushcart Prize Anthology.*

"Fugue on a Line of Amr bin M'ad Yakrib" was shortlisted for the 2012 Forward Prize for a single poem, and appeared in the *Forward Book of Poetry 2013* (Faber & Faber, UK).

"A Stranger's Mirror" was commended for the 2013 Forward Prize for a single poem, and appeared in the *Forward Book of Poetry 2014* (Faber and Faber, UK).

Many of the individual renga in "Syria Renga" and "Râb'ia's Renga" appear in the book *Diaspo/Renga*, a collaboration in renga between

Deema K. Shehabi and Marilyn Hacker, published by Holland Park Press, UK, in 2014.

Claire Malroux's "Grottoes" appeared in *Poetry Review* (UK).

Marie Étienne's "Ochre Night " appeared in *PN Review* (UK).

Vénus Khoury-Ghata's poems beginning "The stones in your garden" and "The faces lined up on the walls" appeared in *Prairie Schooner.*

Emmanuel Moses' "Prelude 1," "Fugue 1," "Prelude 2," and "Fugue 2" appeared in *The Kenyon Review*; "Prelude 3" and "Fugue 3" appeared in *FIELD*. The entire sequence was published in *Modern Poetry in Translation* (UK).

Guy Goffette's "Elegy for a Friend" was published in *Modern Poetry in Translation* (UK).

Jean-Paul de Dadelsen's "The Last Night of the Pharmacist's Wife" appeared in *The Kenyon Review.* "The Bridges of Budapest" appeared in *Crazyhorse* and in *PN Review* (UK).

New Poems

Casting Out Rhymes

Yes, dictionaries opening again
Yes, scorched across her forehead like a stain
Yes, less to say than cognate words contain
Yes, caffeine and butalbital for pain
Yes, ruptured synapses hobble the brain.

No, watched gray water circle down the drain
No, looking out the window of the train
No, not the melody, just the refrain
No, not temerity, no, not disdain
No, stroked across the scar against the grain.

Yes, written in another alphabet
Yes, odor of the night's first cigarette
Yes, hands and face and trousers grimed with soot
Yes, from the freeway, saw the minaret
Yes to the proposed amendment, but . . .

No, the initial syllable was not
No, it was the verb that she forgot
No, his raincoat and his shoes were wet
No, coffee, and a square of chocolate
No, just damp earth still clinging to the root.

Yes, with a clear head after a day's fast
Yes, any fantasy could be the last
Yes, touched the scar that used to be a breast
Yes, eidolon of the iconoclast
Yes, clasped the beaded bracelet on her wrist.

No, taciturn, an unforthcoming guest
No, thought of lapis, thought of amethyst
No, spoke to no one and still felt harassed
No, knew it was futility at best
No, watched the street after the bus had passed.

Yes, blotches that the years splotched on her skin
No, long-term memory inviolate
Yes, twice a day, pre-prandial insulin
No, from the bakery across the street
Yes, furious, implacable and sane.

No, fingers splayed, not clenched into a fist
Yes, classes daily at the institute
No, more like car exhaust than autumn mist
Yes, rainwater, the print of hoof and boot
No, not, in fact, "the lips your lips had kissed."

Pantoum in Wartime

In memory of Adrienne Rich

Were the mountain women sold as slaves
in the city my friend has not written from for two weeks?
One of the Just has given his medal back.
I wake up four times in the night soaked with sweat.

In the city my friend has not written from, for two weeks
there was almost enough electricity.
I wake up four times in the night soaked with sweat
and change my shirt and go to sleep again.

There was almost enough electricity
to heat water, make tea, bathe, write e-mails
and change her shirt and go to sleep again.
Her mother has gallstones. Her sister mourns.

Heat water, make tea, bathe, write e-mails
to Mosul, New York, London, Beirut.
Her sister mourns a teenaged son who died
in a stupid household accident.

To Mosul, Havana, London, Beirut,
I change the greeting, change the alphabet.
War like a stupid household accident
changes the optics of a scene forever.

I change the greeting, change the alphabet:
Hola, morning of light, ya compañera.

Change the optics of a scene forever
present, and always altogether elsewhere.

Morning of roses, kiss you, hasta luego
to all our adolescent revolutions,
present and always altogether elsewhere.
It seemed as if something would change for good tomorrow.

All our adolescent revolutions
gone gray, drink exiles' coffee, if they're lucky .
It seemed as if something would change for good tomorrow.
She was our conscience and she died too early.

The gray exiles drink coffee, if they're lucky.
Gaza's survivors sift through weeping rubble.
She was our conscience, but she died too early,
after she spoke of more than one disaster.

Cursing, weeping, survivors sift through rubble.
One of the Just has given back his medal,
after he spoke of more than one disaster.
How can we sing our songs if we are slaves?

Pantoum

for Fadwa Soleiman

Said the old woman who barely spoke the language:
Freedom is a dream, and we don't know whose.
Said the insurgent who was now an exile:
When I began to write the story I started bleeding.

Freedom is a dream, and we don't know whose—
that man I last saw speaking in front of the clock tower
when I began to write the story? I started bleeding
five years after I knew I'd have no more children.

That man I last saw speaking in front of the clock tower
turned an anonymous corner and disappeared.
Five years after I knew I'd have no more children
my oldest son was called up for the army,

turned an anonymous corner and disappeared.
My nephew, my best friend, my second sister
whose oldest son was called up for the army,
are looking for work now in other countries.

Her nephew, his best friend, his younger sister,
a doctor, an actress, an engineer,
are looking for work now in other countries
stumbling, disillusioned, in a new language.

A doctor, an actress, an engineer
wrestle with the rudiments of grammar

disillusioned, stumbling in a new language,
hating their luck, and knowing they are lucky.

Wrestling with the rudiments of grammar,
the old woman, who barely speaks the language,
hated her luck. I know that I am lucky
said the insurgent who is now an exile.

Room

Sudden reflection,
leaf-shadows cast on the shades
the night bus' headlights,

the dictionary
open on a Qu'ran stand,
notebook and pencils.

Three in the morning
noise in the street, parch of thirst
glass of water, sleep

recedes as heartbeats,
decided jurors, pace in
with avid faces.

Fugue on a Line of Amr bin M'ad Yakrib

Those whom I love have gone
And I remain, like a sword, alone.

Gone, yes, or going, determination hardens
Into a self-destructive stubbornness.

What melody will resonate its presence
If you play the same old self-reflective chord alone?

Someone who wrote, "Never to lose you again"
Moved, sent no message with a new address

And in that memory there is a mountain,
Above it, a reddish hawk that swooped and soared alone.

Who held a sword and said that he resembled
A sword, in his solitude was nothing less.

Between the old man and the steely angel,
A sleep-drunk intern holding down the ward alone.

The word-root's there, you look into the branches'
cadence and contexts you can only guess.

Translating from a slow-emerging language
Resembles dialogue, and I'm less bored, alone.

Though it's a doubled blade to be a weapon
And turn yourself onto your own distress.

Silent among her servants, Balqis riding
Back toward her queendom praised the Lord alone.

If the beloved asked, what would you wish of me?
That without my asking, you would answer "Yes."

The glass of wine not offered to the stranger,
The nightly second glass of wine I poured alone.

Paragraphs

for Rose Marie Dorn

Above the river, old and alien,
a foreigner among the foreigners
as I was among the Syrian
opposition demonstrators
gathered in the Place du Panthéon
under a July sky low-bellied with rain
that dumped down to waterlog
their block-long sheltering flag.
This was your river, Hayden, hills you left
in your strong middle age,
mourning their ruin. Late lust was a gift
of distance, despair, then rage
to know new joy and grief. What grief was hers
remained unuttered, as she turned the page,
a foreigner among the foreigners.

You'd have had no one coming to Vermont
for art camp, to the mill that does not mill
grain any longer, though the bannered air will flaunt
high above the Lamoille
River, blue-green vistas your argument
(despite gimcrack guttings) can't discount.
But you left, and you died,
though I write to you beside
the Gihon flowing from Eden in the rain.
I walked with the one you left
whom I toast solo with Grand Union wine:
she cursed also. Then, as she laughed,
tested my recognition of each scene
you wrote, her history you'd paragraphed
in wind-heft, cord-wood, winter, water, green.

Headaches

Wine again. The downside of any evening's
bright exchanges, scribbled with retribution:
stark awake, a tic throbs in the left temple's
site of bombardment.

Tortured syntax, thorned thoughts, vocabulary
like a forest littered with unexploded
cluster bombs, no exit except explosion
ripping the branches.

Stacks of shadowed books on the bedside table
wall a jar of Tiger Balm. You grope for its
glass netsuke hexagon. Tic stabs, dull pain
supersedes voices,

stills obsessive one-sided conversations.
Turn from mouths you never will kiss, a neck your
fingers will not trace to a golden shoulder.
Think of your elders—

If, in fact, they'd died, the interlocutors
who, alive, recede into incoherence,
you would write the elegy, feel clean grief, still
asking them questions

—though you know it's you who'd provide the answers.
Auden's "Old People's Home," Larkin's "The Old Fools"
are what come to mind, not Yeats. In a not-so
distant past, someone

poured a glass of wine at three in the morning,
laid a foolscap pad on the kitchen table,
mind aspark from the long loquacious dinner
two hours behind her,

and you got a postcard (a Fifties jazz club)
next day across town, where she'd scrawled she'd found the
tail-end of a good Sancerre in the fridge and
finished the chapter.

Now she barely knows her friends when you visit.
Drill and mallet work on your forehead. Basta!
And *it is Margaret you mourn for.* Get up,
go to the bathroom.

You take the drugs. Synapses buzz and click.
You turn the bed lamp on, open a book:
vasoconstrictor and barbiturate
make words in oval light reverberate.
The sky begins to pale at six o'clock.

Syria Renga

Driving a flatbed
truck of sheep alongside the
Qalamoun hills, he

glances at the mountains and
thinks of his brothers who are

still in Kirkuk. Once
borders were porous, work meant
crossings, for those who

are amateur refugees
now, inadvertent exiles.

■ ■ ■

Two hundred miles from
the refugee camp outside
Damascus, Zainab

descends the stone ramp from Baal's
temple, becomes her namesake

Queen Zenobia
who held off Roman legions
Sanuqawimu!

White cotton scarf round her face
like the headdress of a queen.

■ ■ ■

Road signs say *Tadmor*
not *Palmyra*. Prisoners
shackled in a truck

don't see the temple pillars
or the Bedouin kids who

sell you keffiyehs.
They learned about Palmyra
in history class,

what they will learn in Tadmor
is a lesson from elsewhere.

■ ■ ■

Her father will die
without seeing her again.
He's ninety-four now.

Safe in exile, they watch the
insurrection in cafés.

She asks her husband
"But who'll take power after
your revolution?"

—thinks of the old man she loves,
the hills near Latakia.

■ ■ ■

Chams calls his mother
and she talks to him in code:
"It rained yesterday,

a strong cold wind from the mountains
blew down the telephone lines.

Now the power's back."
A YouTube of the demo
on his Apple screen,

but not his younger brother's
face in the tide of faces.

■ ■ ■

Returning at last
she requested a visa
as she'd always done.

The man at the embassy
asked "Are you a journalist?"

Not exactly truth
to declare that she wasn't:
who knew what she'd write?

But she said "No." Nonetheless
she didn't get the visa.

■ ■ ■

The boy's round-faced smile,
and the image of his corpse
returned, thus tortured

have gone viral on YouTube
and on ten thousand posters.

Internet switched off
on Friday, the day of the
week's demonstrations.

Where's Joumana, where's Imân?
Eina Najîb wa Ahmad?

■ ■ ■

The telephone rings
in Reem's apartment. And rings.
Nobody answers.

She's gone to market, or she's
working in the library.

Rings late at night, rings
early in the morning. Still
nobody answers.

She's gone to her family
in the country? She has none.

■ ■ ■

Chams' mother tells him
three soldiers knocked on her door,
asked where her son was.

Military service meant
he was still in the reserves,

called back to duty.
She said he was in England.
Now he'll have to stay.

He hadn't planned to go back,
nor thought he was in exile.

■ ■ ■

His cousin tweets from
"Syrian Revolution,"
needs cell phones, thumb drives,

so he's on Turkish Airlines
toward two days in Istanbul

with his French passport,
appointment with a stranger
in their first language,

feeling like a boy again,
mother tongue and contraband.

■ ■ ■

"I slept in Yarmouk
and I dreamed of Palestine.
When I'm asleep here

I dream I'm sleeping in the
camp, dreaming of Palestine,

but when their troops kill
Syrian children, and bomb
Syrian cities

and say it's for Palestine,
I don't want that Palestine."

■ ■ ■

In a Damascene
pizza parlor they worked on
translations of Plath

stacking up saucers of sweet
thick coffee they drank till dusk.

Two years later one
portable phone's been cut off.
No number to call—

the now-distant friend translates
silence that's not poetry.

October Sestina

Addictive day starts with the lit-up screen
against the backdropped window, while the street's
still dark, the gray slate roofs oily with rain.
Distraction of four newspapers' front pages
clicked on to spiral into distant windows
and distant, virtual and dulled encounter.

Better to go down to the café counter
(above which, on a television screen,
the same heads talk) and watch, beyond the windows
a drizzly morning's intersecting streets
that used to open into day, their pages
etched with the calligraphy of rain.

One more fall day, whose uncertain rain
is the most probable, least vexed encounter.
From left to right, from right to left, on pages
or posters, paravent or movie screen,
spectators are the spectacle, when streets
unwind their bobbins below open windows.

Addicted, then transfixed behind the window's
barrier, slant light, slant fall of rain,
a mystery enacted on the street's
begrimed and glittery parquet will counter-
act the dire pronouncements you can't screen
out, the bad news on the daily pages.

You write what someone wrote on other pages
when lives were flexed and fixed in different windows,

a nightstand's pile of books enough to screen
out anguish. Morning whispers through fine rain.
Upstairs, on the blue-tiled kitchen counter,
the coffeemaker waits. Perhaps the street's

doubling for the discovery of streets
paced briefly or long-viewed on midnight pages.
Elsewhere's tired eyes, an elsewhere you encounter
lowering the shades, turning from the windows
to walk downstairs and veer into the rain,
thrust your borrowed double through the screen.

Out of the rain, a cup chinks on the counter;
out there, the street's, and here, the morning's pages
fold like a screen. Time to open the windows.

Sapphics in Winter

Rain falls till and through the five-thirty nightfall
as this time I walk with you to the métro,
going different places, of course, but on the
same rush-hour subway.

Winter trees and scarves over quilted jackets,
folded headlines cover the small disasters:
Three days' massacres trump a decade's failures,
linked though they might be.

You are young and what I began to tell you
was the wrought-iron grille of a conversation,
curlicues, non sequiturs, screens unfolded
hiding a rag-heap.

You're already elsewhere, redeemed from exile,
blue-green eyes assessing the slack and blemish
as I kiss you back to the intifada—
kissed at a distance.

Three Variations on Fado Themes

TAHRÎR

Through the skein of years, I had nothing to fear from this place.
How final and brief it would be to disappear from this place.

The tangle of driftwood and Coke cans and kelp in the sand
made me think of the muddle that drove us (my dear) from this
place.

An orchard, a vineyard, a stable, a river. A wall.
The impassible distance today once seemed simple and near from this
place.

There was the word *refuge*, there was the word *refugee*
who, confused and disrupted, began to appear from this place.

The silence that lasted for decades, for months or for hours
will sooner or later be broken. You'll hear from this place.

There is a wall, and the words that we write on the wall.
Libertação! Can you make out *Tahrîr* from this place?

From your bedroom window with the sun coming up
I could see dusty jitneys crawl toward the frontier from this place.

My name's rhyme with yours and the things that are done in our
names
in whatever language no longer sound terribly clear from this place.

SAHAR AL-BEITUNIA

She lives in Beitunia
And her name is Sahar
Her name is the hour
Between sunrise and morning.

Her bougainvillea
Overlooks Beitunia
Where a mango-bright bedspread
Hangs over the railing
Lit by first light
That reflects from a wall.

Not the wall of a house
Or her family's orchard.
She can see the graffiti
Ich bin ein Berliner

Marwân had orchards
Al-zaytûn wa-l-'inab
Olive trees, grapevines,
Where they went out to work
Between sunrise and morning.

She is *bint Marwân*
(and also *bint Su'âd*).
She is *ukht Târiq,*
Ukht Mahmûd, ukht Asmâ.

When jeeps and bulldozers
Converged on Beitunia

A hundred and twenty
All walked out at midday
Were chased back with teargas
And rubberized bullets.

Seventeen thousand dunnams
Of orchards and wheatfields
With a wall thrust between them
And the doors of Beitunia.

Her name is Sahar
At dawn in Beitunia
Where the first light reflects
On the wall of a prison.

Ismuhâ Sahar
Bayn al-fajr wa-l-subh
—her name is Sahar,
between sunrise and morning.

BLACK BOAT

If you were there when I woke
With my barbed wire, with my scars
You would avert your green gaze
I would feel the chill of regret

Though you said something else
In sunlight, over wine.

I saw a cross on a tall rock
And a black boat danced on light.
Someone waved, was it you,
A brown arm between white sails.

Old women know
That more go away
Than will ever return
Than the morning has scars.

In the wind as it blows
Wet sand against the panes
On the water that sings
In the fire as it dies
In blue sheets warmed by
Someone sleeping alone
On an empty park bench
When they lock up the square
You are still there

Brown arm green gaze black boat blown sand barbed wire.

Alcaics for a Wedding

You, from the start a child who was ready to
whisk down the slide, then jump off the diving board,
    walk home from school alone or fly to
    Nice "unaccompanied minor," seven,

drove through Ohio snow in a rented truck,
one skittish cat and laptop for company,
    headed from Tucson to Manhattan
    and new exigencies of vocation.

Rode through the jasmine dawn on a motorbike—
Someone was being born; as a glimmer of
    day pearled the green, you came to meet her
    leaving the hills and your night behind you.

Leap into trust, commitment, companionship,
one more departure, this time accompanied.
    Sometimes a diving board, love's an airport
    where all the signs today read ARRIVALS.

Fadwa: The Education of the Poet

When I was made a prisoner in my father's house
my mother turned away, swallowing her words.
The potholed path to school became a vision
of my lost future. It was not even because of a letter—
a boy from the boys' school entrusted a flower,
an almond sprig, for me, to my little brother

that was intercepted by our middle brother.
I was twelve. I was no longer to leave the house.
If that twig in blossom sullied an imagined flower
the metaphor wasn't mine. I had no words
for my ignorance, or innocence. I copied letter
after wobbly letter on my school slate. My vision

was astigmatic, strained. Was their double vision
god's eyes? Six months later, my eldest brother
came back, his studies done, Doctor of Letters.
He shelved his books in the attic room of the house
that became a hive, crawling and buzzing with words
and dripping the honey of words on a carpet of flowers.

My days were dust and water, starch and flour,
chores, silence, shame. I could not envision
the schoolchild I'd been, or what I would be. Words
receded, I spoke so little. I brought my brother
his coffee and shaving water. The rest of the house
was dawn-silent. He was writing a letter—

—no, something else, carefully, hearing each letter
that rounded or plunged like the bud or the root of a flower,

and the vowels' perfume pervaded the house
as he weighed the choices that shaped his revision.
"Do you still care about poetry?" asked my brother,
and began to recite, a mellifluous ladder of words.

"Take this book. Copy this poem. Make a list of the words
you don't know. Mark the small vowel for each letter.
Learn it, and recite it to me," said my brother.
"We are dust and ash, and beauty is brief as a flower.
But a stanza's a room; a single line is the house
you—*you*—can build, throw open the doors to your vision."

I live in my own house now. In my line of vision,
an almond tree flowers. I live by words,
and on each page I write a letter to my brothers.

Dahlia and Fadwa

When I see her come through the orchard toward my house,
I begin composing answers to the words
with which she'll challenge me, her vision
precise and focused, as if every letter
were drawn beforehand. She stops, stoops toward a flower
she doesn't pick, and the word I'm left is "sister"

as if she were coming to visit a younger sister
after a long absence, as if this stranger's house
were where she'd watched my awkward gestures flower
to eloquence, as if I had measured her words
from a wider world each time a letter
arrived, message inhabited by a vision.

But I was the only messenger of my vision.
My father died. There was no brother or sister
but the orphan's clan of branching letters
I taught myself. I moved from house to house
with a dented trunk, a few books, many words,
an acid fruit preceding its own flower.

Am I named for a traitor or for a flower?
Blood and fire have cindered and stained my vision;
I cursed their politics; they praise my words.
What answer can I give my older sister
when I open the door, welcome her into my house,
and she hands me the incriminating letter

while I imagine opening a letter
after scouring the sink, dusting a loaf with flour,

that tells me my house no longer is my house.
Beyond the window and my field of vision
my son is shouting something to his sister—
and it's hard, all at once, to make out their words.

But in my yellow kitchen these are only words
I can't pronounce correctly, in the letters
of my sister's language, language that is sister
to mine. I watched her flame and flower
as boys threw stones—spliced shots on television,
and felt the leaning walls of my bright house.

In that threatened house, at a loss for words
my letter would be subject to revision,
flow or gutter, never reach my sister.

. . . But in Things

A zinc bucket with roses from last week
infused with sugar-water, fading fast.
The summer solstice is a fortnight past,
light on the cusp of evening is oblique.
A photocopied page in Arabic
graffitied with French/English pencil scrawl,
not a testimony, just a tale
that I can translate better than I speak.
Beside a blue placemat from Monoprix,
Le Monde des Livres and the *LRB*,
an earthenware pitcher from Tripoli.
As in that shop bead-curtained from the street
the statement in a shaft of alien light
questions an object and its history.

Question an object and its history:
this dark red silk scarf, lightweight, lightly shirred,
whose tiny label says that it was made
in Halab. Ordinary luxury
to roam the ancient caravanserai
then everybody's market, where I played
at haggling with a wisecracking, yes, almond-eyed
merchant of twenty whose job skill was repartee,
and two girl students loitering near the stall
praised me for speaking Arabic at all.
You've read the same damn headlines. *Ubi sunt*
the multicolored silks, the girl students?
The fourteenth-century wooden arcades
burned first. The boy, what choice on fire, what words?

Chapter

Bending down to gather the scattered Lego
Jan recalled a moment, another childhood,
scooping up a ball in a sunlit courtyard
before her name changed.

Who was she in consonants that her children
mispronounced, or did on the rare occasions
that they tried? Her grandmother used a pet name
almost forgotten.

Hidden child, perpetually a virgin,
learned to make love in an adopted language:
twelve years old, her mother tongue blushed though she herself was a mother.

A Stranger's Mirror

Beside her bookshelves, in his winter coat,
a denim jacket lined with cotton fleece,
and who might not have said to him, "Then stay . . ."
as there was, all at once, a lot to say,
except that was another century's
invitation. Her questions, bilingual jests
came from the creased lips and crepey throat
of a woman in her sixties.
Alone, and with a choice of alphabet
she did not reconstruct the repartee,
at once anodyne and intimate,
nor pause at her stacked desk to contemplate
disaster she might well precipitate
if her neck were smooth. If she had breasts.

When her neck was smooth, when she had breasts,
she thought the body was the least of it,
the site of some desires and appetites
and certain others' ardent interests.
Not beautiful, not scandalous. Requests
like touch and hold, like any intimate
avowal, shocked no one, under any light;
Now, inadvertent archaeologist
she contemplates the ruin of a face
(the downside is quotidian dis-grace,
the upside is invisibility)
and the ravenous mythology
in which she's exiled from her own desire,
reflected strangely, in a stranger's mirror.

Reflected strangely, in a stranger's mirror
the exile's eyes themselves mirror a sky
more clear than when familiarity
abstracted it to gray and azure blur, or
instilled, by means of likenesses, the torpor
of saying the same old dull thing endlessly.
Here is an Elsewhere, all the cues that she
found in a cloud, a wall, a stone are Elsewhere.
At dusk in the street warren near the port
with a witty quadrilingual friend,
distancing the old narrative seems plausible.
Weeks later, after a day much too short,
a white night staggering hours before its end,
the graying woman yawns, sits at her table.

The graying woman yawns, sits at her table,
insomniac after the equinox.
The words she wants are in some padlocked box
whose combination she's incapable
of calling from the incoherent babble
of panic and despair, of dream that shocks
her out of brief and febrile sleep. The lacks,
the slack, the slide, the sunrise above rubble—
is that all, all want, that heat, all need,
that model of unspeakable obsession,
senile in promise, infantile in greed,
horseblindered to the world beyond its skin?
How much despair is clinical depression,
and how much what they still call mortal sin?

How much of what they still call mortal sin
is more like moral, mental masturbation,
slothful, not sexual, the titillation
of knowing what one might, and giving in
to entropy instead. Dead stop. Begin
the stagnant list, the stunning conjugation.
An ice-floe drips, bird drops, an abstract nation
raises its colors to an alien sun.
Neither testosterone nor estrogen
scabs the cut, blocks synapses of hurt in
a mind that spills its seed in solitaire.
Would I pick up the pen, the phone, again,
open the windows to the winter air,
if I were you, and were, as you are, certain?

If I were you and were, as you are, certain
as anyone can be, of pages spread
across long days like crisp sheets on a bed,
and of the bed itself, a blue voile curtain
behind it, and beyond that, light, alert in
a lovers'-morning sky, the book you read
the night before close by, and commented-
upon, in two alphabets, inadvertent
discoveries in margins, I'd agree
(and do) the body is a festival.
Also a house of mourning, and a field
soldiers have fought and camped on, burned and fouled,
and a mote in the absence that we whirl
toward with our metered love-words, almost free.

Untoward, metered love-words, almost free
to mean a thing and still mean its negation
to be avowal and renunciation
in a vexed breath's simultaneity
once had a different utility.
The inadmissible elucidation
is not pronounced, a train that left the station,
one rainy weeknight wolf-hour, half-past three.
There's not one story only, there are threads
of consanguinity and contraband.
A risk that is familiar and remote,
in remembered streets, imagined beds,
shrugs into its sleeves, extends a hand
beside the bookshelves, in a borrowed coat.

Ghazal: Outside the door

for Farkhonda and Bina

Laughter, music, voices singing verses can be heard outside the door.
The little girl is memorizing every word outside the door.

Light in the stairwell, seen through the judas-hole:
is that the visitor you longed for or you feared outside the door?

Long hours in lamplight practicing his scales,
in counterpoint to solfège of a bird outside the door.

The diplomat entering the leader's office
forgets the Copt, the communist, the Kurd outside the door.

Praise for the leader, loyalty till death!
Another imprecation is whispered outside the door.

The first love left, the second packs her bags.
Are those the nervous footsteps of the third outside the door?

Self is a mirror, poster-color bright,
but notice how the colors become blurred outside the door.

The revolutionaries' nameless laundress
wonders "What happens to a dream deferred?" outside the door.

Ghazal: In the wind

for Somaia

The exiled ney's lament that's hanging there in the wind
translates a language we can share in the wind.

Clothed by calligraphers in gold and lapis,
the poet's words flow supple, black and bare in the wind.

The scholars and the tourists and the pilgrims
Crowd past the shrine, then step outside and stare in the wind.

There was no room behind the slats of the partition.
She walked out to a field, knelt down and said her prayer in the wind.

Alone, unwound her headscarf under a lime tree,
ran down the goatpath with her tangled hair in the wind.

You understood there would be peace with justice?
What voice informed you history was fair in the wind?

A fox deleted from another poem
Crouched down beside the entrance to his lair in the wind.

What's all this orientalist palaver?
A vast vocabulary misheard where? In the wind.

The hakawati weaves another story,
a garment that a traveler can wear in the wind.

Ghazal: For it

The moment's motion blurs the pose you hold for it
as if you knew what future were foretold for it.

Is there a course in loss and cutting losses,
and what ought I to do to be enrolled for it?

I could walk to the canal and watch the barges
from the footbridge, but this morning it's too cold for it.

However the diplomat flattered the dictator
or threatened him, ten prisoners were paroled for it.

Who wants to push poems on reluctant readers?
Keep them in notebooks, wait to be cajoled for it.

A lover's hand reaches for the beloved's
hand, as yours would too. Are you too old for it ?

Yâ 'ainy, if revolution shakes your stupor
awake, open your eyes, your arms, be bold for it!

Ghazal: A woman

Across the river, in the orchard on the hill, a woman
said, sometimes a handful of red earth can fulfill a woman.

She remains a speaker, although silent;
remains, although invisible, a woman.

I loved a man, I loved a city, I loved a language.
I loved, make of it what you will, a woman.

No one spoke up against the law forbidding speech,
until a schoolboy, until a monk, until a woman . . .

Who might have thought they'd hesitate to kill a child,
who might have thought they'd hesitate to kill a woman?

Rita shoulders her rifle in front of the looking-glass.
There's more than one way a uniform can thrill a woman.

The hakawati with gray hair and no breasts
writing words and crossing them out is still a woman.

Râbi'a's Renga

Now who will correct
her eccentric spelling in
animal stories

where the wolf talks resistance
to the dog in a café?

After a day when
even an old wolf would weep,
who will tell her that

her astigmatic eyes are
the color of dark honey?

■ ■ ■

Wish for desire back
and there's age, not a promise:
no future in it.

The woman in the proverb
exogamous or exiled,

her foreign mirror
in which she saw her stranger's
unfamiliar face,

her face the face of exile,
turns away from the mirror.

■ ■ ■

She is alive now.
He thought she was too old and
ugly to be raped

—if he had raped her, he would
have thought he had to kill her.

So, just robbery,
two interminable hours
of sequestration,

now, apprehension that does
not rejuvenate her face.

■ ■ ■

This letter, witty,
flirtatious: six names for *horse*
from the Abassids.

This one from elsewhere: they shot
a teacher in the street, left

the body to be
defiled. A border apart,
on her desk today.

She begs her not to despair.
Copies the six words for *horse*.

■ ■ ■

"Come share my meal," said
the poet to the red wolf
in desert firelight.

Fangs glistened in the flames,
and a honed sword set aside.

The wolf was hungry,
accustomed to betrayals.
The man offered food

and a truce. What happened next,
neither wolf nor poet told.

■ ■ ■

A cinnamon clove,
a mollusk sheathing a pearl,
a wild strawberry,

humble hidden savory
engorged with a mother tongue,

not the origin
of the world, hell's mouth, flood-rift
blind spot in the sea.

What are you talking about?
Irrelevant metaphors.

■ ■ ■

The eggplant curry
she made for herself eaten, she
washes the dishes,

turns out the kitchen light and
watches the quieting street.

Is she like Li Po
with her third, last glass of wine,
the midnight window,

the high nail-clipping moon and
the plants that got through winter?

■ ■ ■

Gaza, and Warsaw
where her father was martyred,
though that's not her word—

she weighs the two names as she
drives to the women's prison

on the Hudson. Shoes
off, belt off, full-body scan;
then, in the green room

political prisoners
open their dogeared readers.

■ ■ ■

When the worst recedes,
the night soldiers' wreckage,
the boy with the knife,

there are verbs to conjugate,
grammar books' bland sentences

in which a young man
gets off a train, a schoolgirl
runs to catch a bus,

the ruptured world composes
yumkininî ân. I might.

■ ■ ■

If I got to choose—
she said in the library—
between one long night

with my body replenished,
all the parentheses filled

and a chain of days
for a year or two, weekly,
puzzling through these books

water-pitcher, languages
crossing—I'd forgo that wine.

■ ■ ■

She stands in line with
other asylum seekers, exiles,
in fall rain outside

the Préfecture de Police
for her resident's permit,

after two hours is
scanned, seated to wait some more.
She'd like to tell them

"Write down, I am an Arab!"
this morning of Yom Kippour.

■ ■ ■

Cloud cuneiforms,
the calligraphy of rain,
alphabets of sand

slip through her fingers as the
old woman copies the words

of a couplet by
Abu al-'Atahiyya
"Death is certain yet

I am still joyful, as if
I knew death through denial."

■ ■ ■

The night bus speeds by
without her flagging it down
from the bus-stop bench.

She lights one more cigarette
and looks up at two windows

alight in pocked walls,
where damasked shadows gesture
behind the curtains

under a moon that sums up
what she knows about distance.

Luzumiät: Necessities of what was unnecessary

for Golan Haji and Fady Joudah, because of al-Ma'arri and Amin Rihani

The politicians lie, and having lied
Make grand pronouncements about genocide,
Red lines, civilization and the rest
Elsewhere from anywhere anyone died.

■ ■ ■

Midsummer's lingering azure was misspent.
The morning light is late and different.
A man in Ghouta on a shopping street
Held his son's hand. Now tell me where they went.

■ ■ ■

The left knows: intervention by the West
Would be imperial self-interest.
The teacher came out of the bakery
And took a sniper's bullet in her chest.

■ ■ ■

The children whine and sulk or break things but
The little village school is bolted shut.
One teacher joined a katiba from Homs
A week after the other one was shot.

■ ■ ■

Ghosts in the alleyways, under the eaves,
With knives and hit lists hidden in their sleeves
Who were the grocer, neighbor, carpenter,
At least that is what everyone believes.

■ ■ ■

We talked about war, exile, prosody.
He wrote, and published afterwards, that he
"Could almost see the woman she once was"
Though not the one I am, or hope to be.

■ ■ ■

She was the daughter of a journalist
Who signed a letter that became a list
Of detainees at 'Adra, where he spent
A decade of the childhood that she missed.

■ ■ ■

When I was twenty-three or twenty-four
I'd have some revelation in a bar,
And write it down, and get another drink
Like my friend Omar. No, not *that* Omar!

■ ■ ■

The lucky exiles drink their arak, call
For more mudardara and muttabal.

Newly arrived, the cook's thin artist son
Paints hunger strikers on the café wall.

■ ■ ■

We eat grapes and pistachios while we
Debate the relevance of theory.
She's couch-surfing; she can look for a job
Now she's officially a refugee.

■ ■ ■

As who I once was can't embrace who you are,
Admitting longing is evoking error.
There is no lovely stranger in my bed,
Only the stranger in the bathroom mirror.

■ ■ ■

After his talk on djinns in the Qu'ran,
Requisite drinks with almost everyone,
He Skypes his older sister in Tartous
To ask if she's had word yet from her son.

■ ■ ■

Bright in the pink and green of your indoors,
A screen set with your mother's miniatures
Of village life remembered in Iran,
Her memories, not yours, becoming yours.

■ ■ ■

Tents stretched to the horizon. Nothing green.
No bread, no books, no fruit, no gasoline.
The boy best in his class at sports and math
Has diarrhea in a trench latrine.

■ ■ ■

Her erstwhile colleagues at the institute
Called her a turncoat and a prostitute.
Three bearded country boys invoking God
Got her across the border to Beirut.

■ ■ ■

June Jordan wrote *Directed by Desire*
—for sex, for justice, trees, words, life entire.
The one direction my desire suggests
Is from the frying pan into the fire.

■ ■ ■

The bright and brief parenthesis won't last.
Dawn hesitates, the sky is overcast.
Rain crosshatches its notes on yellow leaves,
Like headlines, like footsteps approaching, fast.

■ ■ ■

Children flock round the tired, balding man
Who summons language that was once his own,
Remembers his grandmother's formula,
And begins the tale: *Kan ya ma kan*. . . .

from *Winter Numbers* (1994)

Against Elegies

James has cancer. Catherine has cancer.
Melvin has AIDS.
Whom will I call, and get no answer?
My old friends, my new friends who are old,
or older, sixty, seventy, take pills
before or after dinner. Arthritis
scourges them. But irremediable night is
farther away from them; they seem to hold
it at bay better than the young-middle-aged
whom something, or another something, kills
before the chapter's finished, the play staged.
The curtains stay down when the light fades.

Morose, unanswerable, the list
of thirty- and forty-year-old suicides
(friends' lovers, friends' daughters) insists
in its lengthening: something's wrong.
The sixty-five-year-olds are splendid, vying
with each other in work hours and wit.
They bring their generosity along,
setting the tone, or not giving a shit.
How well, or how eccentrically, they dress!
Their anecdotes are to the point, or wide
enough to make room for discrepancies.
But their children are dying.

Natalie died by gas in Montpeyroux.
In San Francisco, Ralph died
of lung cancer, AIDS years later, Lew
wrote to me. Lew, who at forty-five,

expected to be dead of drink, who, ten
years on, wasn't, instead survived
a gentle, bright, impatient younger man.
(Cliché: he falls in love with younger men.)
Natalie's father came, and Natalie,
as if she never had been there, was gone.
Michèle closed up their house (where she
was born). She shrouded every glass inside

—mirrors, photographs—with sheets, as Jews
do, though she's not a Jew.
James knows, he thinks, as much as he wants to.
He's been working half-time since November.
They made the diagnosis in July.
Catherine is back in radiotherapy.
Her schoolboy haircut, prematurely gray,
now frames a face aging with other numbers:
"stage two," "stage three" mean more than "fifty-one"
and mean, precisely, nothing, which is why
she stares at nothing: lawn chair, stone,
bird, leaf; brusquely turns off the news.

I hope they will be sixty in ten years
and know I used their names
as flares in a polluted atmosphere,
as private reasons where reason obtains
no quarter. Children in the streets
still die in grandfathers' good wars.
Pregnant women with AIDS, schoolgirls, crack whores,

die faster than men do, in more pain,
are more likely than men to die alone.
What are our statistics, when I meet
the lump in my breast, you phone
the doctor to see if your test results came?

The earth-black woman in the bed beside
Lidia on the AIDS floor—deaf and blind:
I want to know if, no, how, she died.
The husband, who'd stopped visiting, returned?
He brought the little boy, those nursery-
school smiles taped on the walls? She traced
her name on Lidia's face
when one of them needed something. She learned
some Braille that week. Most of the time, she slept.
Nobody knew the baby's HIV
status. Sleeping, awake, she wept.
And I left her name behind.

And Lidia, where's she
who got her act so clean
of rum and Salem Filters and cocaine
after her passing husband passed it on?
As soon as she knew
she phoned and told her mother she had AIDS
but no, she wouldn't come back to San Juan.
Sipping café con leche with dessert,
in a blue robe, thick hair in braids,
she beamed: her life was on the right

track, now. But the cysts hurt
too much to sleep through the night.

No one was promised a shapely life
ending in a tutelary vision.
No one was promised: if
you're a genuinely irreplaceable
grandmother or editor
you will not need to be replaced.
When I die, the death I face
will more than likely be illogical:
Alzheimer's or a milk truck: the absurd.
The Talmud teaches we become impure
when we die, profane dirt, once the word
that spoke this life in us has been withdrawn,

the letter taken from the envelope.
If we believe the letter will be read,
some curiosity, some hope
come with knowing that we die.
But this was another century
in which we made death humanly obscene:
Soweto El Salvador Kurdistan
Armenia Shatila Baghdad Hanoi
Auschwitz. Each one, unique as our lives are,
taints what's left with complicity,
makes everyone living a survivor
who will, or won't, bear witness for the dead.

I can only bear witness for my own
dead and dying, whom I've often failed:
unanswered letters, unattempted phone
calls, against these fictions. A fiction winds
her watch in sunlight, cancer ticking bone
to shards. A fiction looks
at proofs of a too-hastily finished book
that may be published before he goes blind.
The old, who tell good stories, half expect
that what's written in their chromosomes
will come true, that history won't interject
a virus or a siren or a sealed

train to where age is irrelevant.
The old rebbetzin at Ravensbrück
died in the most wrong place, at the wrong time.
What do the young know different?
No partisans are waiting in the woods
to welcome them. Siblings who stayed home
count down doom. Revolution became
a dinner party in a fast-food chain,
a vendetta for an abscessed crime,
a hard-on market for consumer goods.
A living man reads a dead woman's book.
She wrote it; then, he knows, she was turned in.

For every partisan
there are a million gratuitous
deaths from hunger, all-American

mass murders, small wars,
the old diseases and the new.
Who dies well? The privilege
of asking doesn't have to do with age.
For most of us
no question what our deaths, our lives, mean.
At the end, Catherine will know what she knew,
and James will, and Melvin,
and I, in no one's stories, as we are.

Nearly a Valediction

You happened to me. I was happened to
like an abandoned building by a bulldozer, like the van that missed my skull
happened a two-inch gash across my chin.
You were as deep down as I've ever been.
You were inside me like my pulse. A new-
born flailing toward maternal heartbeat through
the shock of cold and glare: when you were gone,
swaddled in strange air I was that alone
again, inventing life left after you.

I don't want to remember you as that
four o'clock in the morning eight months long
after you happened to me like a wrong
number at midnight that blew up the phone
bill to an astronomical unknown
quantity in a foreign currency.
The dollar dived since you happened to me.
You've grown into your skin since then; you've grown
into the space you measure with someone
you can love back without a caveat.

While I love somebody I learn to live
with through the downpulled winter days' routine
wakings and sleepings, half-and-half caffeine-
assisted mornings, laundry, stockpots, dust-
balls in the hallway, lists instead of longing, trust
that what comes next comes after what came first.
She'll never be a story I make up.

You were the one I didn't know where to stop.
If I had blamed you, now I could forgive
you, but what made my cold hand, back in prox-

imity to your hair, your mouth, your mind,
want where it no way ought to be, defined
by where it was, and was and was until
the whole globed swelling liquefied and spilled
through one cheek's nap, a syllable, a tear,
was never blame, whatever I wished it were.
You were the weather in my neighborhood.
You were the epic in the episode.
You were the year poised on the equinox.

Days of 1992

Pray for the souls of the antisemites.

—ALFRED CORN, "SOMERSET ALCAICS"

I spent the morning waxing the furniture,
thick orange beeswax sprayed on a chamois cloth,
    dull glow on what was flat and dusty,
    odor of beeswax, a tinge of honey.

Storm-crumbled plaster, storm-swollen window frames,
work for the bookish Orthodox carpenter,
    then a discussion with the plumber
    who will dismantle the bathtub Thursday.

If it were Sunday, I'd do a market run,
Boulevard Richard-Lenoir, half past nine, I'd
    be filling up a wicker basket,
    bathed in the polyglot cries of vendors.

Chard, eggplant, onions, cherries and strawberries,
ecru pleurottes, their undersides filigreed,
    to be sautéed with a breast of chicken?
    Trout? Or a skate wing (black butter, capers)?

But it's not Sunday, only a workaday
Tuesday, after the bleak anniversary
    fifty years later makes the papers
    those city buses, the baffled captives . . .

One generation now, since the stadium,
since the betrayals, since the internment camps,
    since the returning gaunt survivors
    got off the trains that had damned, then saved them,

just to confront revisionist bureaucrats.
As I live out my frozen diaspora
    watching the clouds and writing letters
    what earthly good is my faceless mourning?

What did my neighbors do when the gendarmes came
Jew-hunting in this Jewish arrondissement?
    I've never asked my next-door neighbor,
    frail centenarian, who was fifty-

two then, a few years older than I am now.
I'm wary of inquiring in memories:
    Maybe she liked Pétain, perhaps she
    told the gendarmes where a man was hiding,

Maybe she knew no Jews, ignored the buses,
maybe she hid a scared Jewish girl in her
    dank Turkish toilet on the landing
    until an aunt with forged papers fetched her.

So I invent her, paint her with politics
past, while she follows soaps on her TV set,
    cleans, totes her bread and wine upstairs, feeds
    sparrows in the Place des Vosges dry bread crumbs,

reads daily papers, rightish and populist.
I wait to hear the Sécu come check on her
    mornings. She's background music to my
    life for eight years here, for how much longer?

She goes, and I go, into our histories
as the century's flame-darkened ending
    silences us if we've stayed silent,
    letting the noise of the street subsume us.

Elysian Fields

"Champs Elysées of Broadway" says the awning
of the café where, every Sunday morning,
young lawyers in old jeans ripped at the knees
do crosswords. Polyglot Lebanese
run it: they've taken on two more shopfronts
and run their banner down all three at once.
Four years ago, their sign "Au Petit Beurre,"
was so discreet that, meeting someone there,
I'd tell her the street corner, not the name.
They were in the right place at the right time.
Meanwhile, the poor are trying hard enough.
Outside, on Broadway, people sell their stuff
laid out on blankets, cardboard cartons, towels.
A stout matron with lacquered auburn curls
circles the viridian throw rug
and painted plaster San Martín to hug
a thinner, darker woman, who hugs her
volubly back, in Spanish, a neighbor,
I guess, and guess they still have houses.
The man with uncut, browned French paperbacks,
the man with two embroidered gypsy blouses
and three pilled pitiful pairs of plaid slacks
folded beside him on the pavement where
there was a Puerto Rican hardware store
that's been a vacant shopfront for three years
may not. There's a young couple down the block
from our corner: she's tall, gaunt, gangly, black;
he's short, quick, voluble, unshaven, white.
They set up shop dry mornings around eight.

I've seen him slap her face, jerking her thin
arm like a rag doll's—a dollar kept from him,
she moves too slow, whore, stupid bitch . . . "She's
my wife," he tells a passing man who stops
and watches. If anyone did call the cops
it would be to prevent them and their stacks
of old *Vogue*s and outdated science texts
from blocking access to the "upscale" bar
where college boys get bellicose on beer.
"Leave him" would I say? Does she have keys
to an apartment, to a room, a door
to close behind her? What we meant by "poor"
when I was twenty was a tenement
with clanking pipes and roaches, what we meant
was up six flights of grimed piss-pungent stairs,
four babies and a baby-faced welfare
worker forbidden to say "birth control."
I was almost her, on the payroll
of New York State Employment Services,
the East 14th Street Branch, whose job it was
to send day workers, mostly black, to clean
other people's houses. Five-fifteen,
and I walked east, south, walked up my own four
flights. Poor was a neighbor, was next door,
is still a door away. The door is mine.
Outside, the poor work Broadway in the rain.
The cappuccino drinkers watch them pass
under the awning from behind the glass.

A Note Downriver

Afternoon of hungover Sunday morning
earned by drinking wine on an empty stomach
after meeting Tom for a bomb on Broadway—
done worse, known better.

I feel muggy-headed and convalescent,
barely push a pen across blue-lined paper,
scowl at envelopes with another country's
stamps, and your letter.

Hilltop house, a river to take you somewhere,
sandwiches at noon with a good companion:
summer's ghost flicked ash from the front porch railing,
looked up, and listened.

I would grouse and growl at you if you called me.
I have made you chamomile tea and rye bread
toast, poured us both orange juice laced with seltzer
similar mornings.

We'll most likely live in each other's houses
like I haunted yours last July, as long as
we hear rivers vacillate downstream. They say
"always," say "never."

An Absent Friend

Perched on a high stool, the auburn sybil
eats Fig Newtons, elbows on the sink,
the other lively hand exhorting. Think
through these words to silences. Think to refill
the teapot. The water's come to a boil.
It's four A.M. Tea steeps. We pour, we drink
it milky, volley talk, talk, link
lines read out loud from some book on the pile
accumulated on the scarred oak table
to some felicity of memory.
The black cat hums like a fridge. The three
daughters are asleep in different rooms.
Through the steamed window on the garden comes
pearl dawnlight, lovely and unremarkable.

Lovely and unremarkable, the clutter
of mugs and books, the almost-empty Fig
Newtons box, thick dishes in a big
tin tray, the knife still standing in the butter,
change like the color of river water
in the delicate shift to day. Thin fog
veils the hedges where a neighbor dog
makes rounds. "Go to bed. It doesn't matter
about the washing-up. Take this book along."
Whatever it was we said that night is gone,
framed like a photograph nobody took.
Stretched out on a camp cot with the book,
I think that we will talk all night again
there, or another where, but I am wrong.

Cleis

She's sixteen, and looks like a full-grown woman,
teenage status hinted at by the acne.
I remember infancy's gold, unblemished
skin. I remember

every time I scolded her, slapped her, wished her
someone whom she wasn't, and let her know it.
Every mother knows she betrays her daughter.
Does she? Well, maybe.

She was not the builder of model airplanes.
She was not the runner I never could be.
She was not the pillager of my bookshelves,
Rimbaud or Brontë.

She was not the heroine of a novel.
She was only eight, with a perfect body
caught above the swimming pool, midair, leaping
into blue water

(snapshot, 1982, Vence—she joined me
Air France Unaccompanied Minor). She's the
basic human integer, brown-skinned, golden,
wingless, but flying.

She has breasts and buttocks to keep her earthbound
now. She rereads children's books in her loft bed:
Little Women, *Anne of Green Gables*, *Robin
Hood* and *Black Beauty*

—dreaming herself back out of adolescence
while she talks of cars and her own apartment.
Sixteen is a waiting room: older, younger,
anything's better.

Every day a little bit more a grown-up
face not known yet superimposed on her face
as it turns, a sunflower, out of childhood
"bright and amazing"

like one of her lullabies (by a poet
ragged, old, incontinent, isolated
in a walk-up cluttered with rocks and papers
now, a flamboyant

balladeer once): cats by the fire in winter,
magic cat-king purring beside the singer,
famine and despair in the cries of scrawny
cats on the pavement.

Years now since I stroked her and sang that to her.
Since her breasts grew, I haven't seen her naked.
Infant sweat's like lavender water, hers is
womanly, pungent.

When I was in love with her, with a lover's
tendency to mythify the beloved
did I know her better than I do now, when
we know our limits?

Now she is a traveler like the others,
blonde braid, man's hat, jeans and a gray tweed blazer,
pushing one old duffle bag on a trolley,
free, in an airport

full of haggard voyagers, coming, going.
She stops, sees me. Under the sign ARRIVALS
we embrace, and heft the old bag up, one strap
each, on our shoulders.

Quai Saint-Bernard

I take my Sunday exercise riverside,
not quite a local, not quite a transient.
  Dutch houseboats, gravel barges nose by
  teenagers tanning in day-glo gym shorts.

Waves slick as seal pelts undulate after, like
sun-dappled, ludic sexual animals
  —if you can ignore the floating garbage
  cast by the strollers and weekend sailors.

Three German students nap on their sleeping bags,
backpacks and water bottles niched next to them,
  up on the slope of lawn beside the
  playground, as safe as suburban puppies,

while, underneath a willow, a family,
blonde woman, man like sun-burnished ebony,
  her mother, almond-golden toddler,
  picnic on Camembert, bread and apples.

I brought a book to sit in my favorite
spot, concrete steps that arc in a half-circle
  out from the water. Sometimes barges
  pull up and tie up beside my elbow.

Shit! someone's standing inches in back of me
with all this space . . . From vision's periphery
  I can just make out that it's a woman,
  so I relax. Then she walks around me

on down the quai—a derelict madwoman,
drugged, drunk or tranced, long hair to her knees, with bare
feet, flowered blouse and filthy trousers,
teetering there like a tightrope walker.

She pauses, kneels down, flinging her copious
brown, half-soaked hair, a blindfold, in front of her
so she can't see where she is going
inches away from the churning water.

Who stops her, leads her farther away from the
edge, even asks her what she was doing there?
I don't although she'd come so close her
serpentine shadow fell on my notebook.

She halts, and sways in front of a sunbathing
young man engrossed in reading a paperback.
We others watch her staring at him,
grateful we aren't the one she's chosen.

No crisis, she traverses the half-circle
stone steps, away from water and audience,
sits in the dust behind a basalt
statue, lies down like exhausted dogs do.

So I dismiss her, turn back away from her.
So does the almost-naked man opposite.
We read, relieved of ever knowing
even what language she might have spoken.

Year's End

for Audre Lorde and Sonny Wainwright

Twice in my quickly disappearing forties
someone called while someone I loved and I were
making love to tell me another woman
had died of cancer.

Seven years apart, and two different lovers:
underneath the numbers, how lives are braided,
how those women's deaths and lives, lived and died, were
interleaved also.

Does lip touch on lip a memento mori?
Does the blood-thrust nipple against its eager
mate recall, through lust, a breast's transformations
sometimes are lethal?

Now or later, what's the enormous difference?
If one day is good, is a day sufficient?
Is it fear of death with which I'm so eager
to live my life out

now and in its possible permutations
with the one I love? (Only four days later,
she was on a plane headed west across the
Atlantic, work-bound.)

Men and women, mortally wounded where we
love and nourish, dying at thirty, forty,

fifty, not on barricades, but in beds of
unfulfilled promise:

tell me, senators, what you call *abnormal*?
Each day's obits read as if there's a war on.
Fifty-eight-year-old poet dead of cancer:
warrior woman

laid down with the other warrior women.
Both times when the telephone rang, I answered,
wanting not to, knowing I had to answer,
go from two bodies'

infinite approach to a crest of pleasure
through the disembodied voice from a distance
saying one loved body was clay, one wave of
mind burst and broken.

Each time we went back to each other's hands and
mouths as to a requiem where the chorus
sings death with irrelevant and amazing
bodily music.

Cancer Winter

for Rafael Campo and for Hayden Carruth

1.

Syllables shaped around the darkening day's
contours. Next to armchairs, on desks, lamps
were switched on. Tires hissed softly on the damp
tar. In my room, a flute concerto played.
Slate roofs glistened in the rain's thin glaze.
I peered out from my cave like a warm bear.
Hall lights flicked on as someone climbed the stairs
across the street, blinked out, a key, a phrase
turned in a lock, and something flew open.
I watched a young man at his window write
at a plank table, one pooled halogen
light on his book, dim shelves behind him, night
falling fraternal on the flux between
the odd and even numbers of the street.

2.

I woke up, and the surgeon said, "You're cured."
Strapped to the gurney, in the cotton gown
and pants I was wearing when they slid me down
onto the table, made new straps secure
while I stared at the hydra-headed O.R.
lamp, I took in the tall, confident, brown-
skinned man, and the ache I couldn't quite call pain
from where my right breast wasn't anymore

to my armpit. A not-yet-talking head,
I bit dry lips. What else could he have said?
And then my love was there in a hospital coat;
then my old love, still young and very scared.
Then I, alone, graphed clock hands' asymptote
to noon, when I would be wheeled back upstairs.

3.

The odd and even numbers of the street
I live on are four thousand miles away
from an Ohio February day
snow-blanketed, roads iced over, with sleet
expected later, where I'm incomplete
as my abbreviated chest. I weigh
less—one breast less—since the Paris-gray
December evening, when a neighbor's feet
coming up ancient stairs, the feet I counted
on paper, were the company I craved.
My calm right breast seethed with a grasping tumor.
The certainty of my returns amounted
to nothing. After terror, being brave
became another form of gallows humor.

4.

At noon, an orderly wheeled me upstairs
via an elevator hung with Season's
Greetings posters, bright and false as treason.

The single room the surgeon let us share
the night before the knife was scrubbed and bare
except for blush-pink roses in a vase on
the dresser. Veering through a morphine haze on
the cranked bed, I was avidly aware
of my own breathing, my thirst, that it was over—
the week that ended on this New Year's Eve.
A known hand held, while I sipped, icewater,
afloat between ache, sleep, lover and lover.
The one who stayed would stay; the one would leave.
The hand that held the cup next was my daughter's.

5.

It's become a form of gallows humor
to reread the elegies I wrote
at that pine table, with their undernote
of cancer as death's leitmotiv, enumer-
ating my dead, the unknown dead, the rumor
of random and pandemic deaths. I thought
I was a witness, a survivor, caught
in a maelstrom and brought forth, who knew more
of pain than some, but learned it loving others.
I need to find another metaphor
while I eat up stories of people's mothers
who had mastectomies: "She's eighty-four
this year, and *fine*!" Cell-shocked, I brace to do
what I can, an unimportant exiled Jew.

6.

The hand that held the cup next was my daughter's
—who would be holding shirts for me to wear,
sleeve out, for my bum arm. She'd wash my hair
(not falling yet), strew teenager's disorder
in the kitchen, help me out of the bathwater.
A dozen times, she looked at the long scar
studded with staples, where I'd suckled her,
and didn't turn. She took me / I brought her
to the surgeon's office, where she'd hold
my hand, while his sure hand, with its neat tool, snipped
the steel, as on a revised manuscript
radically rewritten since my star
turn nursing her without a "nursing bra"
from small firm breasts, a twenty-five-year-old's.

7.

I'm still alive, an unimportant Jew
who lives in exile, voluntarily
or not: Ohio's alien to me;
Death follows me home here, but I pay dues
to stay alive. White cell count under two:
a week's delay in chemotherapy
stretches it out. Ohio till July?
The Nazarenes and Pentacostals who
think drinking wine's a mortal sin would pray
for me to heal, find Jesus, go straight, leave.

But I'm alive and can believe I'll stay
alive a while. Insomniac with terror,
I tell myself, it isn't the worst horror,
it's not Auschwitz, it's not the Vel d'Hiv.

8.

I had "breasts like a twenty-five-year-old,"
and that was why, although a mammogram
was done the day of my year-end exam
in which the doctor found the lump, it told
her nothing: small, firm, dense breasts have and hold
their dirty secrets till their secrets damn
them. Out of the operating room
the tumor was delivered, sectioned, cold-
packed, pickled, to demonstrate to residents
an infiltrative ductal carcinoma
(with others of its kind). I've one small, dense
firm breast left, and cell-killer pills so no more
killer cells grow, no eggs drop. To survive
no body stops dreaming it's twenty-five.

9.

It's not Auschwitz. It's not the Vel d'Hiv.
It's not gang rape in Bosnia or
gang rape and gutting in El Salvador.
My self-betraying body needs to grieve
at how hatreds metastasize. Reprieved

(if I am) what am I living for?
Cancer, gratuitous as a massacre,
answers to nothing, tempts me to retrieve
the white-eyed panic in the mortal night,
my father's silent death at forty-eight,
each numbered, shaved, emaciated Jew
I might have been. They wore the blunt tattoo,
a scar, those who survived, oceans away.
Should I tattoo my scar? What would it say?

10.

No body stops dreaming it's twenty-five,
or twelve, or ten, when what is possible's
a long road poplars curtain against loss, able
to swim the river, hike the culvert, drive
through the open portal, find the gold hive
dripping with liquid sweetness. Risible
fantasy, if, all the while, invisible
entropies block the roads, so you arrive
outside a ruin, where trees bald with blight
wane by a river drained to sluggish mud.
The setting sun looks terribly like blood.
The hovering swarm has nothing to forgive.
Your voice petitions the indifferent night:
"I don't know how to die yet. Let me live."

11.

Should I tattoo my scar? What would it say?
It could say "KJ's Truck Stop" in plain English, highlighted with a nipple ring
(the French version "Chez KJ: Les Routiers").
I won't be wearing falsies, and one day
I'll bake my chest again at Juan-les-Pins,
round side and flat, gynandre/androgyne
close by my love's warm flanks (though she's sun-shy
as I should be: it's a carcinogen
like smoked fish, caffeine, butterfat and wine).
O let me have my life and live it too!
She kissed my breasts, and now one breast she kissed
is dead meat, with its pickled blight on view.
She'll kiss the scar, and then the living breast.

12.

I don't know how to die yet. Let me live!
Did Etty Hillesum think that, or Anne Frank,
or the forty-year-old schoolteacher the bank
robber took hostage, when the cop guns swiv-
eled on them both, or the seropositive
nurse's aide, who, one long-gone payday, drank
too much, fucked whom? or the bag lady who stank
more than I wished when I came closer to give
my meager change? I say it, bargaining
with the Contras in my blood, immune

system bombarded but on guard. Who's gone?
The bookseller who died at thirty-nine,
poet, at fifty-eight, friend, fifty-one,
friend, fifty-five. These numbers do not sing.

13.

She'll kiss the scar, and then the living breast,
and then again, from ribs to pit, the scar,
but only after I've flown back to her
out of the unforgiving Middle West
where my life's strange, and flat disinterest
greets strangers. At Les-Saintes-Maries-de-la-Mer
lust pulsed between us, pulsed in the plum grove where
figs dropped to us like manna to the blessed.
O blight that ate my breast like worms in fruit,
be banished by the daily pesticide
that I ingest. Let me live to praise
her breathing body in my arms, our wide-
branched perennial love, from whose taproot
syllables shape around the lengthening days.

14.

Friends, you died young. These numbers do not sing
your requiems, your elegies, our war
cry: at last, not "Why me?" but "No more
one-in-nine, one-in-three, rogue cells killing
women." You're my companions, traveling

from work to home to the home I left for
work, and the plague, and the poison that might cure.
The late sunlight, the morning rain, will bring
me back to where I started, whole, alone,
with fragrant coffee into which I've poured
steamed milk, book open on the scarred pine table.
I almost forget how close to the bone
my chest's right side is. Unremarkable,
I woke, alive today. Does that mean "cured"?

from *Squares and Courtyards* (2001)

Street Scenes: Sunday Evening

Flowers at the plinth, curbside, said he wouldn't be back—
a bucket of red tulips, with a sign:
he died the night of February 9th
of cold—hand-scrawled, but worded like a plaque
for an assassinated partisan
"shot here in '42 by the SS."
His friends (it said) are invited to Mass
at the Église St-Paul, rue St-Antoine.
He was called Monsieur Guy. The man who lived
there, died under the arches of the Place
des Vosges. We went, remembering him, to the late Mass.
An edgy, bearded man lit white votive
candles, one, then another, dropping five-franc
pieces, his lunch, his breakfast, his next drink,
into the alms box, pacing like a monk
from Mary's candle-lit feet to coin slot. Mink
coats, camel's hair, duffles and anoraks
(on children) clustered in the center pews.
In torn windbreakers and bald running shoes,
the pavement's citizens observed their backs
discreetly. An eighty-year-old woman, whose
face was intelligent as it was clear,
embraced, named, greeted each street person there
to mourn a friend. Couchant beside her, close
to the hem of her neat navy blue
wool coat, a huge Rottweiler cocked black ears
when she (chime-pure) sang, but didn't echo her
responses. Monsieur Guy had a dog too,
a venerable Labrador. Its bed

was blankets in a carton. Mornings, he'd
spit on his handkerchief and swab its eyes.
Thin, lank black hair—both looked wearily old.
"Is that *his* dog?" you asked, meaning the wise
vigilant Rottweiler. Stories I'd told
to you out of the fabric of a day
included them: I'd have passed them on the way
to Sunday market, stood in line behind while he
bought dog food in the late-night Arab grocery.
Once, man and dog reclined against a sun-
warmed wall in August. Monsieur Guy clapped his hands
as an old Japanese couple danced
deft tangoes to a street accordion.

The priest referred to wanderers, not sinners,
and the good works of St. Vincent de Paul.
A red-faced woman teetered in the aisle.
We shared handclasps of greeting with the small
street cohort, the bright-voiced dog owner, whose smile
included us. The Labrador is gone,
his friend dead of exposure in the cold,
the evanescent fellowship of one
evening's community hardly recalled
when pious civic families meet old
souls in the street, ragged and reeking wine.
Beside the altar, diffident as beginners,
beside the bucket, with a fresh bouquet,
young street musicians, scrubbed, still dressed in jeans,
grouped in their student string quartet, which plays

under the arches weekend afternoons,
struck up unfunereal Vivaldi
as recessional music Monsieur Guy
(and I) had heard on numberless Sundays.
We went to our respective fasts and dinners.

Scars on Paper

An unwrapped icon, too potent to touch,
she freed my breasts from the camp Empire dress.
Now one of them's the shadow of a breast
with a lost object's half-life, with as much
life as an anecdotal photograph:
me, Kim and Iva, all stripped to the waist,
hiking near Russian River on June first
'79: Iva's five-and-a-half.
While *she* was almost twenty, wearing black
T-shirts in D.C., where we hadn't met.
You lay your palm, my love, on my flat chest.
In lines alive with what is not regret,
she takes her own path past, doesn't turn back.
Persistently, on paper, we exist.

Persistently, on paper, we exist.
You'd touch me if you could, but you're, in fact,
three thousand miles away. And my intact
body is eighteen months paper: the past
a fragile eighteen months regime of trust
in slash-and-burn, in vitamin pills, backed
by no statistics. Each day I enact
survivor's rituals, blessing the crust
I tear from the warm loaf, blessing the hours
in which I didn't or in which I did
consider my own death. I am not yet
statistically a survivor (that
is sixty months). On paper, someone flowers
and flares alive. I knew her. But she's dead.

She flares alive. I knew her. But she's dead.
I flirted with her, might have been her friend,
but transatlantic schedules intervened.
She wrote a book about her Freedom Ride,
the wary elders whom she taught to read,
—herself half-British, twenty-six, white-blonde,
with thirty years to live.
                                        And I happened
to open up *The Nation* to that bad
news which I otherwise might not have known
(not breast cancer: cancer of the brain).
Words take the absent friend away again.
Alone, I think, she called, alone, upon
her courage, tried in ways she'd not have wished
by pain and fear: her courage, extinguished.

The pain and fear some courage extinguished
at disaster's denouement come back
daily, banal: is that brownish-black
mole the next chapter? Was the ache enmeshed
between my chest and armpit when I washed
rogue cells' new claw, or just a muscle ache?
I'm not yet desperate enough to take
comfort in being predeceased: the anguish
when the Harlem doctor, the Jewish dancer,
die of AIDS, the Boston seminary's
dean succumbs "after brief illness" to cancer.
I like mossed slabs in country cemeteries
with wide-paced dates, candles in jars, whose tallow
glows on summer evenings, desk-lamp yellow.

Aglow in summer evening, a desk-lamp's yellow
moonlight peruses notebooks, houseplants, texts,
while an ageing woman thinks of sex
in the present tense. Desire may follow,
urgent or elegant, cut raw or mellow
with wine and ripe black figs: a proof, the next
course, a simple question, the complex
response, a burning sweetness she will swallow.
The opening mind is sexual and ready
to embrace, incarnate in its prime.
Rippling concentrically from summer's gold
disc, desire's iris expands, steady
with blood-beat. Each time implies the next time.
The ageing woman hopes she will grow old.

The ageing woman hopes she will grow old.
A younger woman has a dazzling vision
of bleeding wrists, her own, the clean incisions
suddenly there, two open mouths. They told
their speechless secrets, witnesses not called
to what occurred with as little volition
of hers as these phantom wounds.

Intense precision
of scars, in flesh, in spirit. I'm enrolled
by mine in ranks where now I'm "being brave"
if I take off my shirt in a hot crowd
sunbathing, or demonstrating for Dyke Pride.
Her bravery counters the kitchen knives'

insinuation that the scars be made.
With, or despite our scars, we stay alive.

"With, or despite our scars, we stayed alive
until the Contras or the Government
or rebel troops came, until we were sent
to "relocation camps," until the archives
burned, until we dug the ditch, the grave
beside the aspen grove where adolescent
boys used to cut class, until we went
to the precinct house, eager to behave
like citizens . . ."

I count my hours and days,
finger for luck the word-scarred table which
is not my witness, shares all innocent
objects' silence: a tin plate, a basement
door, a spade, barbed wire, a ring of keys,
an unwrapped icon, too potent to touch.

The Boy

Is it the boy in me who's looking out
the window, while someone across the street
mends a pillowcase, clouds shift, the gutterspout
pours rain, someone else lights a cigarette?

(Because he flinched, because he didn't whirl
around, face them, because he didn't hurl
the challenge back—*"Fascists?"*—not *"Faggots"*—*"Swine!"*
he briefly wonders—if he were a girl . . .)
He writes a line. He crosses out a line.

I'll never be a man, but there's a boy
crossing out words: the rain, the linen-mender,
are all the homework he will do today.
The absence and the privilege of gender

confound in him, soprano, clumsy, frail.
Not neuter—neutral human, and unmarked,
the younger brother in the fairy tale
except, boys shouted *"Jew!"* across the park

at him when he was coming home from school.
The book that he just read, about the war,
the partisans, is less a terrible
and thrilling story, more a warning, more

a code, and he must puzzle out the code.
He has short hair, a red sweatshirt. They know

something about him—that he should be proud
of? That's shameful if it shows?

That got you killed in 1942.
In his story, do the partisans
have sons? Have grandparents? Is he a Jew
more than he is a boy, who'll be a man

someday? Someone who'll never be a man
looks out the window at the rain he thought
might stop. He reads the sentence he began.
He writes down something that he crosses out.

Days of 1994: Alexandrians

for Edmund White

Lunch: as we close the twentieth century,
death, like a hanger-on or a wannabe
sits with us at the cluttered bistro
table, inflecting the conversation.

Elderly friends take lovers, rent studios,
plan trips to unpronounceable provinces.
Fifty makes the ironic wager
that his biographer will outlive him—

as may the erudite eighty-one-year-old
dandy with whom a squabble is simmering.
His green-eyed architect companion
died in the spring. He is frank about his

grief, as he savors spiced pumpkin soup, and a
sliced rare filet. We'll see the next decade in
or not. This one retains its flavor.
"Her new book . . ." ". . . brilliant!" "She slept with . . ." "*Really!*"

Long arabesques of silver-tipped sentences
drift on the current of our two languages
into the mist of late September
midafternoon, where the dusk is curling

Just thirty-eight: her last chemotherapy
treatment's the same day classes begin again.
    I went through it a year before she
    started; but hers was both breasts, and lymph nodes.

She's always been a lax vegetarian.
Now she has cut out butter and cheese, and she
    never drank wine or beer. What else is
    there to eliminate? Tea and coffee . . .

(Our avocado salads are copious.)
It's easier to talk about politics
    than to allow the terror that shares
    both of our bedrooms to find words. It made

the introduction; it's an acquaintance we've
in common. Trading medical anecdotes
    helps out when conversation lapses.
    We don't discuss Mitterrand and cancer.

Four months (I say) I'll see her, see him again.
(I dream my life; I wake to contingencies.)
    Now I walk home along the river,
    into the wind, as the clouds break open.

Rue Debelleyme

Rain from the channel: wind and rain again
umbrellas jostle on the pavement, crowd
together, move apart. Atlantic rain

south from the British Isles. A monocloud
covers the sky that yesterday was blue
and filled with light, where clement winds allowed

expansive breathing, new air flowing through
a sentence or a ribbon or a song
children sang complicated verses to:

a day I could be grateful to for long
light: although not June, still just July
when no direction was entirely wrong

for finding points to take my bearings by
and walk around the corner of a street
that's always there, a small discovery.

If you've misplaced the key, the door is shut
but every street's a door that opens up,
the narrow gangway to a bannered boat:

run up before it raises anchor, slipping otter-like from moorings. On the dock
hands wave bright scarves, and colored pennants flap.

A bus pulled out, a taxi stopped, a truck
parked curbside, the driver undid a latch,
put down a ramp, rolled out a garment-rack.

Two black girls on boot-skates stopped to watch
dusty from play, homebound at one o'clock
with nectarines and two baguettes for lunch.

(If you've misplaced the key, you're out of luck,
but every window framed another key
A garden past the crossing winked back black,

copper, gold children to their serious play.)
Sisters, from their matching innocent
navy-blue pleats hemmed short above the knee.

(Somewhere in the next arrondissement
women do piecework in small factories,
mostly undocumented immigrants,

Filipina, African, Chinese,
some of whose children become secular
and republican *lycéens*.) Did these

two with neat ribbons in their cornrowed hair
and roller-skating scabs on their bare knees
memorize La Fontaine and Baudelaire,

and did the rack of one cloned summer dress
with lime-green polka dots and large puffed sleeves
remind them of the end-of-term *kermesse*,

the job their mother hardly ever leaves,
or some preadolescent feminine
world I wouldn't recognize, believe,

or, with the best-intentioned will, imagine?
Their futures opening like a painted fan:
hairdresser, film director, *lycéenne*,

they skated off, one with the nectarines
the other with the loaves under her arm
towards a deserved repast of citizens,

(as I imagined going home with them
the driver, padlocking his empty van,
set off an inadvertent car-alarm)

the lost key in some jacket pocket found
as, equally irrelevant, the rain
clouds open out onto the blue of noon.

Again, the River

for Geneviève Pastre

Early summer in what I hope is "midlife,"
and the sunlight makes me its own suggestions
when I take my indolence to the river
and breathe the breeze in.

Years, here, seem to blend into one another.
Houseboats, tugs and barges don't change complexion
drastically (warts, wrinkles) until gestalt-shift
dissolves the difference.

Sentence fragments float on a wave of syntax,
images imprinted in contemplation,
indistinct impressions of conversations
which marked some turning.

Food and drink last night with a friend—we've twelve years'
history of Burgundy and good dinners
and as many books off the press between us
toasted together.

Writing is a difficult form of reading.
Paragraphs that roll away from their moorings
seem like passages to another language
half-comprehended.

Sometimes thought is more like a bad translation
Hazy shapes resistant to sentence structure
intimate—but what do they mean, exactly?
Texture, sound, odor

(dockside, urinous, up on green slopes, roses
in full bloom like elegant girls of forty)
imprint images in aleatoric
absence of order.

Isolated words can unlock a story:
what you ate, she felt when she heard the music,
what's brought back by one broken leaf, whose sticky
sap on a finger

named a green, free season to city children.
Now, daylight's duration is equinoctial:
spring is turning swiftly to summer, summer's
ripeness brings endings.

I can feel a change in the weather coming.
When I catch a glimpse of myself in mirrors,
I see someone middle-aged, with my mother's
sallow complexion.

Who do we write books for—our friends, our daughters?
Last night's dinner companion has two daughters,
women in their thirties with strong opinions.
My child is younger

and might say there won't be books in the "20 . . . s,"
just hard copy downloaded from computers.
Children won't haunt library aisles, as I did,
tracking their futures.

(What about the homeless man reading science
fiction on the steps of St. Paul, a tattered
paperback, a galaxy on the cover
he was approaching?)

Houses are precarious or unsettling.
We who left them young, and applaud our daughters'
rootlessness still scrutinize wind-chapped faces
of pavement dwellers.

My friend's postcard goddesses, morning teapot,
Greek and Latin lexicons, Mac computer
fill the magic cave of a room she works in
which she'll be leaving

when her lease is up (as provincial theater
troupes strike sets, pack trunks) lares and penates
ready to be set on a desk and bookshelves
in closer quarters

where she'll re-establish haphazard suppers
on her Cévennes grandmother's round oak table.
Where will I be? Too many airline tickets
away to answer.

(I lead two lives superimposed upon each
other, on two continents, in two cities,
make believe my citizenship is other
than that blue passport's.)

But today there's wind on the Seine; a tugboat
with embroidered curtains and gardened windows
looks like home as it navigates the river
toward other moorings.

Wednesday I.D. Clinic

for K.J.

Your words are ones the patients said themselves.
You carry them inside yourself, their vessel.

The widowed black man with two half-white children might
have given them up, have given up, this time

next month. But you don't say: that woman; this man.
You know their faces. You tell me a first name,

temperament and age, even a T-cell
count, if I ask, which will probably be less

than it was. Not always. Someone bursts into tears.
Someone drags his chair closer, to stare

at you, as if your eyes, your collar, your lips,
said more than that sentence. He asks for vitamin pills.

She asks for condoms. He asks for simpler words.
She shifts the murmuring baby, lets him drowse

against her breast, bounces him on her knee,
starts, almost imperceptibly, to keen

a lullaby, or is it a lament?
As your heart beats, you rock her, in a mental

mutual embrace (you've hugged her) which allows
you to breathe with her, pause with her, swallow

the hard words. She's with you when you come downtown
later. You could keep it to yourself. You won't.

Squares and Courtyards

Across the Place du Marché Ste-Catherine
the light which frames a building that I see
daily, walking home from the bakery,
white voile in open windows, sudden green
and scarlet window-box geraniums
backlit in cloud-encouraged clarity
against the century-patinaed gray
is such a gift of the quotidian,
a benefice of sight and consciousness,
I sometimes stop, confused with gratitude,
not knowing what to thank or whom to bless,
break off an end of seven-grain baguette
as if my orchestrated senses could
confirm the day. It's fragrant. I eat it.

Confirm the day's fragrance. I eat, bit
by bit, the buttery *pain aux raisins*
shell-coiled beside my steaming afternoon
tea. It's the hour for a schoolchild's treat,
munched down, warm in waxed paper, on the street,
or picked at on chipped earthenware (like mine)
beside books marked with homework to be done
while the street's sunlit, dusk-lit, lamplit.
She sucks her pencil, window-framed. I sip
nostalgia for a childhood not my own
Bronx kitchen table, with a fire escape
in the alley shaded by sumac trees
which filtered out the other languages
I heard the airshaft's crosscurrents intone.

I heard the airshaft's crosscurrents intone
below the minyan davening morning offices.
A childish rasp that slurred and sputtered was
the Polish janitor's red-knuckled son
helping his father empty garbage cans.
His voice was why I thought him rough (as is
English when spoken by its novices)
a voice I never heard speaking its own
language. His name was Joseph. He was six.
Other syllables connected news
from gutted Europe to the dusty motes
of Sabbath morning. Ash settled on bricks,
spun up the shaft with voices of old Jews,
was drawn down garrulous chain-smokers' throats.

Drawn up from garrulous chain-smokers' throats
at round tin tables on wet cobblestones
just hosed down by a green-clad African
street cleaner: strikes, prices, who still votes
Left, sex, a sick child. Hands unbutton coats
halfway. The wind's mild, but it looks like rain
above the Place du Marché Ste-Catherine
where charcoal-bellied clouds converge like boats
in the mutable blue harbor sky.
Another coffee, another *blanc sec*—
as if events were ours to rearrange
with words, as if dailiness forestalled change,
as if we didn't grow old (or not) and die
as long as someone listened when we spoke.

As long as someone listened when I spoke
especially someone walking a dog—
I'd launch into juvenile monologue:
Greek myths, canine behavior—and could I stroke
the Lab or spaniel? Speech and touch invoked
my grandmother, the bookkeeper from Prague,
who died as I emerged out of the fog
of infancy, while lives dispersed in smoke
above the camps (and Dresden, and Japan)
and with them, someone else I might have been
if memory braided with history.
I pressed my face into the dog's warm fur
whose heat and smell I learned by heart, while she
receded into words I found for her.

Receding into words I found for her
delight, someone was dispossessed of her own
story (she thought) by mine.
Receding in-
to words, the frail and early-rising neighbor
who died during my cancer-treatment year
is not summed up by "centenarian."
Her century requires a lexicon.
I wrote a girl on paper when I bore
a child, whose photocopied life became
letters tattooed across a watermark,
a woman's in the world, who shares her name.
And Gísela, who took me to the park,
for whom I pieced together sentences
—it's all the words she said to me I miss.

It's all the words she said to me I miss,
down to unechoed accents. Did she speak
Yiddish to me? With whom did she speak Czech?
German was what my father spoke till his
sixth year, first grade (when did he tell me this?)
—his parents' common tongue. And did they make
love in their second language? The air's thick
with cognates, questions and parentheses
she'll scribble down once she's back in her room,
chewing her braid, tracing our labyrinthine
fragments. She zips her anorak
and shifts the heavy satchel on her back
watching low clouds gather as she walks home
across the Place du Marché Ste-Catherine.

Not knowing what to thank or whom to bless,
the schoolgirl at the window, whom I'm not,
hums cadences it soothes her to repeat
which open into other languages
in which she'll piece together sentences
while I imagine her across the street
as late light shifts, sunlit, dusk-lit, lamplit.
Is there a yellow star sewed to her dress
as she exults, confused with gratitude,
her century requires a lexicon
of memory braided with history
she'll have reflective decades to write down?
Not thinking, she'll get old (or not) and die;
thinking: she can, if anybody could.

from Paragraphs from a Daybook

for Hayden Carruth

Filthiest of cold mornings, with the crumbs
of my breakfast *tartine* and the dregs of tea,
to clear away. On the market street the bums,
long-term jobless, stateless, *sans-abri*—
meaning, those without shelter—
crouch on cardboard, wrapped in frayed woolens, filter
out the wind as best they can, discreetly beg:
a plastic bowl beside rag-swaddled legs.
They all are white, and half of them are women.
I talk with one: tall, stained teeth, arched nose and cheekbones
like Norman gentry. She's soft-spoken
as a fifth-grade teacher, who'd have shown
me fluvial maps, and pointed out the human
scale of geography. She huddles down
on the florist's doorstep in the rue Saint-Antoine.

Her friend camps daily on the métro stairs,
a tiny skinny woman with blue eyes.
I gave her my old gloves and a blue mohair
scarf when it was five below. Despite her size
and lack of an ounce of fatty insulation,
she vaunts her indomitable constitution
to layered housewives who pass the time of day
with laden caddies, homebound before noon.
In summer, they more or less live on the Quai
Saint-Bernard. The little one strips for the sun
to shorts and a tank top, turning crinkled coffee-brown
around her aster eyes, and looks even thinner,
while her friend tucks a print skirt over her knees and relaxes.
Close to midnight, I sometimes see them sharing dinner
on a plastic plate, on the steps of the Bureau of Taxes.

The topic was "love," and I thought about bound feet;
"how writers invent love with words"—somewhere there is
a trove of "lotus-foot" poems. But how do girls complete
a thought without a word for "clitoris"?
—though there probably is a word meaning "what is cut,"
semantically akin to venom or shit
used when five-year-olds are maimed
with razor blades: that once, it's named.
We think about the things for which we've words;
words tell us what they think of
us, and the paragraph fast-forwards
to a trampled patch of bloody turf
or a kaleidoscope of bright imagination
in which it is possible to focus "love"
without envisaging some mutilation.

In winter, the produce on the stalls
is rufous roots, dark leaves, luminous tubers,
as if earth voided jewels from its bowels
for my neighbors'
Sunday stew pots. Concurrent raucous calls
and odors waft among the vegetables:
merguez sizzles in a skillet, fowl
turn on a row of spits. Damp dogs prowl
between wool-stockinged calves and corduroys
Tissue-wrapped clementines
from Morocco (gold from old colonies)
salt fish from Portugal and Spain's
olives and oil; cauliflower from Brittany,
also the channel-crossing mist of rain
down from the northwest coast since yesterday

I almost gushed to my friend about a movie
I'd just seen: the son of a concentration-
camp survivor's homage. Mother tells son
the volumes she remembers. Now she's seventy-
something, tangoes in high-heeled elegance
over the abyss of memory.
But we were balancing fine points of translation
with forkfuls of ratatouille in a café
the freezing afternoon of New Year's Eve,
and both of us had other things to say.
Our plates were cleared. With habitual diffidence,
she handed a new manuscript to me
and took (to the Ladies') momentary leave.
I turned a page and read the dedication
to her father, who died at Bergen-Belsen.

for Muriel Rukeyser

Was a woman. Was quixotically prolific.
Was a Jew. Died too young.
Owned up to violence to be pacific.
Unmothered by the words shaped on her tongue,
she chose a child, borne in the aftershock
of her own youth, awakened in world wars.
Her first lover was Republican Spain;
her brothers, the strip miners with brown-lung
disease, brothers for whom
she bore witness, from whom she learned to listen.
Square and dark, she keened outside locked doors,
Homer and Hecuba, woman as epic.
The river of her body was the Hudson,
to which all other rivers brought her back,
an important Jew who died at home.

However well I speak, I have an accent
tagging my origins: that teflon fist,
that hog-wallow of investment
that hegemonic televangelist's
zeal to dumb the world down to its virulent
cartoon contours, with the world's consent:
your heads of state, in cowboy suits
will lick our leader's lizard boots.
My link to that imperial vulgarity
is a diasporic accident:
pogroms in Austria, in Hungary,
the quota, the boat, the apartment
up six cabbagey flights, overtime in the garment
trade, the children fiercely intent
on speaking well, without an accent.

Thought thrusts up, homely as a hyacinth
wrapped in its bulb like a root-vegetable,
a ninth-month
belly, while the green indelible
pattern's inscribed into the labyrinth.
Lanced into light, it's air's inhabitant
with light and air as food and drink.
A hyacinth, tumescent pink
on the low wooden Mexican chest
confronts the wintry dusk
with informed self-interest.
Leaf-spears extravagantly ask
what idea, still gnarled up in a knot
of ganglions, will break through the husk
shaped at last, recognizable as thought

Trace, on a city map, trajectories
of partially forgotten words
along the river's arteries,
volatile substance of a sentient world.
Mauve heather crowds the window-grille. The light
lingers a little later, with a slight
vernal inflection. In a moon-glazed vase
bloom yellow freesias, like some rainy day's
brook bank, in someone else's memory.
Small whirlpools of perception widen, ring
an infant's numinous discoveries
of syllables for animals, toys, trees:
a Lab's thick coat, the dusty birds
in Claremont Park each tardy urban spring,
a stuffed pink leather horse with button eyes.

A question mark in yellow overalls,
I could read. I was three.
I slept with that pink horse. My one doll's
name was V.J. She'd been given to me
to celebrate the Victory
over Japan, that is to say, the Bomb
I'd spend my schooldays taking shelter from.
I couldn't tie my shoes. But Reddy the fox,
Tootle the engine who jumped off the tracks,
spelled me their stories on my mother's lap
despite weak eyes and poor small-motor skills.
My grandparents were dead: not in pogroms,
not in the camps—of strokes and heart attacks,
merely immigrants, not deportees.
"When you die, does everything just stop?"

Death is the scandal we wake up to, Hayden:
that flash in childhood, then every blue day.
Once conscious of desire, we're laden
with its accountability.
Death and the singer; death and the maiden:
duets you've taken both voices, and played into measured words, their numbers cast-
out lines which lured a shape out past
the lovely bodies which it mimed and praised.
Now the dazzling shape's your own
daughter's in the dance, appalled, amazed,
as soundwaves track disaster in the bone.
Futile and gorgeous gestures: words employed
tracking the inconscient revolution
of the wheel whose spokes revolve around a void.

I'm four, in itchy woolen leggings,
the day that I can't recognize the man
down at the park entrance, waving,
as my father. He has ten
more years to live, that spring. Dapper and balding
he walks toward me; then I run toward him, calling
him, flustered by my flawed vision.
Underfoot, the maples' green-
winged seeds splay on mica-specked octagons.
His round face, thin nose, moustache silvered gray
at thirty-eight look (I think now) Hungarian.
I like his wood-smell of two packs a day
as he swings me up to his shoulder
and I say, things look blurry far away
—one Saturday, two years after the war.

Grief, pain and sorrow all are *"la douleur,"*
while *"le bonheur"* is simple happiness
which we savored in the hour
seized as the solstice passed
across the heather-misted calendar
whose olive-brown hillocks' December blur
was pierced by the setting sun
as we meandered, *vigneron*
to *vigneron*, well-spring to orchard, stopped
for *Le Monde*, for the view,
pleasure both cumulative and abrupt:
sudden suave vista; beauty we knew
(mist imperceptibly becoming rain)
well enough to recall, while going through
the nuances of sorrow, grief and pain.

Cherry-ripe: dark sweet *burlats*, scarlet *reverchons*
firm-fleshed and tart in the mouth
bigarreaux, peach-and-white *napoléons*
as the harvest moves north
from Provence to the banks of the Yonne
(they grow *napoléons* in Washington
State now). Before that, *garriguettes*,
from Perigord, in wooden punnets
afterwards, peaches: yellow-fleshed, white,
moss-skinned ruby *pêches de vigne.*
The vendors cry out "Taste," my appetite
does, too. Birdsong, from an unseen
source on this street-island, too close for the trees:
it's a young woman with a tin basin
of plastic whistles moulded like canaries—

—which children warbled on in Claremont Park
one spring day in my third year. Gísela
my father's mother, took
me there. I spent the days with her
now that my mother had gone back to work.
In a brocade satchel, crochet-work, a picture book
for me. But overnight the yellow bird
whistles had appeared
and I wanted one passionately.
Watching big girls play hopscotch at curb's edge
or telling stories to V.J.
under the shiny leaves of privet hedge
were pale pastimes compared to my desire
Did I hector one of the privileged
warblers to tell us where they were acquired?

—the candy store on Tremont Avenue
Of course I don't call her *Gísela*.
I call her Grandma. "Grandma will buy it for you,"
—does she add *"mammele"*
not letting her annoyance filter through
as an old-world friend moves into view?
The toddler and the stout
gray-haired woman walk out
of the park oval toward the shopping streets
into a present tense
where what's ineffaceable repeats
itself. Accidents.
I dash ahead, new whistle in my hand
She runs behind. The car. The almost-silent
thud. Gísela, prone, also silent, on the ground.

Death is the scandal that was always hidden.
I never saw my grandmother again.
Who took me home? Somebody did. In
the next few days (because that afternoon
and night are blank) I don't think I cried, I didn't
know what to ask (I wasn't three), and then I did, and
"She's gone to live in Florida" they said
and I knew she was dead.
A black woman, to whom I wasn't nice,
was hired to look after me.
Her name was Josephine—and that made twice
I'd heard that name: my grandmother's park crony
was Josephine. Where was Grandma, where was Gísela?
she called me to her bench to ask one day.
I say, "She's gone to live in Florida."

for Claire Malroux

On a beechwood sideboard, there sat in state
an object whose functional equivalent
would be, in American, a trivet,
but "trivet" originally meant
something three-legged—no, that isn't it.
A recollection I cannot translate:
carved wood, a blue ceramic square,
chimes which a child with short brown hair
released into the air, turning a key,
on a noon-shuttered kitchen's red-tiled floor.
The still heat of the estival Midi
exhaled, leonine, beyond the door
as the child, bare-legged and barefoot,
made up verses for
the tune she'd conjured out of the hot plate—

—if that's the word for it.
A gray June afternoon outside Auxerre,
the last few tables of a flea market:
on one of them, box-like, carved wood, a square
tile, with fin-de-siècle bathers, set
in it, a key between its four squat feet
which I turn. *"Für Elise"*
chimes in the dusty marketplace.
And somehow I participate
in a midsummer memory
of a cool moment, a still neutral date.
The thin child, a large scab on her right knee,
stands in the shuttered midday darkness, while
I hold what's entered my own history:
music; carved wood, a blue ceramic tile.

from *Desesperanto*
(2003)

Elegy for a Soldier

June Jordan, 1936–2002

I.

The city where I knew you was swift.
A lover cabbed to Brooklyn
(broke, but so what) after the night shift
in a Second Avenue
diner. The lover was a Quaker,
a poet, an anti-war
activist. Was blonde, was twenty-four.
Wet snow fell on the access
road to the Manhattan Bridge. I was
neither lover, slept uptown.
But the arteries, streetlights, headlines,
phonelines, feminine plural
links ran silver through the night city
as dawn and the yellow cab
passed on the frost-blurred bridge, headed for
that day's last or first coffee.

The city where I knew you was rich
in bookshops, potlucks, ad hoc
debates, demos, parades and picnics.
There were walks I liked to take.
I was on good terms with two rivers.
You turned, burned, flame-wheel of words
lighting the page, good neighbor on your
homely street in Park Slope, whose
Russian zaydes, Jamaican grocers,

dyke vegetarians, young
gifted everyone, claimed some changes
—at least a new food co-op.
In the laundromat, ordinary
women talked revolution.
We knew we wouldn't live forever
but it seemed as if we could.

The city where I knew you was yours
and mine by birthright: Harlem,
the Bronx. Separately we left it
and came separately back.
There's no afterlife for dialogue,
divergences we never
teased apart to weave back together.
Death slams down in the midst of
all your unfinshed conversations.
Whom do I address when I
address you, larger than life as you
always were, not alive now?
Words are not you, poems are not you,
ashes on the Pacific
tide, you least of all. I talk to myself to keep the line open.

The city where I knew you is gone.
Pink icing roses spelled out
PASSION on a book-shaped chocolate cake.
The bookshop's a sushi bar

now, and PASSION is long out of print.
Would you know the changed street that
cab swerved down toward you through cold white mist?
We have a Republican
mayor. Threats keep citizens in line:
anthrax; suicide attacks.
A scar festers where towers once were;
dissent festers unexpressed.
You are dead of a woman's disease.
Who gets to choose what battle
takes her down? Down to the ocean, friends
mourn you, with no time to mourn.

II.

You, who stood alone in the tall bay window
of a Brooklyn brownstone, conjuring morning
with free-flying words, knew the power, terror
in words, in flying;

knew the high of solitude while the early
light prowled Seventh Avenue, lupine, hungry
like you, your spoils raisins and almonds, ballpoint
pen, yellow foolscap.

You, who stood alone in your courage, never
hesitant to underline the connections
(between rape, exclusion and occupation . . .)
and separations

were alone and were not alone when morning
blotted the last spark of you out, around you
voices you no longer had voice to answer,
eyes you were blind to.

All your loves were singular: you scorned labels.
Claimed *black*; *woman*, and for the rest eluded
limits, quicksilver (Caribbean), staked out
self-definition.

Now your death, as if it were “yours”: your house, your
dog, your friends, your son, your serial lovers.
Death’s not “yours,” what’s yours are a thousand poems
alive on paper.

You, at once an optimist, a Cassandra,
Lilith in the wilderness of her lyric,
were a black American, born in Harlem,
citizen soldier.

If you had to die—and I don’t admit it—
who dared “What if, each time they kill a black man /
we kill a cop?” couldn’t you take down with you
a few prime villains

in the capitol, who are also mortal?
June, you should be living, the states are bleeding.
Leaden words like “Homeland” translate abandoned
dissident discourse.

Twenty years ago, you denounced the war crimes
still in progress now, as Jenin, Ramallah
dominate, then disappear from the headlines.
Palestine: your war.

"To each nation, its Jews," wrote Primo Levi.
"Palestinians are Jews to Israelis."
Afterwards, he died in despair, or so we
infer, despairing.

To each nation its Jews, its blacks, its Arabs,
Palestinians, immigrants, its women.
From each nation, its poets: Mahmoud Darwish,
Kavanagh, Sháhid

(who, beloved witness for silenced Kashmir,
cautioned, shift the accent, and he was "martyr"),
Audre Lorde, Neruda, Amichai, Senghor,
and you, June Jordan.

Crepuscule with Muriel

Instead of a cup of tea, instead of a milk-
silk whelk of a cup, of a cup of nearly six-
o'clock teatime, cup of a stumbling block,
cup of an afternoon unredeemed by talk,
cup of a cut brown loaf, of a slice, a lack
of butter, blueberry jam that's almost black,
instead of tannin seeping into the cracks
of a pot, the void of an hour seeps out, infects
the slit of a cut I haven't the wit to fix
with a surgeon's needle threaded with fine-gauge silk
as a key would thread the cylinder of a lock.
But no key threads the cylinder of the lock.
Late afternoon light, transitory, licks
the place of the absent cup with its rough tongue, flicks
itself out beneath the wheel's revolving spoke.
Taut thought's gone, with a blink of attention, slack,
a vision of "death and distance in the mix"
(she lost her words and how did she get them back
when the corridor of a day was a lurching deck?
The dream-life logic encodes in nervous tics
she translated to a syntax which connects
intense and unfashionable politics
with morning coffee, Hudson sunsets, sex;
then the short-circuit of the final stroke,
the end toward which all lines looped out, then broke).
What a gaze out the window interjects:
on the southeast corner, a black Lab balks
tugged as the light clicks green toward a late-day walk
by a plump brown girl in a purple anorak.

The Bronx-bound local comes rumbling up the tracks
out of the tunnel, over west Harlem blocks
whose windows gleam on the animal warmth of bricks
rouged by the fluvial light of six o'clock.

Alto Solo

Dear one, it's a while since you turned the lights out
on the porch: a decade of separate summers
passed and cast shed leaves on whatever river
carried our letters.

Merely out of habit, I sometimes tell you
when I've learned a word, made a friend, discovered
some small park where old men debate the headlines,
heard some good music

—it's like jazz, which, even at its most abstract
has the blues in it, has that long saudade
like a memory of what didn't happen
someplace that might be

inlaid with mosaics of recollection
which, in fact's a street corner of the utmost
ordinariness, though the late light steeps it
in such nostalgia

I can hear a saxophone in the background
wail an elegy for the revolution
as someone diminishes in the distance
and the film's over.

Now you know there won't be another love scene.
Do those shadows presage undreamt-of war years?
Twenty, thirty pass, and there's still a sound track
behind the credits:

Cecil Taylor's complex riffs on the keyboard
which a prep-school blonde, seventeen, named Julie
sneaked me into the Blue Note for, because she
knew how to listen—

or it could be Janis packing the Fillmore
West with heartbreak, when I knew that I'd see her
playing pool again at Gino and Carlo's
some weekday midnight.

This is not about you at all: you could be
anybody who died too young, who went to
live in São Paolo or back to Warsaw
or just stopped calling.

(Why did Alice Coltrane stop cutting records?
—think of Pharaoh Sanders being your sideman!—
Lapidary grief: was its consolation
all stone, all silence?)

Now it's morning, gray, and at last a storm came
after midnight, breaking the week-long dog days.
Though I woke at three with a splitting headache,
I lay and listened

to the rain, forgave myself some omissions
as the rain forgave and erased some squalor
It was still too early for trucks and hoses.
A thud of papers

dropped outside the news agent's metal shutters.
Am I glad we didn't last out the winter?
You, the street I made believe that I lived on
have a new address.

You've become—and I never would have wished it—
something like a metaphor of the passage
(time, a cobbled alley between two streets which
diverge, a tune that

re-emerges out of the permutations
rung on it by saxophone, bass and piano,
then takes one more plunge so its resolution's
all transformation).

Someone's always walking away; the music
changes key, the moving men pack the boxes.
There the river goes with its bundled cargo:
unanswered letters.

Rue des Écouffes

for Marie-Geneviève Havel

RUE DES ÉCOUFFES

The street is narrow, and it just extends
rue de Rivoli/rue des Rosiers
a street from which the children went away
clutching their mothers, looking for their friends—
on city buses used for other ends
one not-yet-humid morning in July.
Now kosher butchers coexist with gay
boutiques, not gaily. Smooth-cheeked ephebes hold hands.
Small boys with forelocks trail after bearded men—
and I have dragged that story in again
and will inevitably next compare
the curtains of the creaky balcony
smelling of female exile, exhaled prayer
with the discreet shutters of the women's bar.

LES SCANDALEUSES

Hung on the exposed brown stone of the bar's
back wall, words and collage on aquarelle
metaphor a landscape or a well-
traveled sky, thigh, eye with a view of stars:
her latest work. A child between two wars,
she learned her own vision from the salty squall
of Norman winters, learned what she couldn't tell
except with brush, chalk, pencil, engraver's

stylus and blade, with ink spilled on a stone
as the sea spills up and over the stones when the tide
comes in. Leather jacket, cap, she stands, briefly alone
at the bar with a glass of wine, her Celtic moon-
stone eyes as light and dark as the shapes she made
while the night's first women come in out of the rain.

LES SCANDALEUSES II

The night's first women come in out of the rain:
two couples who arrived, enlaced, astride
two motorcycles, pulled up just outside
the door, doff helmets and leather, order gin/
tonic, beer, beer, a kir. From the bar, they crane
their necks toward the row of dreams, mindscapes, implied
back roads they're too young to have traveled; slide
closer together, wanting things to begin.
Watching, she doesn't envy them their youth,
their way of being in a pack, in pairs
(wounds inflicted, in the name of "truth,"
on friends, near-infidelities on stairs).
But the lacework beginning near that one's mouth
is elegant: Engraver's grooves. Soft dares.

NULLE PART

The elegant engraver's grooves: soft dares
to follow down to the glass-roofed quai, embark

on the last train's last car hurtling through the dark
tunnel irregularly blazed with flares
alizarin, viridian. Lit by the glare's
a silhouette, androgynous, at work
setting (in Paris? London? Prague? New York?)
mosaic tiles. She leads you up spiral stairs
into the blue explosion of the air's
matinal brilliance. But she disappears—
avid flesh, mercurial avatar
desire or imagination sends?
And then you know exactly where you are:
the street is narrow; you see where it ends.

Ghazal on Half a Line by Adrienne Rich

In a familiar town, she waits for certain letters,
working out the confusion and the hurt in letters.

Whatever you didn't get—the job, the girl—
rejections are inevitably curt in letters.

This is a country with a post office
where one can still make oneself heard in letters.

(Her one-street-over neighbor's Mme de Sévigné
who almost always had the last word in letters.)

Was the disaster pendant from a tongue
one she might have been able to avert in letters?

Still, acrimony, envy, lust, disdain
are land mines the unconscious can insert in letters.

Sometimes more rage clings to a page than she would claim—
it's necessary to remain alert in letters

(an estranged friend donated to a library
three decades of her dishing out the dirt in letters)

and words which resonate and turn within
the mind can lie there flattened and inert in letters.

The tightest-laced precisely-spoken celibate
may inadvertently shrug off her shirt in letters.

Ex-lovers who won't lie down naked again
still permit themselves to flirt in letters.

What does Anonymous compose, unsigned
at night, after she draws the curtain? Letters.

Grief

Grief walks miles beside the polluted river,
grief counts days sucked into the winter solstice,
grief receives exuberant schoolyard voices
as flung despisals.

It will always be the first of September.
There will be Dominican boys whose soccer
game provides an innocent conversation
for the two people

drinking coffee, coatless. There will be sunset
roselight on the river like a cathedral.
There will be a rusty, amusing tugboat
pushing a barge home.

Did she think she knew what her friend intended?
Did she think her brother rejoiced to see her?
Did she think she'd sleep one more time till sunrise
holding her lover?

Grief has got no brother, sister or lover.
Grief finds friendship elsewhere. Grief, in the darkened
hours and hours before light flicks in one window
holds grief, a mirror.

Brother? He was dead, in a war-drained city.
Grief was shelling peas, with cold water running
in the sink; a harpsichord trilled Corelli
until the phone rang.

And when grief came home from a post-op nightwatch
two small girls looked reticent over homework.
Half the closet, half the drawers were empty.
Who was gone this time?

Grief is isolationist, short-viewed. Grief lacks
empathy, compassion, imagination;
reads accounts of massacres, floods and earthquakes
mired in one story.

Grief is individual, bourgeois, common
and banal, two women's exchange in Sunday
market: "Le mari de Germaine est mort." They
fill bags with apples.

Grief is primagravida, in her fifth month.
Now she knows the fetus has died inside her.
Now she crosses shopping-streets on a sun-shot
mid-winter morning.

Winter licks the marrow from streets that open
onto parks and boulevards, rivers, river-
parallel parkways, arteries to bridges,
interstates, airports.

Grief daubs kohl on middle-aged burning eyelids.
Grief drives miles not noticing if the highway
runs beside an ocean, abandoned buildings
or blackened wheatfields

—and, in fact, she's indoors. Although her height is
average, massive furniture blocks and crowds her:
oak and pine, warm gold in their grain she thought would
ransom her season.

Workmen clear a path to repair the windows,
not with panes of light on their backs, no message-
bearers these. Still stubbornly green, a street leads
back to the river.

Fourteen years drained into the fifteen minutes
that it took a late-summer sun to douse its
light behind the opposite bank, the boys to
call their match over.

Explication de texte

Plusieurs réponses sont possibles, mais montrez comment la ville
peut se lire comme un substitut de l'objet désiré.

—TEXT ON APOLLINAIRE FOR LYCÉENS PREPARING THE BACCALAURÉAT

Paris nights, drunk on gin,
aflame with electrical fire.
Trolleys with green-lit spines
sing their long route down wire
and rail, deranged machines.

—GUILLAUME APOLLINAIRE
TRANSLATED BY M.H.

Paris is wintry gray.
The small rain spits and sputters.
Before the break of day
when green trucks hose the gutters
lights go on in the bakery.

The days go on, routine
light lingers on the clocks
Yellow and red and green
crowd in the window box
impermanent and benign.

The tiny *sans-abri*
and her more substantial friend
arrive from a night on the quay
at their avenue, extend
their hands to earn their pay,

each on her opposite side.
They've been on the street together
for over a decade
while others jettisoned other
partners and promises made.

Bickering all the way
but punctual at their labors
weekday and holiday
they are my long-term neighbors
with Mme de Sévigné

The days go on, routine.
I would be happy never
to board another plane.
My feet, crossing the river,
and the La Défense/Vincennes

line, or Balard-Créteil
are forms of transportation
quite adequate for me.
Other communication
failed: well, let it be.

Sorrow becomes a sink
and loss becomes a drain.
The drain begins to stink.
Call the plumber again.
Remember how to think.

The poet who wrote and longed
for a woman he barely knew
by whom he thought he'd been wronged
gave Paris new verses to
her electrical torch-song:

the weedy, lovelorn merman's
complaint to pitiless sirens,
some similes, some sermons,
Montmartre and environs
—he even included Germans.

When friends say what they mean
companionship illumines
nights that unroll, routine
in being scaled for humans
choosing their food and wine.

We ordered a house *pichet*
and argued down to the wire at
a smoke-stained brown café:
my friend looked more like the pirate
than the pirate's fiancée.

Poached salmon followed soup
while another loquacious friend
talked such amazing shop
we left the "Vagenende"
when the waiters were cleaning up.

Days and nights, routine
as unambiguous words:
accompanied, alone,
the hours are not like swords,
strike gently, like the rain.

I have two pairs of glasses:
for the peopled world beyond
the panes; for the small world this is
where I eat, and read *Le Monde*,
and drink, and the evening passes.

I grill my trout. I drink
three glasses of Brouilly
or some adequate Southwest plonk.
A Mozart symphony
drowns out the screech and honk

of buses, bikes and vans
and the selfsame garbage truck,
manned by green-clad Africans
come back at nine o'clock
to empty the big green cans.

Paris, elegant gray
godmother, consolation,
heartbroken lullaby,
smell of the métro station,
you won't abandon me.

A hot bath; Couperin:
the hours are not like swords,
strike gently, like the rain,
notes on a harpsichord
impermanent, benign.

Chanson de la mal aimée

December fog condensed above the Seine.
Though it was not the season to atone
for sins, for my sins (unknown) the tears began
again. Unknown, as another mind's unknown
till written, shouted, sung, spray-painted: spoken.
Perhaps (why we all cry) unknown even then.
I walked toward my own bed that I slept and woke in
across that river, or another one.

Between harp-cables humming toward Brooklyn
we were, we thought, descendants of Hart Crane,
emancipated, nocturnal, nineteen.
The city's nightwatch glistened on either side.
Streetlights were haloed by the damp and dawn.
Shadows beneath implied what they implied.
Once we arrived, we'd just turn back again
across that river, or another one.

I crossed the Thames in taxis for a man
who lived quite well without me in South London.
Later there was a girl who heard a train
head north, beneath red leaves, along the Hudson.
The Joyce Bridge on the Liffey one damp June
morning stretched forward like a conversation
I'd no reason to think would not go on
across that river, or another one.

A fresh breeze from the arm of the Malvan
fingered across the terrace of the stone

house where a card table sat in the sun
at which I wrote, bare-chested, dripping sweat.
A Gauloise smouldered out in the Cinzan-
o ashtray. Soft hiss, like the cigarette:
a bird rushed up through oak leaves and was gone
across that river or another one.

The Seine descends from sources in the Yonne;
The children of Vincelles and Vincelottes
launched lanterns cradling candles, let them float
downriver, to begin the village fête.
There would be fireworks, in the misty rain.
A couple on the terrace of the inn
mourned someone as the fireflies blinked out
across that river, or another one.

The Hudson saw my heart break. The Hudson
took it, discarded garbage on its swells.
I'm leaving you. There is nobody else.
She lied she lied she lied she lied she lied.
Walk away from the river, shaking, stunned
as you once came back to it, glad bride,
found child, proud friend. Sewage seeped and spread
across that river, or another one.

There is no heaven and it has no Queen
(There is no God and Mary is His Mother).
I have one life and one is all I get;
it will be "same" unless I make it "other."

The workings of the wind, so intricate,
augmented to a night of devastation.
Water bloated the banks, with bridges down
across that river or another one.

Betrayal isn't torture, cancer, rape.
Authorities don't gas abandoned wives;
deplore the ones who put their heads in stoves
(still, don't suggest it to the Taliban).
Betrayal is a dull stereotype.
"The friends who met here and embraced are gone
each to his own mistake . . ." (that's wartime Auden)
across that river or another one.

I crossed the Pont Sully, above the Seine,
below a small, bright, distant winter moon
tasting the conversation and the room,
glad that my lungs were clear again, and wine
had flavor, that I wanted to walk home
at midnight, that walking home alone
at midnight was a privilege of mine
across that river or another one.

Paragraph for Hayden

Quadruple bypass: yes, he had it.
What happens next is anybody's guess.
After the surgeon's pre-op visit
he pulled the tubes and needles out, got dressed
and stalked outside to smoke a cigarette.
The surgeon threatened not to operate.
Old heart, old curmudgeon,
old genius, terrified old man
who more than anyone knows form
is one rampart of sanity,
your mind is ringing like a fire alarm
and you still smoke three packs a day.
Not lover, barely friend, from this distance
I break your rule and say,
stay in the present tense. Stay in the present tense.

Quoi de neuf sur la guerre?

(Café Le Diplomate, Turenne/Saint Claude)

Five old men
dissect last week's election.
Jacques' student granddaughter bought
a studio apartment

—bigger than
the three rooms that he lived in
with his two brothers, parents,
in the rue du Pont-aux-Choux . . .

(two streets up).
Glasses folded on his cap,
Maurice fishes for a not-
quite-lost riposte in Yiddish.

(His accent
is a familiar garment
on a neighbor, here or in
Strauss Park on upper Broadway.)

The senior
four worked here before the war.
Now they're back in the rag trade.
An eleven o'clock break

—tradition:
black coffee and discussion,
the *cheder* relived later.
The one two decades younger,

Victor, will
at last bring up Israel
—sixtyish son asking his
elders what ought to be done.

And Maurice,
the pouches around his eyes
creased deep in a sad smile, says,
having known wars, not much peace,

(a schoolboy
in Krakow in 1930)
"A solution? There is just
one. The final solution."

Does he mean
the British had a plan in
'48: Arabs could finish
Hitler's job in the new state?

Does he mean
genocide in Palestine
to be practiced by "our own"?
Victor changes the subject.

The waitress
interrupts exegesis:
Please pay, her shift is over.
The watchdog of the café,

a boxer,
trails his young boss, stops at her
trim heels. He scowls, sniffs the floor
and gets sawdust on his jowls.

Ghazal

She took what wasn't hers to take: desire
for all that's not her, for what might awake desire.

With it, the day's a quest, a question, answered where-
ever eye, mind lights. Desire seeks, but one can't seek desire.

A frayed wire, a proof, a flame, a drop of globed hot wax,
a riddle solved or not by William Blake: desire.

Erase the film with light, delete the files,
re-reel the story, will all that unmake desire?

For peace or cash, lovers and whores feign lust or climaxes.
A solitary can evoke, but cannot fake desire.

Crave nothing, accept the morning's washed and proffered air
brushing blued eyelids with an oblique desire.

There was an other, an answer, there was a Thou
or there were mutilations suffered for your sake, desire.

Without you, there is no poet, only some nameless hack
lacking a voice without your voice to speak desire.

Omelette

You can't break eggs without making an omelette
—That's what they tell the eggs.

—RANDALL JARRELL, "A WAR"

First, chop an onion and sauté it separately
in melted butter, unsalted, preferably.
    Add mushrooms (add girolles in autumn)
    Stir until golden and gently wilted.

Then, break the eggs as neatly as possible,
crack! on the copper lip of the mixing bowl;
    beat, frothing yolks and whites together,
    thread with a filet of cream. You've melted

more butter in a scrupulous seven-inch
iron skillet: pour the mixture in swiftly, keep
    flame high as edges puff and whiten.
    Lower the flame to a reminiscence.

When I was twenty, living near Avenue
D, there were Sunday brunches at four o'clock.
    Eggs were the necessary protein
    hangovers (bourbon and pot) demanded.

Style: that's what faggots (that's what they called themselves)
used to make dreary illness and poverty
    glitter. Not scrambled eggs, not fried eggs:
    Jamesian omelettes, skill and gesture.

Soon after, "illness" wouldn't mean hangovers.
How many of those glamorous headachy
    chefs sliding perfect crescents onto
    disparate platters are middle-aged now?

Up, flame, and push the edges in carefully:
egg, liquid, flows out toward the perimeter.
    Now, when the center bubbles thickly
    spoon in the mushroom and onion mixture—

though the Platonic ideal omelette
has only hot, loose egg at its heart, with fresh
    herbs, like the one that Lambert Strether
    lunched on, and fell for that lost French lady.

Those were the lunchtime omelettes Claire and I
(three decades after the alphabet avenue
    brunch) savored at the women's bookshop/
    salon de thé, our manila folders

waiting for coffee—Emily Dickinson's
rare tenses and amphibious metaphors.
    Browned, molten gold ran on the platter:
    a homely lyric, with salad garland.

Outside, it rained in June, or was spring for a
brief February thaw. Now the bookshop's one
    more Left Bank restaurant, with books for
    "atmosphere": omelettes aren't served there . . .

With (you've been using it all along) a wood
spatula, flip one half of the omelette
    over the girolle-garnished other.
    Eat it with somebody you'll remember.

For the 6th of April

for Marie Ponsot

Eden is
pots and tubs on the terrace.
Tenacious seeds root, wind-strewn,
to bloom around the ficus.

Light and shade
from this and every decade
cross and dapple the notebook
you hold open on your lap.

Eighty? Well,
forty, too, and twenty: still
no one's fool, a canny heart,
spirit joyously at school.

Precocious
child, you run ahead of us
aging enfants terribles of
a later generation

Slim mother
of a brood of boys, you were
(seemed) all honed will, clear mind, like
a boy, hermit, young sybil

while the day-
to-day life of the body
which needs food on the table,
an orderly neighborhood

and wages
worked through you. You filled pages
nonetheless: fables, lines, rhymes,
hints from all your languages:

how to live
well on bread and wine, forgive
old enemies and lovers
so that full days pass in peace.

Is it luck
no one gets her old life back?
What you regret you redress
if you can; use; don't forget.

Your daughter,
one more city gardener,
tends your best cuttings in pots
in pale sun a half-block west.

Your desk looks
out on your trees (past the books).
Thick thumbs of amaryllis
work their way up and spring comes.

Jean-Michel Galibert, épicier à Saint-Jean-de-Fos

for Guy Goffette

Reconstitute a sense to make of absence
in the still heat of noon, south, summer
where spindled years unravel and unwind.
A hound bays behind a fence. An old white van
beached beneath oleander in a yard
rusts where it ran down, where something came to grief.

Some summers, joy illuminated grief
and solitude was savory. Then, absence
was a prelude, then stiff, starched, flag-striped yards
of sheet on a clothesline flapped in a sudden summer
gust, like the curtains on a caravan
parked in the town square, billowing with wind,

while children anticipated drumrolls, wind
instruments, brasses, florid joy and grief
mimed close to home. From the striped awning of a van
whiffs of *merguez* fried with onions, smell whose absence
would be a small, real rift in the stuff of summer.
Would have been. The dog paces in his three square yards

of territory, the paved part of a yard
where jasmine and oleander wind
their ribboned leaves like schoolgirls starting summer
vacation. Decline "departure," decline "grief,"
compose an essay illustrating absence
using, for instance, the abandoned van

that used to be, let's say, the grocer's van
which parked on Wednesdays opposite the schoolyard
and the children who were present, who were absent.
Women came up in print dresses, cardigans, windbreakers, seasons changing, even grief
fading like the painted sign in summer

sun, winter rain. After a few winters, springs, summers
the bright sign was illegible, the van
rusted, someone had grown into grief.
The van is parked in the grocer's son's back yard,
its windows shattered, spiderwebbed. The wind
blows through it, marks itself present in that absence.

The grocer's son sat in the van each summer
morning that first year. Even grief was absent
as the wind unwound the streamers in his yard.

A Sunday After Easter

Ah! que le monde est grand à la clarté des lampes!
Aux yeux du souvenir, que le monde est petit.

—BAUDELAIRE, "LE VOYAGE"

A child who thought departure would be sweet,
I roam the borders of my neighborhood,
dominical, diminished. Young gay men
their elbows brushing, Sunday-stroll, in pairs
headed for the weekend flea market
on the boulevard Richard-Lenoir
at Oberkampf. I sit in a café
nursing a decaf. A small Chinese boy
(or girl) in sweats stands on tiptoes to reach
the flippers of the "Space Pirates" machine.
I want to find some left turn into dream
or story, the next chapter, memory.
not saturated with regret, into
a vision as unlikely as the mare
with sweat-soaked roan flanks and a tangled mane,
dragon's breath steaming from her flared nostrils
onto a wind too sharp to call a breeze,
cantering riderless across the square
opposite, between the children in
the sandbox and the old men arguing
on benches, in French, in Mandarin,
in Arabic, Yiddish and Portuguese,
despite the afternoon's dour, bone-deep chill,
early February in late April,
except for punctual persistent green

(Fringy and still-fragile auburn fronds
burst from the rhododendron's rubbery
green leaves, red swellings globe the tips
of cherry boughs, japonica blooms
in double yellow stars on bamboo stalks.
The orange crocuses are past their prime
on the lawn, but now the purple ones
emerge, and pansies, with mascaraed petals
in their beds are gold and purple too.
Three early roses, peach-dappled white,
stand out on bushes nearly bare, studded
with sparse, furled, also reddish nascent leaves.)

of trees and shrubs.
                                        An afternoon when sun
is as unlikely as a riderless
horse to cross the square.
                                                    Imagine that
it were given back to me to be
the child who knew departure would be sweet,
the boy who drew square-rigged ships, the girl who knew
truck routes from Ottawa to Mexico,
the one who found a door in Latin verse
and made a map out of hexameters.

A young Moroccan or Tunisian
with a thick, kinky auburn pony tail,
vastly pregnant, in an oversized
sweater and cargo pants, a toddler and

an almond-eyed five-year-old in tow,
sits with an older blonde in camel's hair
coat and tailleur, who orders a Sancerre
(the name of the café is "Le Sancerre").
The young mother gets three Oranginas
but the Chinese child and her older son
have found each other (school friends) and begun
to play a giggling round of hide-and-seek
under the two empty tables between us.
The little one, whose name is Dominique,
slurps Orangina through a straw, and sways
to the loudspeaker's Motown Muzak (O
bars of my girlhood. Ô saisons, ô châteaux),
then slaps his snowpants and begins to howl
"Je veux ma doudoune! Je veux ma cagoule!"
(a child already eager to depart).
"On dit" she tells him gently, "*je voudrais.*"

She reaches in her pants pocket to pay
but the tall blonde has (she's whom to the young mother?).
Bise-bise, the ritual, they leave each other
at the door; the girl, in a long dark-
blue greatcoat crosses the street to the park,
a child tugging each hand; the woman turns
the corner into the rue des Archives,
becomes a shape the falling light discerns.
(One would postulate brown, not nascent leaves,
the color of absinthe, but innocent,
the color of a world renewed, present,

where absence has become a habit and
occasionally less significant
than wind turning a corner, than the frond
from which a bloom breaks, than the old storm-bent
willow skimming the rain-swollen pond.)

The Chinese child is crying silently
but finds the seated couple he or she
belongs to (both are French, white, not-quite-young).
The tear-tracked child and father re-install
themselves at "Space Pirates" pinball.
The roan mare pauses, thrusts her head among
the rhododendron bushes, nibbles the tender shoots,
unseen beast nourished on unlikely fruits,
turning her copper head in fits and starts.
And what is riderless in me departs
around the corner, into the next street,
into the afternoon, holding its light
later in each day's cloud-leaded sky.
Or stays, doglike, between the wrought-iron feet
of the small table, ears at the alert,
actively silent, having learned to wait.

Desesperanto

after Joseph Roth

Son service est plus propre à un estat trouble et malade
comme est le nostre présent: vous dirieiz souvent qu'il nous
peinct et qu'il nous pinse.

—MONTAIGNE, "DE L'ART DE CONFÉRER"

The dream's forfeit was a night in jail
and now the slant light is crepuscular.
Papers or not, you are a foreigner
whose name is always difficult to spell.
You pack your one valise. You ring the bell.
Might it not be prudent to disappear
beneath that mauve-blue sky above the square
fronting your cosmopolitan hotel?
You know two shortcuts to the train station
which could get you there, on foot, in time.
The person who's apprised of your intention
and seems to be your traveling companion
is merely the detritus of a dream.
You cross the lobby and go out alone.

You crossed the lobby and went out alone
through the square, where two red-headed girls played
hopscotch on a chalk grid, now in the shade,
of a broad-leafed plane tree, now in the sun.
The lively, lovely, widowed afternoon
disarmed, uncoupled, shuffled and disarrayed
itself; despite itself, dismayed
you with your certainties, your visa, gone

from your breast-pocket, or perhaps expired.
At the reception desk, no one inquired
if you'd be returning. Now you wonder why.
When the stout conductor comes down the aisle,
mustached, red-faced, at first jovial,
and asks for your passport, what will you say?

When they ask for your passport, will you say
that town's name they'd find unpronounceable
which resonates, when uttered, like a bell
in your mind's tower, as it did the day
you carried your green schoolbag down the gray
fog-cobbled street, past church, bakery, *shul*,
past farm women setting up market stalls
it was so early. "I am on my way
to school in ______." You were part of the town
now, not the furnished rooms you shared
with Mutti, since the others disappeared.
Your knees were red with cold; your itchy wool
socks had inched down, so you stooped to pull
them up, a student and a citizen.

You are a student and a citizen
of whatever state is transient.
You are no more or less the resident
of a hotel than you were of that town
whose borders were disputed and redrawn.
A prince conceded to a president.
Another language became relevant

to merchants on that street a child walked down
whom you remember, in the corridors
of cities you inhabit, polyglot
as the distinguished scholar you were not
to be. A slight accent sets you apart,
but it would mark you on that peddlers'-cart
street now. Which language, after all, is yours?

Which language, after all these streets, is yours,
and why are you here, waiting for a train?
You could have run a hot bath, read Montaigne.
But would footsteps beyond the bathroom door's
bolt have disturbed the nondescript interior's
familiarity, shadowed the plain
blue draperies? You reflect, you know no one
who would, of you, echo your author's
"Because it was he; because it was I,"
as a unique friendship's non sequitur.
No footsteps and no friend: that makes you free.
The train approaches, wreathed in smoke like fur
around the shoulders of a dowager
with no time for sentimentality.

With no time for sentimentality,
mulling a twice-postponed book review,
you take an empty seat. Opposite you
a voluble immigrant family
is already unwrapping garlicky
sausages—an unshaven man and his two

red-eared sons.

You once wrote: it is true,
awful, and unimportant, finally,
that if the opportunity occurs
some of the exiles become storm-troopers;
and you try, culpably, to project these three
into some torch-lit future, filtering out
their wrangling (one of your languages) about
the next canto in their short odyssey.

The next canto in your short odyssey
will open, you know this, in yet another
hotel room. They have become your mother
country: benevolent anonymity
of rough starched sheets, dim lamp, rickety
escritoire, one window. Your neighbors gather
up their crusts and rinds. Out of a leather
satchel, the man takes their frayed identity
cards, examines them. The sons watch, pale
and less talkative. A border, passport control,
draw near: rubber stamp or interrogation?
You hope the customs officer lunched well;
reflect on the recurrent implication
of the dream's forfeit. One night in jail?

Canzone

Late afternoon, a work-table four stories
above the rain-slick January street
—and words begin to slide into a story
someone told once. Repeating well-known stories
with new inflections, does the teller add
a nuance or a chapter to the story—
the teller's own, or a recounted story—
so that it takes an unexpected turn
and doesn't, like a child from school, return
at the same time, to the same place? History
cycles over in place, unless we learn
something from the cycle—learn to unlearn

what's overdetermined. The child learns how to learn
from listening to, embroidering on stories
repeated to delight, to soothe. She learns
from delight, from repetition, learns
syntactic play, learns courtesies the street
exacts (accepts, rewards when they're well learned),
learns over time how much there's still to learn.
At eight, eighteen, you promise that you'll add
a word to your lexicon each day, add
a book to your bedside reading, start to learn
a language. Now, like a trip, you plan returning to a book read once, think how you'll turn

that page down, give the writer one more turn
to teach what you were not prepared to learn
in adolescence, stubborn, taciturn

inclined to shut the book, mentally turn
on your heel, exit the uncongenial story
which did not give your *idée fixe* a turn
to play the diva. Less inclined to turn
on Flaubert, having walked down the street
Mme Moreau lived in, you know your street
is also paved with stories. If you could turn
doors and windows back like pages, had
a listener's wit, there'd be nothing to add.

But even a silent interlocutor adds
something to a narrative, which turns
in spirals, auricular labyrinths, to add
conjunction and conjecture. (The teller adds
specifics, so the listener will learn
extreme attention.) Remember how you had
smiled and hummed the line you knew on the ad
in the métro, history and a story
Clément wrote, Montand sang, and, one more history:
the passage from commune to commuter. Add
the station's name, a grassy path, a street
whose western limit is your own home street.

Life hums, a wire pulled taut between that street
and one across an ocean. Stretch back, add
East Fifth, East Sixth, East Tenth, Henry Street,
Perine Place, Natoma Street, Paddington Street.
In dream-labyrinth nights, I turn
a corner, one street becomes another street

in another country, yet on that street
doorway flows into hallway: no need to learn
my way; I know the way. Awake, I earn
the daily recognition of the streets
I live on, dual, counterpoint, their stories
enunciate a cautious history.

Now and from memory's clerestory,
my vision of that palimpsest, a street,
(as fading daylight, gold on velvet, adds
textured layer) turns outward as streetlights turn
on, lights cut out lives, limits: What can I learn?

Respite in a Minor Key

I would like an unending stretch of drizzly
weekday afternoons, in a moulting season:
nowhere else to go but across the street for
bread, and the paper.

Later, faces, voices across a table,
or an autumn fricassee, cèpes and shallots,
sipping Gigondas as I dice and hum to
Charpentier's vespers.

No one's waiting for me across an ocean.
What I can't understand or change is distant.
War is a debate, or at worst, a headlined
nightmare. But waking

it will be there still, and one morning closer
to my implication in what I never
chose, elected, as my natal sky rains down
civilian ashes.

Morning News

Spring wafts up the smell of bus exhaust, of bread
and fried potatoes, tips green on the branches,
repeats old news: arrogance, ignorance, war.
A cinder-block wall shared by two houses
is new rubble. On one side was a kitchen
sink and a cupboard, on the other was
a bed, a bookshelf, three framed photographs.

Glass is shattered across the photographs;
two half-circles of hardened pocket bread
sit on the cupboard. There provisionally was
shelter, a plastic truck under the branches
of a fig tree. A knife flashed in the kitchen,
merely dicing garlic. Engines of war
move inexorably toward certain houses

while citizens sit safe in other houses
reading the newspaper, whose photographs
make sanitized excuses for the war.
There are innumerable kinds of bread
brought up from bakeries, baked in the kitchen:
the date, the latitude, tell which one was
dropped by a child beneath the bloodied branches.

The uncontrolled and multifurcate branches
of possibility infiltrate houses'
walls, window frames, ceilings. Where there was
a tower, a town: ash and burnt wires, a graph

on a distant computer screen. Elsewhere, a kitchen
table's setting gapes, where children bred
to branch into new lives were culled for war.

Who wore this starched smocked cotton dress? Who wore
this jersey blazoned for the local branch
of the district soccer team? Who left this black bread
and this flat gold bread in their abandoned houses?
Whose father begged for mercy in the kitchen?
Whose memory will frame the photograph
and use the memory for what it was

never meant for by this girl, that old man, who was
caught on a ball field, near a window: war,
exhorted through the grief a photograph
revives. (Or was the team a covert branch
of a banned group; were maps drawn in the kitchen,
a bomb thrust in a hollowed loaf of bread?)
What did the old men pray for in their houses

of prayer, the teachers teach in schoolhouses
between blackouts and blasts, when each word was
flensed by new censure, books exchanged for bread,
both hostage to the happenstance of war?
Sometimes the only schoolroom is a kitchen.
Outside the window, black strokes on a graph
of broken glass, birds line up on bare branches.

“This letter curves, this one spreads its branches
like friends holding hands outside their houses.”
Was the lesson stopped by gunfire? Was
there panic, silence? Does a torn photograph
still gather children in the teacher’s kitchen?
Are they there meticulously learning wartime lessons with the signs for house, book, bread?

Essay on Departure

And when you leave, and no one's left behind,
do you leave a cluttered room, a window framing
a zinc roof, other mansard windows? Do you
leave a row of sycamores, a river
that flows in your nocturnal pulse, a moon
sailing late-risen through clouds silvered by
the lights flung up from bridges? Do you leave
the wicker chairs the café owner stacks
at half-past-midnight while the last small clutch
of two girls and a boy smoke and discuss
what twenty-year-olds in cafés discuss
past midnight, with no war on here? You leave
the one and then the other, the all-night
eight-aisles-of-sundries with a pharmacy
cloned six times in one mile on upper Broadway.
Everywhere you're leaving something, leaving
no one, leaving as a season fades,
leaving the crisp anticipation of
the new, before its gold drops on the rain-
slick crossings to the walkways over bridges,
the schoolyard's newly painted porte-cochère:
remembered details. You're no longer there.
What's left when you have left, when what is left is
coins on the table and an empty cup?
An August lapse begins; the shutters drop
and lock, whatever follows is conjecture.
The sound feels final, punitive, a trap
shutting its jaws, though when the selfsame structure
was rolled up mornings, it was hopeful noise,

a reprieve from insomnia, a day's
presence opening possibility.
As you leave the place, you bring the time
you spent there to a closed parenthesis.
Now it is part of that amorphous past
parceled into flashes, slide-vignettes.
You'll never know if just what you forget's
the numinous and right detail, the key—
but to a door that is no longer yours,
glimpse of a morning-lit interior's
awakening silhouette, with the good blue
sky reflected on the tall blue walls,
then shadow swallows what was/wasn't true,
shutters the windows, sheathes the shelves in dust,
retains a sour taste and discards the kiss,
clings to the mood stripped of its narrative.
You take the present tense along. The place
you're leaving stops, dissolves into a past
in which it may have been, or it may not
have been (corroborate, but it's still gone)
the place you were, the moment that you leave.

Some Translations

Claire Malroux

GROTTOES

I

Without knowing us the dog rushes to greet us
from the path at the morning's most luscious moment when the sky
leans on the church's slate roof

Imperiously she leads us to the enchanted spots
of her dog's life

We must roll with her in the fields sniff the horse-droppings
shake ourselves off in the stream which erases time's borders
like animal tracks

A bridge to our human joy so close
to her domain
and necessary to her happiness
as if a hint of eternity guided her by its smell

When we retrace our steps she'll hurl herself on us again
with grand gestures of gratitude

Swallows have no such fraternities
Barely curious, the horses will have turned away to scratch themselves
and embrace, cheek to cheek

Echo of a group the ancestor's eye gathered on a rocky promontory
not so long ago at the heart of the grotto

II

At the grotto's mouth she forgot the spring
the grass's whispers the stridencies the shiver and thunder
of the branches

shook off the sun's weight to penetrate its silence

Now she is no more than an arm of shadow a snake's sloughed skin
in the stone

Men have crawled into her body with torches
and flints

Europa Eurydice Persephone Beatrice

Their drawings destroy and beget themselves
horse's belly bison's hump and mammoth's chest
doe's head in a crotch

One reads: the god is closest to me in my enemy
With my sharpened lump of clay I hold him in my power
I am embodied in him
Or: you who pass by here help me to escape the stone trap

Some unique artist has left a signature: human
slender, sexless, future pastor of the catacombs

Another is it the same the big-assed female carrying
her clitoris in front of her like a Périgordine's bag
when she goes to market on the village square
at the hour when the sun gnaws the last bones of snow

III

Alone in her grotto where nothing except a rarer air asks questions

The branches up there grinding the dead on their way to a cloud-
eden

Rarer matter than dew on a rose, shadow on a wall, the shiver
of skin stretched over the chasm

For the hermit the days' exhalations the leaves' prodigality the
greenness of rain

What is a day after so many days?

A stone (sometimes white), a marker placed graciously but without
indulgence on the path

Mask the dusks, fire off huge bouquets of dawns

Let your afternoons play at rolling down the slopes
as yesterday you slid your days on the shaft of the abacus
back and forth, without counting them

Where today they are impaled one by one in slow torture

Marie Étienne

OCHRE NIGHT

I take the train the way I always do
Once at my destination, I forget
My overcoat in the train compartment.
I go back for it, dawdle
The train leaves
But the next station isn't very far
It's like Brussels, where the train turns in-
To the subway
It's the month of August
The summer is familial we rent
A villa that looks out over the beach
One day, suddenly, I cry "Where's Clara?"

One day, suddenly, I cry "Where's Clara?"
My sister's expression seems to imply
It's all my fault
Clara comes back though I
Feel immense joy I can no longer stop
Banging up against my anguish.
When I
Go back to my furnished room in Paris
The landlady is disagreeable
"The door of the armoire was boarded up,"
She says
From now on how can I get dressed
I know my clothing has been visited

I know my clothing has been visited
Amputees—each with an arm or a leg
Ending in a swelling of whitish wood
And a steel dart that looked like the one on
Top of the metal helmets which Prussian
Soldiers used to wear—had indulged themselves
Among the folds of fabric cavities
Of sleeves the gaps opened by buttonholes
In practices on which I do not dare
Reflect they so disgust me
Even now

Reflect they so disgust me
Even now
My armoire has an insupportable
Odor and my clothes which are inside it
Dresses skirts cloaks overcoats silk
Dressing-gowns hang there soiled with whitish stains
And ripped
While I'm still standing there stiffened
In front of the obscene wardrobe closet
I witness a visitor's arrival
Whose prosthesis was at least the same size
As the rest and I hear the landlady

As the rest and I hear the landlady
Ask me to be considerate toward him
And caring. "It's his birthday," she proclaims
All I want is to change so I can go

To work and I complain to her that I've
Been delayed already
                                        She goes out
With a ferocious air to the next room
Of palatial size but dark and dirty
The furniture is covered up with dust
I stop there on the threshold not knowing

I stop there on the threshold not knowing
What she is looking for while she warns me
Not to come further as she undertakes
Furiously to dust the family
Of armchairs weighted down in the darkness
So I turn and go back to my armoire
To take out a pair of woven sandals
Whose straps as soon as I've slipped my feet in
To them begin to bleed all over me
Profusely as if they were animals

Profusely as if they were animals
That someone had run over
                                                  I leave, thus
Shod and grieve while I'm walking
                                                            Later on
Another night perhaps, Florence and I
Lie down side by side we are holding each
Other's hands—silently on a bed
Which is rolling through space like the car of

A train
All at once I raise myself up
To speak almost fall but steady myself
On the guardrail she also lifts herself

On the guardrail she also lifts herself
To answer me
"Be careful!" I warn her
But she is already falling falling
Interminably my mouth opens in
A long silent cry
When I pass through their
Villages there are men who cry out too
They all have thick black beards which they swallow
They overturn the tables where seasoned
Animals' bodies waited for them braised
Over coals in kitchens
They overflow

Over coals in kitchens
They overflow
Without waiting through the funnels of doors
I finally arrive at journey's end
In a sandy region which resembles
The sea
The dunes rise up set in the midst
Of them and buried in the sand up to
The waist are figures apparently made
Of the same stuff and which are all facing

My way

        Their torsos are rigid and they
Seem to have been dropped there by the hot wind

Seem to have been dropped there by the hot wind
One can't help but notice a strange thing
The night continues to appear golden
Because of rays of light which emanate
From within the dunes themselves

                                        It is this
Which gives to those silhouettes petrified
Against the sky their indeterminate
Color

        I don't know where I am because
The silkiness of those forms, those ochres
Is as peaceful as it is menacing

Vénus Khoury-Ghata

Water down your anger when the river squats on your doormat
a river can't be wrung out
or kicked to chase it away
a river isn't a dog
don't block its path when it steps across your threshhold
its water will erase the footprints of the winds come to beg at your
door
quench the thirst of your hearth
and make your daughters' knees shine like pebbles in a stream

Shame on your wife who has it drink from a bowl riddled with holes
it knows her secrets
knows the cleft flesh of her stock-pot and the hairy skin of her hemp
shame on you for taking stones from the mourning walls of a
collapsed house to build elsewhere

The stones in your garden speak louder than the people passing by
they claim an ancestry that goes back to the first cave
when two flintstones controlled fire
and a pauper wind swept the brambles of an alphabet gone deaf

Things being what they were
you had only to grasp a stone in your hand to feel the planet's
vibrations
sense a volcano's insurrection
the cry of a mountain collapsed by an ant

Hold back your hand when the sunset draws its last circle on your
wall
the sun is not a drum
and the discussion between darkness and asphalt doesn't concern you
while your shadow follows you by a finger and an eye
You walk and your destinations print themselves on your feet

Everything speaks to you of departure
words squirm around on the page
the man walking under a rainstorm brushes you with his thoughts
you feel the dampness of his feet in your calves
you are you and all the passers-by at once
the keys of their houses jangle in your pockets
you open walls and doors
none of them are like yours that grew old while you went off
following these strangers
their rain-slicked hair accuses you of negligence
if you were more responsible you'd have lined them up on your
windowsill
dried their feet alongside the herbs from your garden
and played out the scene in the opposite sense
inching your way among the people walking to make up for your
own inertia

Three handfuls of red earth for the shoulder that protected you from
 the storm
three handfuls of blue earth for the one that bore your sorrows
a last one for the sparrow that announces your burial to the trees in
 the forest

Deprived of any cermony
your chair and your bed still have something to say about the way you
 drove a nail into bare wood
joined the angles and smoothed the wood by hand the way you'd
 stroke a pet

This hole in the ground is incomprehensible to your possessions, they
 are convinced that everything that breathes and sweats ends up on
 the garbage dump

You watch for the moment when dusk makes its way to your table to
note down its tall tales
Dust unto dust, the words it dictates to you
disappointment weighs down your shoulders
Why were you called to this earth if your fingers dilute the messages
meant for you
Slumped on the suddenly darkened page
you tell yourself that writing is the invention of a mute alphabet,
a petition to a tomb

The faces lined up on the walls invite you to join them
their comments make the plaster sweat
the beggar must take off his shoes before crossing the threshhold
the hunter must hang his game-bag on the walnut tree
and the women hold back their tears that would catch fire touching
    your forehead

when you think about it
your death has not aged your house or whitened the hair of the
    walnut tree that always looks straight into the sun
despite its bitter fruit and its nest's frayed straw
a beating of wings like applause
that's all that you take away
only the sooty-winged lark will wear mourning for you

Guy Goffette

ELEGY FOR A FRIEND

for Paul de Roux

I

When life was strong and when we'd walk
as if in a dream, slipping from the métro to Dante's
hell without changing faces or
pace; when love

carried us like a torch from strands
to waves of hair, crackling flames of promises
cindered swiftly by the wind; when nights
stayed white and

turned our faces to the wall, moving
the moon's reflection quarter to quarter beneath
our lids, it was already dancing there
and strong and white, that shadow

that burns all shadows while it waits for us.

II

Always, still, tomorrow, these paltry
words, thrown off in passing, overflow us.
They pile up in the margins of our lives
smooth, unfevered

sand, to which no one pays attention
until the heart suddenly beats
its wings and begins to count its steps,
because everything has been said,

everything, and all that's left is to shut the door.
But suddenly it resists and creaks like
memory before a mountain of forgetting:
this pile of sand, this

silence that takes up all the space and screams.

III

Perhaps that sudden rain was needed
on the dying roses and summer roofs
to make the sky gray once again
with a dreamer's eyes

and lead slowly up from the depths the figure
of the absent one to his third-story window,
rue Poliveau, when the generous plane trees
still had the wherewithal to return

the poet's greeting, and his breath, and
colors to his room, lightening the grip
of living and his double question
in the raw mirror: who

am I, who? and *my life where have you gone?*

IV

If we had known this, that the wind consoles
more than it batters—and the sorrowful
heart goes to sleep in its hand
like a child

whose tears keep flowing,
or that the forest splits suddenly in two
to let the sky's breath pass through
and drink its fill

of the naked light that woodcutters
grope for like a nest of silence
beyond the chainsaws—if we had known
that, would we have stayed

sitting so long in our afflicted bedrooms?

V

It’s the same story always and we blame ourselves
afterwards for having in the heat of words
and wine allowed dark clouds to rise
on the friend’s brow

reading nothing beyond his eyes, nothing
of the desert of life and the man’s thirst
as he wrestled his shadows, when he kept
turning his uneasy

head toward the lights in the street
as if he already saw *spring*
making its way up the branches and already feared
that he lacked the strength

to climb up with it like a cat in summer.

VI

One day we must depart, no longer knowing
anything of what was at the source
of the fire, nor how nor why
things all at once

began to go wrong
and the fire went out, the rosebush changed
to thorns, love to scorched earth
and what remains with

our footsteps echoing where the heart once was
is paltry: some words on paper that
no longer say anything except that they were
written, read and reread

by a blind man dancing in the fire.

Emmanuel Moses

from *PRELUDES AND FUGUES*

PRELUDE 1

Knight of the deep wound
I see that there is an ash tree
when the universe has foundered
I will find shelter there
I, son of one of the virgins
I know that there is a tree
the first one to lose its leaves
they call it the tree of the world
its foot is sunk deep into ice
the tips of its branches decked with mist
I will find shelter there
with the other men of the middle world
with the leaves trembling on the heavenly path
they say that it is watered by a fountain of tears
that Hecuba and Andromache could be seen there
their proud shadows
but also Achilles and David mourning their companions
I know that there is an ash tree called the tree of the world
it rises under the dew
moaning
how console men otherwise?
I will find shelter there
beneath the sun's wings and the millstone of black clouds
there is a tree of the world
o men of the middle world

FUGUE 1

We are all those trees of which I said:

I know there is a tree

we embrace three worlds

we are the tree of the world

the living and the dead touch in us

the dead rise toward the living

the living descend toward the dead

we are the middle

o world of men

which is the world of the gods

the mortals rise toward the immortals

the immortals descend toward the mortals

who lives at the top of the ash tree if not the most joyous of the gods?
Or perhaps it was wind in the boughs?
In us the dead rise toward the immortals
the immortals lean down toward the dwelling-place of the dead
if the ash tree is green before the oak
if the oak is green before the ash
we will or will not watch raindrops on the windowpane
blur the view it offers of the sky
we are the tree of the world

the dead nourish the gods

the gods grow with the living

I know that there is a tree

PRELUDE 2

The sun was their gold
who remembers that they took so long to start again?
Dogs and doves sought the branches' shadow
but no serpent ever lingered there
someone had fled to Norway on a cutter
a woman pregnant by a foreigner
she passed there amidst the saplings
that sheltered her in their leaves
when they had set sail
she shed tears for her father, her brothers and handmaidens
once out of the forest she never returned
who would have told her where the tree of the world can be found?
But she had seen the great vegetal procession!
Later she would often pick up her lute
and sing of sylvan beauty
where she had sought cool and calm
where the sun was her gold
old betrayals smoldering like brushfire . . .
in a castle in Norway
a woman would pick up her lute
and call on the tree of the world

FUGUE 2

Or perhaps she died
as the wives of fallen soldiers sometimes do
her son will bear the saddest name
and will become a hero in his turn
sung by poets from court to court
or perhaps she took up her lute to intone a lament
between the walls of her nun's cell
or she could have woven a tapestry for her beloved
which shows two armies before their confrontation
knights in armor with painted shields and plumed helmets
foot soldiers armed with halberds banners flying
and in the background bouquets of trees at the foot of blue hills
one of them spiked with a pink fortress
where she would imagine herself seated at the window
fearfully waiting
for the outcome of the fray
a chaffinch would be perched on the branch of a walnut tree
that defines the world
and builds it
it could be said the tree kept it balanced
recognizable from far off and always solitary
the woman would pass long hours depicting it in the landscape

tree-world with packets of dry leaves
hanging over it a sun sending out long rays
which will be their gold

PRELUDE 3

I will sing in bird language
or like the organs of Dresden
there was an old woman who still remembered them
beneath the syringas of exile
(and now she sleeps in a sad suburban cemetery)
a linden used to shade her grave that I'm told was cut down
words of smoke will emerge from my mouth
when I sing in the tongue of the bird of paradise!
They'll have called me guilty, a vagabond
till I raise my song toward the evening's blue clouds
how strong the wind is at nightfall
it unfurls flags of smoke between the gallows
each one was the tree of knowledge
lit by a guilty sun
I will sing between brick walls
dragging my vagabond feet from hut to hut
each rag drying in the wind will be a shroud for me
we drank down great mouthfuls of gall!
There were days when the wind exulted
even the sun was guilty—
o vagabonds huddled in ditches in vacant lots
o sleepers in eternity's hot morning
two by two the shadows withdraw toward the bend in the road
the world's marvels will have to step aside
with the streaming light

FUGUE 3

Here rose a bravely woven song
—the assassin's whistling in back alleys—
bread that was delivered gnawed by rats
or at any rate spotted with mold
o unscrupulous bakers of the dismal years!
Mothers you kept your little girls close by your skirts
—the ogres whistling in back alleys—
mornings were pale and hard as bone
life rotted in foul-smelling kitchens
no one drank wine in leafy bowers
at most a beer or two between factory walls
we lived in city centers under the purr of machines
sirens ripped the morning fog
sometimes we thought we were on the road to the sun!
Men and women had exchanged their roles
they drowned the wound as best they could
sailors left on the dock sang with peasants beneath the earth
they died beside the factory
without sacraments
cursing those who cursed them and kicked them aside
now we can see the oblivion where an insane chimney burgeons
what became of the songs and wisps of smoke?
Which god spared the last tree in the world?

Jean-Paul de Dadelsen

THE LAST NIGHT OF THE PHARMACIST'S WIFE

The wind above the glaciers that rushed here from the desert
comes barely cooled to torment the tall pine tree's branches.
When everything is in labor, how can you sleep, how
can you die?

On the slow smooth waters, the flat boats,
the black boats are, like the soul
almost permanently moored.

The year grows long before it brings back
the distracted daughter, the son loved from afar.
The children who laughed in her arms, on her breasts
rarely write to the pharmacist's wife
even to ask for remedies.

Beauty is cheap, except
as a last appeal that will no longer be heard. O captive
between the seasons, the barrel of fresh cabbages
cut in the cellar in October's first frosts,
when with the swallows all at once departed, you
wake in the first silence of late fall.

Odile, the plain is merciless. At night
frogs at a loss to reproduce complain.
The stork plunges its long lecherous beak down other chimneys.

The clock with heavy wooden shoes, the heart with its heavy steps
measure the night which barely drifts. How hard it is
to break loose from the moorings! How long it takes
for the water's traction to tear loose
the chain that for so long grasped the riverbank!

The heart in its heavy clogs paces the nocturnal prairies,
stands shifting its feet on the shore of the water
which very soon it must cross.

The Bridges of Budapest

They hanged me for having wanted to live.
They hanged me for not having killed.
They—it's not the same ones every day—hanged me
for having believed what the others foretold
in their textbooks at the night school for retarded adults. They
hanged me
for nothing. To forget their fear. To strangle their shame.

Hear them, on the bridges of Budapest, coexist,
the hanged men of every catechism and cosmogony.
Once the bad moment is over, we keep each other company
the more hanged men there are, the more conversation,
where we're at now, we can laugh all the more.
The wind on the blue Danube fills our pockets now emptied forever
of grenades,
frost stiffens our bodies' excrescences. For six days
I slaved away; the seventh day I rested, I saw.

Strange mandrakes will be born on the roads
when the tanks, when the dogs, when the overflowing sewers
will have spread into all the veins of the earth, into all
its wombs this jism of hanged men, this blood
spurting in equatorial rain on the slimy trees
these scraps of membrane and bone and nail of thirteen-year-old girls
for precocious weddings bedecked in grenades
slipping under the tanks to blow themselves up with them.

Against, in the great stupid scale—against
the plain where words pile up that mean nothing and everything,

the words that don't make bread, the words that don't make love,
the words made of wind gathered from the long-rotten beards
of professors dressed in long-johns for a revolution in slippers,
against the words which kill without seeing, without looking at
anyone,
against the people who live by keeping others from living
against the people who revenge themselves sixty long years for their
sad childhoods,
against: delivery boys, ironmongers, cesspool emptiers,
typographers, milkmen, little telegraph boys,
a few girls thirteen, twelve, ten years old
suddenly pubescent when it's a question of slipping, to strangle him,
into the butcher's alcove of metal and fire.

We watered, ploughed, seeded the esplanades,
over the asphalt we passed the harrow and the hoe,
we reaped. Your turn, Ivan, to harvest!
Ivan, O son of a sow, O son of a Christian woman,
child of a piglet, child of a Siberian convict, Ivan of a thousand faces,
Ivan of a single poverty, it's against you, it's with you,
it's beside you, it's also for you I fought
against your brother Ivan, against my brother Janos.
The wind makes us waltz on the same lamppost.
From the highest gas lamp, ohé, Janos, you who fart on our heads,
do you see the American tanks coming? do you see them descend
with parachutes, the Titoist, progressive, libertarian, humanist
volunteers?
You wanted to play haughty, Janos. Not like us
who for so many years, in so much night, are waiting,

in so much ice, in so much death, waiting,
in ever-more-ridiculous and more necessary hope, waiting,
when our kids, come home from the idolatrous school, try to teach us
how to make a fire, a roof, bread, how
you slaughter the pig (when you have one), how the wolf seeks its food,
how every spring the river enormously shatters its prison,
how you sell your vegetables, how you feed your old mother,
how you make babies,
how you die.

You who wanted a frank and fraternal world, you've got it,
you who always hoped to see, on your people of drunkards and
  layabouts,
a sudden Pentecost pour down where everyone
would embrace speaking Russian among fiery doves,
you've succeeded.
Around the hanged men a circle of lost children dances, the spirits
of the dead in the most ancient massacres are dancing. Obviously
when you give a party like that it attracts people, you're a success
while waiting for the bloody carnivals of Ukraine, White Russia,
Baltic, Caucasian, Turkestan, Siberian, here are colleagues
come from the human family's little celebrations, suppression
of the Great Mutiny, march to the west, village near Tipiza liquidated
  by bombs
the year of the Liberation, conquest of the Congo, pacification of
  Zulu villages,
Bantu, Viet, Malaysian, Javanese, Philippine, Manchu, Mau Mau,
  tutti quanti.
Come, colleagues, make yourselves at home.

You speak well, Ivan, you always liked to talk. As for us,
here, now, we've brought in this particular harvest. We rest,
we have a look. And however outrageous, useless, it might be,
as for us, what we've done, here, now, as it is,
it pleases us.

All translations in this section are from the French.

from Names (2010)

Ghazal: In Summer

for Mimi Khalvati

The air thickens, already more than half in summer.
At the corner café, girls in T-shirts laugh in summer.

The city streets, crowded with possibility
under spring rain, thin out, don't promise enough in summer.

That urge to write one's life instead of living it
makes sentences slip limply off-the-cuff in summer.

Slipped in a drawer under an expired passport,
curly-head in an orchard smiles for a photograph in summer.

Going downstairs early for bread: two winos snore on the landing,
"Can't they make do with sleeping in the rough in summer?"

Hard-case on the street, teacher out of class both harbor
a lowgrade fever and productive cough in summer.

Espresso winter, springtime of Juliénas:
black tea with honey's what I'll quaff in summer.

Despite my wall of books and Bach's geometries,
some scent wafts from the street to call my bluff in summer.

Not in a tank but a golf cart rides the oligarch:
however, he does not dismiss his staff in summer.

Let them not, in Maryam's name or Marilyn's,
blot any cindered city off a graph in summer.

Lettera amorosa

Where's the "you" to whom I might write a letter?
There are dozens, none is the "thou" I never
did engage in dialogue: was I talking
to myself, loudly?

You, your lion's mane and your pale-rimmed glasses,
buccaneering over the swells of panic,
settled now, with a child, career and spouse in
lower Manhattan?

You, who picked the riverbank where we often
told our days with coffee at almost-twilight
to announce irrevocable departure?
You're gone, you said so.

You, but who are you, if I never met you,
man or woman, mother tongue French or English
(maybe Arabic), you're a word, a nameless
presence, night-fancied.

Letter, then, to light, which is open-ended,
folds, expands, but even on winter mornings
faithfully attends to the correspondence,
answers the question.

Glose: O Caravels

A child, I knew how sweet departure was
from never having left the skiff
of hills, split open any horizon
but the rain's when it closed off the morning

—GUY GOFFETTE, "O CARAVELS"
TRANSLATED BY M.H.

Staying put provides the solidest
comfort as daylight diminishes at four:
the street becomes, again, a palimpsest
of hours, days, months and years that came before
and what is better was, and what is best
will be its distillation. In the pause
when blinds are drawn, when tea is brewed, when fast-
falling evening makes lamplight seem more
private and privileged, I can be still because
a child, I knew how sweet departure was

and planned, extravagantly, voyages,
encounters, divagations, chronicles
of travel, unpronounced truths, bright lies.
Imagined stonework of façades, and smells
not of tinned soup or ink. Gratuitous
enormities could be enacted, if . . .
Without constructing model caravels
of balsa-wood or plastic, I saw skies
between the masts, inferred a different life
from never having left the skiff

moored at a dock of dark mahogany
claw feet of overstuffed postwar club chairs
whose own piled or brocade upholstery
recalled cities that were not anywhere
inscribed in that apartment's memory.
There was only one window, opening on
an alley, garbage cans, one tree, a square
of sky on which the day's calligraphy
scribbled in slate-gray rain, anticipation
of hills, split open any horizon.

Now, even tawdry dreams turn polyglot,
suggestion of the wished-for western wind.
Two syntaxes, more tenses, alternate
merging reflections that are less than kind
(and more than kin) into a better plot
that has to do with transformation. Waking
was easy. The street outside was wet: it rained
all night. The sky's washed clean, and I am not
anticipating any leave-taking
but the rain's when it closed off the morning.

from "O Caravels"

Guy Goffette

On ne part pas.

—RIMBAUD

A child, I knew how sweet departure was
from having never left the skiff
of hills, split open any horizon
but the rain's when it closed off the morning

and that I'd have to find at any cost
the right light so that I could fix the seas
in their places on the map and not
overflow the lines. I was ten and

had more voyages in my pockets
than the great explorers, and if
I agreed to trade Sierra

Leone for Yakoutia, it's really
because the snowy frame of lace
around the stamp was sturdier.

1953: The Bus to Menton

for Mavis Gallant

Her own displacement seemed easy in comparison.
She had been a reporter. She would be a novelist
and her country (she'd write about it) seemed provincial.
The war was over. Near the roadside, sheaves
were tied. Gossip behind her, a new dialect. She listened.
Beyond sprawled olive-terraces, unlike the farms

of home, whose outbuildings circled like a garrison.
Her notebook's lined page waited to be kissed.
The noon heat condensed into a mortal chill
up her spine. A blondish man with rolled-up sleeves
had pushed the bus window open. Sunlight glistened
on the long number tattooed on one of his sunburned arms.

Glose: Storm

Blood's risks, its hollows, its flames
Exchanged for the pull of that song
Bone-colored road, bone-colored sky
Through the white days of the storm.

—CLAIRE MALROUX, "*STORM*"
TRANSLATED BY M.H.

Once out of the grip of desire,
or, if you prefer, its embrace,
free to do nothing more than admire
the sculptural planes of a face
(are you gay, straight or bi, are you *queer*?)
you still tell your old chaplet of names
which were numinous once, you replace
them with adjectives: witty, severe,
trilingual; abstracting blood's claims,
blood's risks, its hollows, its flames.

No craving, no yearning, no doubt,
no repulsion that follows release,
no presence you can't do without,
no absence an hour can't erase:
the conviction no reason could rout
of being essentially wrong
is dispelled. What feels oddly like peace
now fills space you had blathered about
where the nights were too short or too long,
exhanged for the pull of that song.

But peace requires more than one creature
released from the habit of craving
on a planet that's mortgaged its future
to the lot who are plotting and raving.
There are rifts which no surgeon can suture
overhead, in the street, undersea.
The bleak plain from which you are waving,
mapped by no wise, benevolent teacher
is not a delight to the eye:
bone-colored road, bone-colored sky.

You know that the weather has changed,
yet do not know what to expect,
with relevant figures expunged
and predictions at best incorrect.
Who knows on what line you'll be ranged
and who, in what cause, you will harm?
What cabal or junta or sect
has doctored the headlines, arranged
for perpetual cries of alarm
through the white days of the storm?

Storm

Claire Malroux

Through the white days of the storm
Bone-colored road, bone-colored sky
High vessels, swaying in place
With flanks open wide to the foe
The perfidious Piper—the same
One who drew young leaves out with his flute
From their seeping, motherly jail
In his wake, flowers and fruits,
Blackbirds, canticles, prophecies
Duets and duels of the sun and moon
The snow's caress, fur of forgetfulness
And the children circling the masts
Plunging entranced toward the routs
Blood's risks, its hollows, its flames
Exchanged for the pull of that song
Bone-colored road, bone-colored sky
Through the white days of the storm.

For Kateb Yacine

(Algerian playwright, novelist, poet and activist, 1929–1989)

A moment jumps the interval; the next
second, a sudden dissonance swells up,
a crack down the smooth surface of the cup,
a dialogue with mistranslated text,
a tense the narrative poises, perplexed,
upon. The dancers and the singer stop,
swirled, each, in shadow like a velvet cape,
potential, and ambiguously sexed.
A gender and a nationality
implicit in the ululation rise
from a long throat to claim or compromise
privilege; responsibility
in texture, in that wound of sound, that vexed
surface, which could detonate, could drop.

Could drop into the anonymity
of headlines: war and fear and fear of war
and war abetted by ambient fear
honed to a hunger by publicity.
But there is a room above the street, a three-
o'clock winter sun, the nuanced, nearly translucent voice of a counter-tenor
threading a cantata by Scarlatti.
There were the exile's words in Arabic
anathematizing any deity
if slaughter is sanctified in its name;
the voice, the struggle from which words became

corporeally transformed to music.
There is the emblematic cup of tea.

There is the emblematic pot of tea
steaming on a wooden bench between
antagonists engaged in conversation
halfway to official enmity,
halfway to some compromise they can agree
upon, and not lose face. A city drones
and screeches in the crepuscule beyond
the room, in contrapuntal energy.
They keep sentences moving, savor the way
to pluck the pertinent or flabby phrase
and skin and gut it, twisting in the air:
a game they magisterially play
like diplomats, not gray-sweatered, gray-haired
exiles filling the breach of winter days.

Exiles filling the breach of winter days
with rhetoric have nothing, but have time
for rhetoric as logical as rhyme.
Meanwhile a speechwriter drafts the ukase
which, broadcast to a military base,
sends children and their city up in flames.
Meanwhile, an editor collects our names
and texts in protest: we can only guess
who else is keeping tabs, who else will be
pilloried in an op-ed in the *Times*,
distracted by brief notoriety

or told a passport will not be renewed.
Imagined exiles, with what gratitude
I'd follow your riposted paradigms.

To make of his riposte a paradigm,
he conjured Nedjma from the wilderness
behind his distance-mined electric face.
Kahina, Nedjma, Ummi, a woman's name
ejaculated to a stadium:
a heroine, a first lover unseen
for decades, a mother who mimed, silence upon
women's silence screamed past millennium.
Another silence, the interlocutor
who argued over wine with Paul Celan
until the words were not German or French
in the cold hypothesis of the Seine,
no longer comforter, companion, tutor
to the last Jew on the November bench.

If the last Jew on the November bench
shivered, rose, walked to the rue de Tournon
and ordered a Rémy and a *ballon*
de rouge with Roth, could it blot out the stench
of ash and lies for both? (Over the ranch
in Texas what smoke rises in premon-
itory pillar?) The gaunt Algerian
asynchronous, among them, needs to quench
an equal thirst. We all had pseudonyms,
code-names, pet-names, pen-names: *noms de guerre*,

simple transliterations, unfamiliar
diphthongs in rote order, palindromes
and puns patched on the untranslatable
(unuttered, anguished) root of a syllable.

Unuttered, anguished, roots of a syllable
in her first language threaded the page she'd sign
(written at night, in a strange town, hidden
among strangers, once betrayed) "Nicole
Sauvage."
                    Sun gilds the roof of the town hall,
its bridal parties gone, too cold, too late.
The February sky is celibate,
precipitated towards a funeral.
Yes, war will come and we will demonstrate;
war will come and reams of contraband
reportage posted on the Internet
will flesh out censored stories, second-hand.
Tire-treads lumbering towards its already-fixed
moment jump the interval: this war, the next.

("Nicole Sauvage" was the pseudonym used by the writer Nathalie Sarraute during the Nazi occupation of France during World War II, in the village where she went into hiding with her daughters after being denounced as a Jew by neighbors in another village.)

Ghazal: Waiting

What follows when imagination's not inspired by waiting,
body and spirit rendered sick and tired by waiting?

Wrinkles, stock market losses, abcessed teeth, rejection slips:
some of the benefits acquired by waiting.

Taught from childhood that patience is a virtue,
she thought that she could get what she desired by waiting.

History, a child at the chapter's cusp
will only find out what transpired by waiting.

Does anyone escape alienated labor's
cycle of being hired, exploited, made redundant, fired, by waiting?

He rolls a pen like a chess-piece between thumb and forefinger:
he won't emerge from the morass in which he's mired by waiting.

If poetry's imagination's daughter,
didn't someone say that she was sired by waiting?

She raised her children, wrote at dawn, ignored the factions,
arrived at being read, remembered and admired by waiting.

Once a pair of lovers downed shots in a Chelsea bar,
their nerves and fantasies hot-wired by waiting.

Sweating, shackled and blindfolded in a basement,
will I get out, the hostage (of whom) inquired, by waiting?

Glose: Willow

And I grew up in patterned tranquility
In the cool nursery of the new century.
And the voice of man was not dear to me,
But the voice of the wind I could understand.

—ANNA AKHMATOVA, "WILLOW"
TRANSLATED BY JUDITH HEMSCHMEYER

A sibilant wind presaged a latish spring.
Bare birches leaned and whispered over the gravel path.
Only the river ever left. Still, someone would bring
back a new sailor middy to wear in the photograph
of the four of us. Sit still, stop *fidgeting.*
—Like the still-leafless trees with their facility
for lyric prologue and its gossipy aftermath.
I liked to make up stories. I liked to sing:
I was encouraged to cultivate that ability.
And I grew up in patterned tranquility.

In the single room, with a greasy stain like a scar
from the gas-fire's fumes, when any guest might be a threat
(and any threat was a guest—from the past or the future)
at any hour of the night, I would put the tea things out
though there were scrap-leaves of tea, but no sugar,
or a lump or two of sugar but no tea.
Two matches, a hoarded cigarette:
my day's page ashed on its bier in a bed-sitter.
No godmother had presaged such white nights to me
in the cool nursery of the young century.

The human voice distorted itself in speeches,
a rhetoric that locked locks and ticked off losses.
Our words were bare as that stand of winter birches
while poetasters sugared the party bosses'
edicts (the only sugar they could purchase)
with servile metaphor and simile.
The effects were mortal, however complex the causes.
When they beat their child beyond this thin wall, his screeches,
wails and pleas were the gibberish of history,
and the voice of man was not dear to me.

Men *and* women, I mean. Those high-pitched voices—
how I wanted them to shut up. They sound too much
like me. Little machines for evading choices,
little animals, selling their minds for touch.
The young widow's voice is just hers, as she memorizes
the words we read and burn, nights when we read and
burn with the words unsaid, hers and mine, as we watch
and are watched, and the river reflects what spies. Is
the winter trees' rustling a code to the winter land?
But the voice of the wind I could understand.

For Anna Akhmatova

Who had been in love with her that summer? Did it matter?
The incidental willow is what she would remember,
bare like a silver brooch on a sky of foxfur
during the winters of famine and deportations.
She wished she had something more cheerful to show them:
a list of the flowering shrubs in a city park,
lovers and toddlers asprawl behind rosebushes;
workers with mallets indulging in horseplay
while knocking partitions of sheetrock to splinters:
energy's avatars, feminine, masculine.
Forehead against the cold pane, she would always be
ten-and-a-half years older than the century.

■ ■ ■

She remembered Mother reading them Nekrasov
as they ate sardines with white cheese and tomatoes
while sun set late on the same seacoast where Tomis
had sheltered and repelled an exiled poet.
She would eat the same briny cheese in the heat of Tashkent
waiting for news from re-named Leningrad.

■ ■ ■

It had pleasured her, a language which incised
choreographed chance encounters, almost-uttered
words, eye-contact, electricity
of an evaded touch: she wrote about
brief summers, solitude's inebriation
in the dusk that fell at almost midnight.

(Louise Labé in a less clement climate,
with electricity and indoor plumbing.)
She and her friends and lovers chiselled lyrics
until the decade (what did they think of revolution?)
caught up with them, the elegant companions,
and set them to a different exercise.

■ ■ ■

(Which travelling companions would you pick?
Who would have chosen to endure Céline?
Pasternak wrote a paean to Stalin;
Donne, for Pascal, would be a heretic.)

■ ■ ■

Something held her back from choosing exile
when the exacting enterprise went rotten
Russia was not her motherland: it was St. Petersburg,
the birch-lined corridors of Tsarskoye Selo—
but she was not retained by bark-scales spreading
up her limbs, with a god's breath in her ears: there
were her threatened friends, her son in prison
(who would not understand her coded letters
or what had held her back from choosing exile
after she did the paperwork to place him
in the Russian Gymnasium in Paris
the year his father met a firing squad;
and Marina—who would not live long—

wrote, she would meet them at the train station).
Tinned fish, gas rings, staggering armchairs, stained toilets
—mass graves of compromising manuscripts.
Was her exigent Muse the despised dictator
who censored, exiled, starved, imprisoned, murdered,
hurting the prodigy of birch and willow
into her late genius of debridement?
"Submissive to you? You must be out of your mind . . ."

■ ■ ■

How could she imagine, the *"gay little sinner,"* up
daybreak to dawn, the exactions of history?
City rerouted for transit to labor camps,
first husband shot in prison, their son in prison,
then in a labor camp, on the front, then still in prison.
She, over fifty, grown aquiline, vigilant,
larger than life, *"casta diva,"* her arias
camouflaged witness, evoking the dailiness
veiled in translation or foreign geographies.
Can you, yourself, in your eyrie, imagine it
while an empire's gearshifts creak behind you?

■ ■ ■

She made her despair the Virgin's or Cleopatra's
—under the circumstances, not outrageous.
She would write in praise of peace brought by the tyrant
if her lines might evoke an adjective passed down

from underling to underling until
some hungry guard unlocked a door . . . It didn't
happen. Her son called her superficial.
Larger than life, with all her flaws apparent
she rolled on the floor and howled in indignation,
more like the peasant she had come to resemble
than Anna Comnena or Cleopatra
or the ikon of words who was asked by other women
at the prison wall *"Can you describe this?"*

■ ■ ■

Once, in a youthful funk, she had made a poem of
her son (then just four) at her churchyard graveside
unable to resurrect his flighty mother
except to the balance sheet of her defections.
She was alone, and he was alive, in prison.
The impatient butterfly of Tsarskoye Selo
a solid matron, stood below the frozen
walls, with her permitted package, like the others—
whether they had been doting or neglectful mothers.

Glose: Jerusalem

The rampart behind the leprosarium:
That also is Jerusalem.
Blue brooks cross the fields,
Light silver-leafs a stocky tree.

—EMMANUEL MOSES,
"THE YEAR OF THE DRAGON"
TRANSLATED BY M.H.

Sunday noon haze on the fruit-stalls of Belleville,
a clochard's clothesline under the Pont des Arts,
the last Alsatian deli in the rue de Tourtille,
the second kosher couscous in the rue Saint-Maur.
The Northern Line at midnight back from Stockwell
via Charing Cross, since no one, not even a cab, had come.
The Black Mountains lurching past a drunken car,
a mail-van threading the Col de Vence in lunar
dawn when the town's enceinte is a colombarium.
The rampart behind the leprosarium.

An equinoctial dusk wrapping the Square
du Temple; a hangnail moon glimpsed through light rain
on the Pont Sully; the 96 bus trapped by parked
motorcycles outside the Royal Turenne,
honking, while truck fumes mount, and the bus driver
shouts at the *motards* what he thinks of them
somewhat distracting his stalled passengers
(the cyclists are pertinently not there);
the glass of water the waiter brings to him:
that also is Jerusalem.

Methods of crossing borders are diverse:
sixty years passed, and trains are innocent
again. Cream-colored cattle kneel; a lone horse
in a barnyard cocks a gray ear to the wind.
The sibilance of riverbanks, the terse
monosyllables a billboard holds
aloft above the tracks, a jet-trail's spent
calligraphy: their messages disperse
in the breached air whistling as it yields.
Blue brooks cross the fields.

In a vision of the perfected past,
a cindered path's circumference of vines
measures the play of words and breath, at last
conjoined in a few salvageable lines:
all of the hour's trajectory not lost
in burnt-out synapses of memory.
Yet some insight bestowed on aliens
inscribes the vineyard on a palimpsest
of city, valley, hills, a different city.
Light silver-leafs a stocky tree.

from "The Year of the Dragon"

Emmanuel Moses

The rampart behind the leprosarium:
That also is Jerusalem.
Blue brooks cross the fields,
Light silver-leafs a stocky tree.
In precious books which slip between our fingers
Each page tells a different story.
I also like to sit with you in that little café
Near the Rohin where the minutes are marked off
By the clanking of the streetcars
In February when the cold bites down
Into the porous flesh of Amsterdam's bricks
The dead rise up with the provisionally
Living and say each in turn
"How we have escaped."

Letter to Mimi Khalvati

Dear, how I hate the overblown diction of
lines for occasions: festschrifts, like elegies
  making a banal birthday seem to
  signpost a passage to unmapped wasteland,

when thoughts and smiles are fresh as they've ever been
—at least my brief years given the privilege
  of bantering across some table,
  words made more fluent by cakes or curries

or by the short time left for exchanging them:
train in an hour, espresso in Styrofoam
  cups. Ciao! I wish . . . I'll tell you next time.
  Bus to the Eurostar, airport taxi.

I'll never see the light of your memories
(joy can be shared, but losses are separate)
  though we're a lucky pair of outcasts,
  free to embellish or keep our stories.

Yours, Mimi, silver's brilliance on velvety
shapes in the no-man's-land between alphabets
  you were obliged to cross and cross to
  write in the white ink of exiled childhood.

Whose children *did* we talk about, smoking and
sipping red wine (an Indian family
  toasting some milestone near us) in the
  restaurant tucked behind Euston Station?

Two women, poised for middle-aged liberty,
still have our fledgling burdens to anchor us,
    wish they were soaring, independent,
    glad when they ground us with tea and gossip.

Think of the friendships lost to geography,
or lost to language, sex, or its absence . . . I
    send, crossing fingers, crossing water,
    bright thoughts, bright Maryam: happy birthday.

Letter to Alfred Corn

Alfred, we both know there's little dactylic pentameter
that can be spotted and quoted from classic anthologies
(although Hephaestion's *Handbook on Meters* cites "Atthis I
loved you once long ago" as an example, without much on
Sappho, but still, could a presence be much more felicitous?)
so this epistle is, much like good friendship, unorthodox,
framed both by Sappho *and* schoolmasters, and, overseeing the
words of itinerants, Wystan? Jean-Arthur? Elizabeth?
Aimé and Léopold lighting the Left Bank with Négritude?
August has shut down the shops and cafés on my market street;
when they re-open, *la rentrée*, fresh start, it will be without
me. I'll be back in New York, feeling ten times more alien
than where the polyglot boulevards intersect, linking up
11e and 20e, Maghreb, punk chic, kashruth, chinoiserie.
Once one could say that Manhattan was barely America,
which—in Manhattan—was meant as an insider's compliment.
Now it's as flag-ridden as the Republican "heartland," where
you behave better than Ovid in exile, not whimpering,
making the best of a stint as a scholar-in-residence.
For two good weeks we were neighbors, and living our parallel
lives, you at liberty, sampling the fruits of the capital—
notably, joys of the eye, its museums and boulevards,
while I attempted to pilot a relay of immigrant
artisans (David and Mario, Portuguese masons and
Nicolas, Serbian, plumber, and Sokli, Tunisian,
painter, Jérôme, electrician, from Sénégal, none of whose
papers I'd swear were in order, no more than my own are, all
working unsupervised for an unscrupulous contractor,
sleazy, incompetent, straight out of some New Wave gangster film,

who extracts money with threats from me, lets *them* go weeks
unpaid)
—if I was lucky and someone showed up on the work-site where
shelter was knocked down to gougings in plaster, precarious
walls spouting naked bouquets of distressed electricity.
That was the place I'd returned to with certainty suddenly
gone, when my life was made moot by disease, when companionate
passion had turned to disdain's acrimonious grievances
and I was left a late-quinquagenarian celibate
—still, when I mounted the spiral stairs slant with three centuries'
steps (like the furrows defining the smile of a laureate)
I was home safe in the cave that I'd made for the possible.
(Who, though is safe, from the "shocking disease," from the
bulldozer
wreaking revenge for the sins of the sons on their families,
or from the dynamite-bookbag in the cafeteria,
or from the tinderbox arrogant ignorance lights for us,
turgid with power, and willing to offer up holocaust?)
Ambulant scholar uprooted from tenure-track coteries
(who dared rip open the envelopes of the academy)
you are an expert at wanderings, vagaries, pilgrimage,
sometimes volitional, sometimes compelled by necessity.
Paris was part of your salad days and your apprenticeships:
Benjamin, Baudelaire, bars, Henry James, baroque opera.
When you returned to it, changed, had it changed for you? Every
street I walk down with one friend, then alone, then with somebody
else is three streets; is a new glyph incised on a palimpsest
someone, a painter, a novelist, poet or essayist
also inflected by naming, exulting or suffering

there, what she noticed, he turned from, we commented on as we
strolled, in no hurry, towards something convivial.

(We could have
stayed on the rue de Belleville to the Buttes-Chaumont, following
Breton and Aragon, seen if the statue was voluble . . .)
Late afternoon of Assumption, a holiday layered with
faces and pages and facets of (largely) this *Hexagone*:
nuns in white habits who sang a cappella, Le Thoronet's
ruins behind them sky-domed, or the nine-o'clock mass which was
washed in the filtered light Matisse's windows poured down the
pews,
dramas in Vence, below, held for post-prandial gossip-fests.
This year it's solo, the 3e, and who'll intercede for us
itinerants while the world goes to hell in a handbasket
(floats like a discarded dinner-dress down to the aqueduct)?
Still, the square's peaceable kingdom of Chinese and African
toddlers, mixed couples, clochards with guitar singing Dylan songs
—slurring the lyrics a bottle of plonk helps approximate—
stays what it's been, just as friendship seems durable. Let me at
least for your birthday, just past, be, though cautious, an optimist,
who loves the world both despite and because of its disarray,
planning new flâneries shared in our mutable capital.
("Atthis, I loved you then long ago when you were
a scruffy ungracious child" was my neophyte's take on it.)

Ghazal: *dar al-harb*

I might wish, like any citizen to celebrate my country
but millions have reason to fear and hate my country.

I might wish to write, like Virginia: as a woman, I have none,
but women and men are crushed beneath its weight: my country.

As English is my only mother tongue,
it's in English I must excoriate my country.

The good ideas of Marx or Benjamin Franklin
don't excuse the gulags, or vindicate my country.

Who trained the interrogators, bought the bulldozers?
—the paper trails all indicate my country.

It used to be enough to cross an ocean
and view, as a bemused expatriate, my country.

The June blue sky, the river's inviting meanders:
then a letter, a headline make me contemplate my country.

Is my only choice the stupid lies of empire
or the sophistry of apartheid: my country?

Walter Benjamin died in despair of a visa
permitting him to integrate my country.

Exiles, at least, have clarity of purpose:
can say my town, my mother and my fate, my country.

There used to be a face that looked like home,
my interlocutor or my mate, my country.

Plan your resistance, friends, I'll join you in the street,
but watch your backs: don't underestimate my country.

Where will justice and peace get the forged passports
it seems they'll need to infiltrate my country?

Eggplant and peppers, shallots, garlic and cumin:
let them be, married on my plate, my country.

Names

1.

A giant poplar shades the summer square.
Breakfast shift done, Reem smooths her kinky mass
of auburn curls, walks outside, her leaf-print dress
green shadow on post-millennial bright air.
It's almost noon. I smell of sweat. I smell
despite bain-moussant and deodorant,
crumpled and aging, while recognizant
of luck, to be, today, perennial
appreciating trees. The sky is clear
as this in Gaza and Guantanamo
about which I know just enough to mourn
yesterday's dead. The elegies get worn
away, attrition crumbles them into
chasm or quicklime of a turning year.

2.

Be mindful of names. They'll etch themselves
like daily specials on the window glass
in a delible medium. They'll pass
transformed, erased, a cloud the wind dissolves
above the ruckus of the under-twelves
on the slide, the toddlers on the grass,
the ragged skinny guy taking a piss
in the bushes, a matron tanning her calves
on a bench, skirt tucked around her knees.
A sparrow lands in the japonica;
as if it were a signal, all at once
massed pigeons rush up from adjacent trees,
wingbeats intrusive and symphonic—a
near total silence is the clear response.

3.

The actress reading sonnets from “The Quest”
made Auden seem, as far as I could tell,
less than Péguy, more than Marie Noël.
The Belgian poet preened. His last, or latest
girlfriend, my neighbor, not quite at her best
in crushed green velvet and a paisley shawl
was looking at once lovely and unwell.
I had to go. I could deduce the rest,
and there was a dinner in the dixième.
I walked up the rue du Temple in the fog,
not a mist of exile and erasure,
but one from which memory and nomenclature
engage (Thank you, Wystan) in a dialogue
with dark streets redolent of almost-home.

4.

Four firelit mirrors lining the Corsican
restaurant's walls reflected divergencies—
Palestinian, Syrian, Lebanese,
expat Russian, expat Jewish American.
A new war had begun that afternoon;
The shrinking world shrieked its emergencies
well beyond our capabilities
if not to understand, to intervene,
though Mourad, who practices medicine,
has made of intervention a career.
Khaled spent decades studying history
in the jaws, shall we say, of an emergency.
Start another bottle of rough-tongued wine,
that sanguine glitter in the midnight mirror.

5.

Edinburgh airport seems provincial when
you're headed back to CDG/Roissy
in dusty sunlight of a mid-July
midday. I had an hour. But there was Hind
(we'd been at the same conference all weekend)
who had three connections: Heathrow, Cairo,
Beirut, where the runways had been bombed,
to Damascus. With airport Starbucks, we brainstormed
the thesis-in-progress she'll have to write
in English if she's going to publish it:
Lesbian writers from the Arab world.
Boarding call. I don't know if she got home.
I e-mailed her. I haven't heard from her.
The war had started five days earlier.

6.

Noura is writing about women also: women
and war. She sends an e-mail from Mosul:
The books arrived, and they are beautiful.
I know, of course, the work of Fadwa Touqan
but since the invasion and the occupation
it is hard to find books, even in Arabic.
Attached is the synopsis of my post-doc
proposal and the draft of a translation.
I cannot visit my old teacher in Baghdad:
because I am Sunni and from Mosul
I would be immediately slain.
Through the cracked prism of Al-Andalus
we witness, mourning what we never had.
(The war goes on and on and on and on.)

7.

A waxing moon, tailwind of a return,
but to what? Life on the telephone,
letters typed on a computer screen
which no one needs to file or hide or burn
at the storm-center of emergency
where there is no coherent narrative.
With no accounting of my hours to give
black holes gape open in my memory.
If there's some story here, it isn't mine,
but one I can imperfectly discern
from what can be imperfectly expressed
by third parties in second languages.
The shots, far off, the power cut, the line
interrupted, the fact I did not learn.

8.

The names have been changed. Nobody's sister
will be gunned down because her brother
shook hands with one politician or another
or because a well-meaning woman activist kissed her
father on both cheeks the way we do
here (thinking, we're all Mediterranean
after all). Nobody's J-1
visa will be revoked because of the conference she went to
in Caracas, or, worse, Teheran.
We have an almost-fiction with mnemonic
cues, which could be proper names, or dates.
Sipping another empire's bitter tonic
an inadvertent exile contemplates
Harvard Square's night-lights on Ramadan.

Ghazal: *min al-hobbi ma qatal*

for Deema K. Shehabi

You, old friend, leave, but who releases me from the love that kills?
Can you tell the love that sets you free from the love that kills?

No mail again this morning. The retired diplomat
stifles in the day's complacency from the love that kills.

What once was home is across what once was a border
which exiles gaze at longingly from the love that kills.

The all-night dancer, the mother of four, the tired young doctor
all contracted HIV from the love that kills.

There is pleasure, too, in writing easy, dishonest verses.
Nothing protects your poetry from the love that kills.

The coloratura keens a triumphant swan-song
as if she sipped an elixir of glee from the love that kills.

We learn the maxim: "So fine the thread,
so sharp the necessity" from the love that kills.

The calligrapher went blind from his precision
and yet he claims he learned to see from the love that kills.

Spare me, she prays, from dreams of the town I grew up in,
from involuntary memory, from the love that kills.

Homesick soldier, do you sweat in the glare of this checkpoint
to guard the homesick refugee from the love that kills?

Paragraphs for Hayden

I'd want to talk to you about desire,
Hayden, the letter I could have written
on a subject you'd never tire
of turning in a glass, smitten
by a song, an argument, long sorrel hair,
profile of a glazed clay icon in the river,
while your knees needled and breathing
hurt, two packs a day bequeathing
what didn't, in fact, kill you in the end.
Was it a distraction
from the inexorable fear, my friend,
its five A.M. gut-contraction?
But who, of your critics or cortège, pretends
that expense of spirit, lust in action,
didn't earn you magnificent dividends?

The week they told me my genetic code
was flawed, I ricocheted, desire and fear
like sun and clouds, a mood-
swing reason had no reason for
(but reason's calibrated in the blood).
Terror. Tumescence. Cloudbursts. Solitude.
No diagnosis, no beloved: balance . . .
I write, not to you; to silence.
By anybody's reckoning, now I'm "old,"
and you, an occasion instead
of an interlocutor. Aura of beaten gold
in a winter of cast lead.
Will the scale tip to the side of pleasure
when a taut cord plucked across the grid
invites, vibrates according to your measure?

A taut-tuned string asserts: the girl in green,
a six-year-old in an oversized sweatshirt
in Gaza City, on a computer-screen
video, not dead, not hurt
but furious. *This is what they've done*
to our house! Our clothes smell of gas! I never wore the sun-
glasses my father gave me
or the earrings my grandmother gave me!
She tosses dark curls, speaks, a pasionaria
in front of a charred wall.
Arching her brows, she orchestrates her aria
with swift hands that rise and fall
while she forgets about fear
even as she ransacks the empty cradle
of its burnt blankets. That baby's—where?

Not like “upstate,” our January freeze
still killed my window-box geraniums.
Beyond that ragged khaki frieze
of dead plants, Sunday hums
up to my windows. I count each of these
hours, respite, respite, from broken treaties
uprooted orchards, shattered concrete.
Eight years later, still on the street
eight years older, two women squabble
and survive improbably.
A dark-haired boy, pale, imperturbable
sits in front of Monoprix,
wrapped in blankets, stroking a silvery cat.
Your voice begins to slip away from me.
Life is like that. Death is like that.

A glass of red wine spills on the grammar book—
the pupil and the teacher gasp, then laugh.
Their voices branch into the baroque
logic of the paragraph.
Does the Brouilly birthmark presage luck
learning the rudiments of Arabic?
This classroom desk is a kitchen table,
but the street outside is peaceful.
Schoolchildren with satchels weave among
shoppers, construction workers, dogs.
No one here is speaking their mother tongue:
perhaps several dialogues
are contradicting contrapuntally.
Two girls in hijab with computer bags
go hand in hand into the library.

Ghazal: Begin

The energy is mounting, something will, again, begin.
You will yourself to, know you will—but when—begin.

Remember anger. Remember indignation. Remember desire.
Feel that deceptive surge of adrenaline begin.

Select a rhyme, trust syntax for reason,
let rash conjunctions of the page and pen begin.

After the last postcard, last phone call from home,
messages from somewhere beyond your ken begin.

Mathilde was eighteen. Arthur was seventeen.
One could see trouble in the ménage Verlaine begin.

Nights of champagne/cocaine pale into dawn.
Mornings of mint tea and ibuprofen begin.

The doctor waking in a refugee camp
heard the keened lesssons of the gaunt children begin.

Lock Bush and Cheney up with Milosevic,
then let the trial of Saddam Hussein begin.

My brave friend's gone; our leaders are blindered bastards:
Thus might an evening's reading of Montaigne begin.

As I poured that glass of wine, I thought of her
and felt the needlings of this damned migraine begin.

On the list I'm writing down of my addictions,
I'll let the oldest one, to oxygen, begin.

There where the fox was too hungry, baffled, tired,
the tracks that led the hunters to his den begin.

Perhaps it will happen if you close your eyes
and count—but very slowly—backwards from ten. Begin.

Sháhid, if my name were Witness, I would sign it.
I leave when the tired jokes about "Marilyn" begin.

A Braid of Garlic

Aging women mourn while they go to market,
buy fish, figs, tomatoes, enough *today* to
feed the wolf asleep underneath the table
who wakes from what dream?

What but loss comes round with the changing season?
He is dead whom, daring, I called a brother
with that leftover life perched on his shoulder
cawing departure.

He made one last roll of the dice. He met his
last, best interlocutor days before he
lay down for the surgery that might/might not
extend the gamble.

What they said belongs to them. Now a son writes
elegies, though he has a living father.
One loves sage tea, one gave the world the scent of
his mother's coffee.

Light has shrunk back to what it was in April,
incrementally will shrink back to winter.
I can't call my peregrinations "exile,"
but count the mornings.

In a basket hung from the wall, its handle
festooned with cloth flowers from chocolate boxes,
mottled purple shallots, and looped beside it,
a braid of garlic.

I remember, ten days after a birthday
(counterpoint and candlelight in the wine-glass)
how the woman radiologist's fingers
probed, not caressing.

So, reprise (what wasn't called a "recurrence")
of a fifteen-years-ago rite of passage:
I arrived, encumbered with excess baggage,
scarred, on the threshold.

Through the mild winter sun in February,
two or three times weekly to Gobelins, the
geriatric hospital where my friend was
getting her nerve back.

At the end of elegant proofs and lyric,
incoherent furious trolls in diapers.
Fragile and ephemeral as all beauty:
the human spirit—

while the former journalist watched, took notes and
shocked, regaled her visitors with dispatches
from the war zone in which she was embedded,
biding her time there.

Now in our own leftover lives, we toast our
memories and continence. I have scars where
breasts were, her gnarled fingers, these days, can hardly
hold the pen steady.

Thousands mourn him, while in the hush and hum of
life-support for multiple organ failure,
utter solitude, poise of scarlet wings that
flutter, and vanish.

In memory of Mahmoud Darwish and Mavis Gallant

INDEX OF TITLES AND FIRST LINES